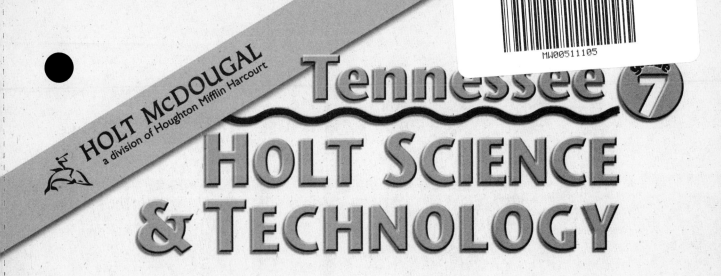

HOLT McDOUGAL
a division of Houghton Mifflin Harcourt

Tennessee 7
HOLT SCIENCE
& TECHNOLOGY

Interactive Reader
and Study Guide

HOLT McDOUGAL
a division of Houghton Mifflin Harcourt

Copyright © Holt McDougal, a division of Houghton Mifflin Harcourt Publishing Company. All rights reserved

Warning: No part of this publication may be reproduced or transmitted in any form or by any means, electronic or mechanical, including photocopy, and recording, or by any information storage or retrieval system without the prior written permission of Holt McDougal unless such copying is expressly permitted by federal copyright law.

Teachers using HOLT SCIENCE & TECHNOLOGY may photocopy complete pages in sufficient quantities for classroom use only and not for resale.

HOLT MCDOUGAL is a trademark of Houghton Mifflin Harcourt Publishing Company.

Printed in the United States of America

If you have received these materials as examination copies free of charge, Holt McDougal retains title to the materials and they may not be resold. Resale of examination copies is strictly prohibited.

Possession of this publication in print format does not entitle users to convert this publication, or any portion of it, into electronic format.

ISBN-13: 978-0-55-401781-5
ISBN-10: 0-55-401781-4

6 7 8 9 0982 12 11
4500310394

Contents

Copyright © by Holt, Rinehart and Winston; a Division of Houghton Mifflin Harcourt Publishing Company. All rights reserved.

Copyright © by Holt, Rinehart and Winston; a Division of Houghton Mifflin Harcourt Publishing Company. All rights reserved.

Copyright © by Holt, Rinehart and Winston; a Division of Houghton Mifflin Harcourt Publishing Company. All rights reserved.

CHAPTER 15 Plate Tectonics

CHAPTER 16 Earthquakes

CHAPTER 17 Volcanoes

CHAPTER 18 Energy Resources

Copyright © by Holt, Rinehart and Winston; a Division of Houghton Mifflin Harcourt Publishing Company. All rights reserved.

UNIT 3 Physical Science

Copyright © by Holt, Rinehart and Winston; a Division of Houghton Mifflin Harcourt Publishing Company. All rights reserved.

Copyright © by Holt, Rinehart and Winston; a Division of Houghton Mifflin Harcourt Publishing Company. All rights reserved.

SECTION 1
Science and Scientists

BEFORE YOU READ

After you read this section, you should be able to answer these questions:

• What is science?

• How does science affect our lives?

• Where do scientists work?

Tennessee Science Standards

GLE 0707.Inq.2
GLE 0707.Inq.4
GLE 0707.Inq.5
GLE 0707.T/E.1

What Is Science?

You probably do not know it, but there are ways you think and act like a scientist. You think like a scientist any time you ask a question about the world around you. For example, you might wonder why leaves change color in the fall.

You act like a scientist any time you investigate a question you ask. For example, you may investigate how your reflection changes in different mirrors. Asking questions and searching for answers are part of a process of gathering knowledge about the natural world. This process is called **science**. ☑

The world around you is full of amazing things that can lead you to ask questions and search for answers. The boy shown below is thinking like a scientist because he is asking questions about things he has observed in the world around him.

STUDY TIP

Outline As you read, make an outline of this section. Use the headings from the section in your outline.

✓ READING CHECK

1. Define What is science?

_____

_____

_____

Why do leaves change color in the fall?

Why did the dinosaurs die out?

How do birds know where to go when they migrate?

Copyright © by Holt, Rinehart and Winston; a Division of Houghton Mifflin Harcourt Publishing Company. All rights reserved.

OBSERVATION

Anywhere you look you may see things that make you wonder. You could look in your home, neighborhood, city, or a nearby forest or beach. You could even look beyond Earth to the moon, sun, and other objects in our universe. Obviously, there are many things that can cause people to think of a question.

Once you ask a question, you can start looking for an answer. There are several different methods you can use to start your search for an answer. You may find the answer by looking around and making observations.

For example, if you want to know how spiders spin their webs, look for a web. When you find one, observe the spider as it spins its web. Sometimes, what you think you might see affects what you do observe. This can introduce error into your investigations. This error, or mistake, is called bias because you might think you already know the correct answer. Be careful to have an open mind when making observations.

RESEARCH

In addition to making observations, you can also research what scientists already know about your question. There are many places to do research and find information. The boy shown below is working in a library, a good place to conduct research.

At a library, you can search for information in books, in magazines, and on the Internet. Always be sure to use information that comes from a reliable source, whether it is a book, magazine article, or a site on the World Wide Web. Ask your teacher or the librarian to find out if a source is reliable. You can also research information by speaking with someone who knows a lot about your question.☑

Critical Thinking

2. Describe Give an example of how making observations helped you answer a question.

_____

_____

_____

_____

 **READING CHECK**

3. Identify List three places to research information in a library.

_____

_____

_____

Copyright © by Holt, Rinehart and Winston; a Division of Houghton Mifflin Harcourt Publishing Company. All rights reserved.

EXPERIMENTATION

You may also answer your question by doing an experiment. Many experiments are easy to perform and use materials you can find at home. However, other experiments are more difficult to perform and may require materials found only in a science laboratory.

All experiments must be carefully planned. You should research similar experiments that have been done in the past. Before performing any experiment, always discuss with an adult what you plan to do. Follow safety precautions, and ask an adult to be present whenever you perform an experiment. ☑

This student is doing an experiment to find out whether this type of plant grows better in shade or in direct sunlight.

✔ **READING CHECK**

4. Describe What should you do to prepare for an experiment?

_____

_____

_____

_____

When you perform an experiment, make careful observations and record them thoroughly. You may observe something unusual or unexpected. You may need to research more, or you may have to perform a different experiment. You may even have to repeat the same experiment several times. In other words, finding an answer to a question may involve several steps, including making observations, doing research, and performing an experiment. ☑

What Role Does Science Play in Our Lives?

Scientists cannot answer every question that they ask. Yet, they continue to ask more questions and search for answers. The answers they find often affect people in their daily lives. For example, scientists are always searching for new medicines, ways to make machines more efficient, and more accurate ways to predict the weather. Their discoveries have resulted in ways to save lives, resources, and the environment.

✔ **READING CHECK**

5. Describe What could you do if you observed something unexpected during an experiment?

_____

_____

_____

Copyright © by Holt, Rinehart and Winston; a Division of Houghton Mifflin Harcourt Publishing Company. All rights reserved.

SECTION 1 Science and Scientists *continued*

SAVING LIVES

One question that many scientists ask is, "How can people be protected from diseases?" Polio is a disease that affects the brain and nerves. In 1955 and 1956, scientists developed vaccines that eliminated polio in the United States. Today, scientists are searching for cures for diseases such as mad cow disease, tuberculosis, and acquired immune deficiency syndrome (AIDS). The figure below shows a scientist learning more about AIDS, a disease that kills millions of people every year.

TAKE A LOOK
6. Infer Why is the scientist wearing gloves?

_____

_____

_____

_____

SAVING RESOURCES

Another question scientists have asked is "How can we make resources last longer?" One answer is to recycle. Scientists have developed more efficient ways to recycle steel, aluminum, paper, glass, and even some plastics. In this way, science helps make resources last longer. The figure below shows how much is saved by recycling one metric ton of aluminum.

Critical Thinking

7. Predict What might happen to our natural resources if we could not recycle?

_____

_____

_____

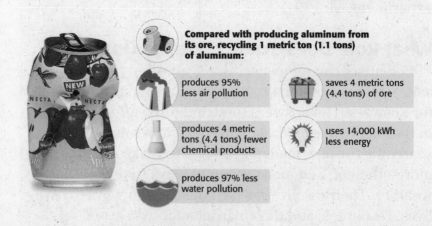

Compared with producing aluminum from its ore, recycling 1 metric ton (1.1 tons) of aluminum:

produces 95% less air pollution

produces 4 metric tons (4.4 tons) fewer chemical products

produces 97% less water pollution

saves 4 metric tons (4.4 tons) of ore

uses 14,000 kWh less energy

Copyright © by Holt, Rinehart and Winston; a Division of Houghton Mifflin Harcourt Publishing Company. All rights reserved.

SECTION 1 Science and Scientists *continued*

ANSWERING SOCIETY'S QUESTIONS

Still another question scientists have asked is, "How to reduce air pollution?" At one time, this question did not have any obvious, reasonable answers. As the problem of air pollution became more important to people, scientists developed technologies to address it.

For example, one source of air pollution is exhaust from cars. However, the millions of people who depend on their cars cannot just stop driving. Using the knowledge gained by science, people have developed cleaner-burning gasoline and new ways to clean up exhaust before it leaves a car's tailpipe! ☑

Where Can You Find a Scientist at Work?

Scientists work in many different places including laboratories, forests, offices, oceans, and space. The following table summarizes some jobs that use science.

Job title	Description
Meteorologist	person who studies the atmosphere and forecasts the weather
Geochemist	person who studies the chemistry of rock, minerals, and soil
Oceanographer	person who studies the ocean
Volcanologist	person who studies volcanoes and their eruptions
Mechanic	person who works on machines
Zoologist	person who studies the lives of animals

A geochemist may work outdoors when collecting rock samples from the field. Then, he may work inside as he analyzes the samples in the laboratory.

✓ READING CHECK

8. Describe How do scientists help reduce air pollution?

_____

_____

_____

TAKE A LOOK
9. Compare How is a meteorologist different from a zoologist?

_____

_____

_____

Copyright © by Holt, Rinehart and Winston; a Division of Houghton Mifflin Harcourt Publishing Company. All rights reserved.

Section 1 Review

GLE 0707.Inq.2, GLE 0707.Inq.4, GLE 0707.Inq.5, GLE 0707.T/E.1 TN

SECTION VOCABULARY

science the knowledge obtained by observing natural events and conditions in order to discover facts and formulate laws or principles that can be verified or tested	

1. Define In your own words, write a definition for the term *science*.

_____

_____

_____

2. Identify List three processes that can be used to answer a question a scientist might ask.

_____

_____

_____

3. Describe How might a meteorologist use science in his or her work to save lives?

_____

_____

_____

_____

4. Calculate Complete the following table.

Resources saved	Amount saved by recycling 1 metric ton of aluminum	Amount saved by recycling 3 metric tons of aluminum
energy	14,000 kWh	
ore	4 metric tons	

5. Explain Name one way that science has played a role in your life today.

_____

_____

_____

Copyright © by Holt, Rinehart and Winston; a Division of Houghton Mifflin Harcourt Publishing Company. All rights reserved.

CHAPTER 1 | Science in Our World

SECTION 2 | # Scientific Methods

 **Tennessee Science Standards**
GLE 0707.Inq.2
GLE 0707.Inq.3
GLE 0707.Inq.4
GLE 0707.Inq.5

BEFORE YOU READ

After you read this section, you should be able to answer these questions:

• What are scientific methods?

• What is a hypothesis?

• How do scientists test a hypothesis?

What Are Scientific Methods?

A group of students in Minnesota went on a field trip to a wildlife refuge. They noticed that some of the frogs they saw looked strange. For example, some of the frogs had too many legs or eyes. The frogs were *deformed*. The students wondered what made the frogs deformed. They decided to carry out an investigation to learn what happened to the frogs.

The students observed the frogs and asked questions about the frogs. By doing these things, the students were using scientific methods. **Scientific methods** are the ways in which scientists follow steps to answer questions and solve problems. The figure below shows the steps in scientific methods. ☑

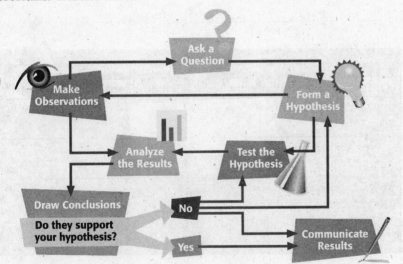

As you can see, the order of steps in scientific methods can vary. Scientists may use all the steps or just some of the steps during an investigation. They may even repeat some of the steps. The order depends on what works best to answer the question.

STUDY TIP

Outline As you read this section, make a chart showing the different steps in scientific methods. In the chart, describe how the students in Minnesota used each step to investigate the deformed frogs.

✓ READING CHECK

1. Define What are scientific methods?

_____

_____

_____

TAKE A LOOK

2. Use Models Starting with "Ask a question," trace two different paths through the figure to "Communicate results." Use a colored pen or marker to trace your paths.

Copyright © by Holt, Rinehart and Winston; a Division of Houghton Mifflin Harcourt Publishing Company. All rights reserved.

Why Is It Important to Ask a Question?

Asking a question helps scientists focus their research on the most important things they want to learn. In many cases, an observation leads to a question. For example, the students in Minnesota observed that some of the frogs were deformed. Then they asked the question, "Why are some of the frogs deformed?" Answering questions often involves making more observations. ☑

☑ **READING CHECK**

3. Explain Why do scientists ask questions?

_____

_____

How Do Scientists Make Observations?

The students in Minnesota made careful observations to help them answer their question. The students caught many frogs. They counted how many normal and deformed frogs they caught. They photographed, measured, and described each frog. They also tested the water the frogs were living in. The students were careful to record their observations accurately.

Like the students, scientists make many different kinds of observations. They may measure length, volume, time, or speed. They may describe the color or shape of an organism. They may also describe how an organism behaves. When scientists make and record their observations, they are careful to be accurate. Observations are useful only if they are accurate.

Critical Thinking

4. Explain Why is it important for observations to be accurate?

_____

_____

_____

_____

TAKE A LOOK

5. Identify Give three kinds of observations that can be made with the tools in the picture.

_____

_____

Scientists use many different tools to make observations. These include microscopes, rulers, and thermometers

Copyright © by Holt, Rinehart and Winston; a Division of Houghton Mifflin Harcourt Publishing Company. All rights reserved.

What Is a Hypothesis?

After asking questions and making observations, scientists may form a hypothesis. A **hypothesis** is a possible explanation or answer to a question. It is based on observations and information a scientist already knows. A good hypothesis can be tested. ☑

Scientists may form more than one hypothesis for a single problem. The students in Minnesota learned about different things that can cause frogs to be deformed. Then, they used this information to form three hypotheses. Were any of these explanations correct? To find out, the students would have to test their hypotheses.

Hypothesis 1:
The deformities were caused by one or more chemical pollutants in the water.

Hypothesis 2:
The deformities were caused by attacks from parasites or other frogs.

Hypothesis 3:
The deformities were caused by an increase in exposure to ultraviolet light from the sun.

☑ **READING CHECK**

6. Define What is a hypothesis?

_____

_____

Say It

Discuss In a group, talk about some other possible hypotheses that the students could have come up with.

TAKE A LOOK

7. Describe What are two things that all the hypotheses have in common?

_____

_____

_____

Copyright © by Holt, Rinehart and Winston; a Division of Houghton Mifflin Harcourt Publishing Company. All rights reserved.

SECTION 2 Scientific Methods *continued*

Critical Thinking

8. Make Connections What is the connection between predictions and tests in an investigation?

_____

_____

_____

_____

_____

PREDICTIONS

Before a scientist tests a hypothesis, the scientist may make a prediction. A *prediction* is a statement that tells the future outcome of an action or event. Predictions are usually stated as *if-then statements*, as shown in the figure below. A prediction can be used to set up a test of a hypothesis. More than one prediction may be made for a hypothesis.

Scientists can perform experiments to test their predictions. In many cases, the results from the experiments match a prediction. In other cases, the results may not match any of the predictions. When this happens, the scientist must make a new hypothesis and perform more tests.

Hypothesis 1:
Prediction: **If** a substance in the pond water is causing the deformities, **then** the water from the ponds that have deformed frogs will be different from the water from ponds in which no abnormal frogs have been found.

Prediction: **If** a substance in the pond water is causing the deformities, **then** some tadpoles will develop deformities when they are raised in pond water collected from ponds that have deformed frogs.

TAKE A LOOK

9. Explain What kind of tests could the students do to test the prediction for hypothesis 2?

_____

_____

_____

_____

Hypothesis 2:
Prediction: **If** a parasite is causing the deformities, **then** this parasite will be found more often in frogs that have deformities.

Hypothesis 3:
Prediction: **If** an increase in exposure to ultraviolet light is causing the deformities, **then** some frog eggs exposed to ultraviolet light in a laboratory will develope into deformed frogs.

Copyright © by Holt, Rinehart and Winston; a Division of Houghton Mifflin Harcourt Publishing Company. All rights reserved.

SECTION 2 Scientific Methods *continued*

How Do Scientists Test a Hypothesis?

One way to test a hypothesis is to do a controlled experiment. A **controlled experiment** is an experiment in which only one factor changes at a time. A *factor* is anything in an experiment that can affect the experiment's outcome. The factor that changes in a controlled experiment is called the **variable**. By changing only the variable, scientists can see the results of just that one change. ☑

In order to test only one variable at a time, scientists use a control group and an experimental group. In the control group, the variable does not change. The control group is used as the standard to compare to changes in the experimental group. In the experimental group, the variable changes for each test. Therefore, any differences in test results are probably caused by the variable. ☑

DESIGNING AN EXPERIMENT

Experiments must be carefully planned. Every factor should be considered when designing an experiment. Remember the prediction for Hypothesis 3: *If an increase in exposure to ultraviolet light is causing the deformities, then some frog eggs exposed to ultraviolet light in a laboratory will develop into deformed frogs.* An experiment to test this hypothesis is summarized in the table below.

This experiment has one control group and two experimental groups. All the factors between these groups are the same except for the amount of UV light exposure. Therefore, UV light exposure is the variable.

The control group receives no UV light. One experimental group is exposed to 15 days of UV light. The other experimental group is exposed to 24 days of UV light.

| Group | Control Factors | | | Variable |
	Kind of frog	Number of eggs	Temperature of water (°C)	UV light exposure (days)
#1	leopard frog	100	25	0
#2	leopard frog	100	25	15
#3	leopard frog	100	25	24

✔ **READING CHECK**

10. Compare What is the difference between a factor and a variable?

_____

_____

_____

_____

✔ **READING CHECK**

11. Compare How are control groups and experimental groups different?

_____

_____

_____

TAKE A LOOK

12. Apply Concepts Which group is the control group? Explain your answer.

_____

_____

_____

Copyright © by Holt, Rinehart and Winston; a Division of Houghton Mifflin Harcourt Publishing Company. All rights reserved.

SECTION 2 Scientific Methods *continued*

COLLECTING DATA

Data are pieces of information gathered from an experiment. When collecting data on living organisms, scientists often try to test many individuals. The more individuals are tested, the smaller the effect of a difference between individuals will be.

For example, in the UV light experiment, a total of 300 frogs were tested. This way, any differences in the results were due to the variable, not to differences in individual frogs. ☑

Scientists often repeat an experiment to see if it produces the same results each time. If an experiment gives the same results each time, scientists can be more certain that the results are true.

The figure below shows the setup of the UV light experiment. It also shows the results of the experiment.

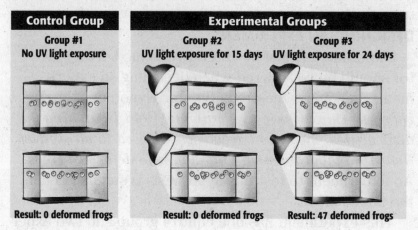

Control Group	Experimental Groups	
Group #1 No UV light exposure	Group #2 UV light exposure for 15 days	Group #3 UV light exposure for 24 days
Result: 0 deformed frogs	Result: 0 deformed frogs	Result: 47 deformed frogs

How Do Scientists Analyze Results?

After an experiment, scientists must organize and analyze their data. When scientists analyze their data, they interpret what the data mean.

Scientists often organize data into tables and graphs. This makes it easier to see the effect of the variable on the experiment. The data the students collected from the UV light experiment are shown in the table below.

Number of days of UV exposure	Number of deformed frogs
0	0
15	0
24	48

READING CHECK

13. Explain Why do scientists try to use many individuals in their experiments?

_____

_____

_____

Math Focus

14. Make a Graph Use the information in the table to fill in the bar graph below.

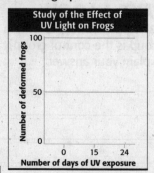

Study of the Effect of UV Light on Frogs

Number of deformed frogs
100
50
0
0   15   24
Number of days of UV exposure

Copyright © by Holt, Rinehart and Winston; a Division of Houghton Mifflin Harcourt Publishing Company. All rights reserved.

What Are Conclusions?

After scientists have analyzed their data, they must decide if the results support the hypotheses. This is called *drawing conclusions*.

The table on the previous page showed that frogs that were exposed to 24 days of UV light developed deformities. Therefore, the results support the hypothesis. However, this does not mean that UV light definitely caused the frog deformities in Minnesota. It only means it may have caused the deformities. Many other factors may also affect the frogs, such as parasites or polluted water.

Sometimes, the results do not support the hypothesis. Finding out that a hypothesis is not true can be as valuable as finding out that a hypothesis is true. When this happens, scientists may repeat the investigation to check for mistakes. Scientists may repeat experiments hundreds of times. Or, they may ask another question and make a new hypothesis.

Questions as complicated as why the frogs were deformed are rarely solved with a single experiment. The search for an answer may continue for many years. Even finding an answer doesn't always end an investigation. In many cases, the answer begins another investigation. In this way, scientists continue to build knowledge.

Why Do Scientists Share Their Results?

After finishing a study, scientists share their results with others. They may write reports and give presentations. They can also put their results on the Internet.

Sharing results also allows other scientists to continue the investigation, ask more questions, and find more answers.

Critical Thinking

15. Infer How can finding out that a hypothesis is not true be useful for a scientist?

_____

_____

_____

_____

_____

TAKE A LOOK
16. Describe Why is it important for scientists to share their results?

_____

_____

_____

Sharing the results of experiments is an important step in scientific methods.

Copyright © by Holt, Rinehart and Winston; a Division of Houghton Mifflin Harcourt Publishing Company. All rights reserved.

Section 2 Review

GLE 0707.Inq.2, GLE 0707.Inq.3, GLE 0707.Inq.4, GLE 0707.Inq.5 TN

SECTION VOCABULARY

controlled experiment an experiment that tests only one factor at a time by using a comparison of a control group with an experimental group	**scientific methods** a series of steps followed to solve problems
hypothesis a testable idea or explanation that leads to scientific investigation	**variable** a factor that changes in an experiment in order to test a hypothesis

1. Describe In a controlled experiment, how are the control and experimental groups the same? How are they different?

_____

_____

_____

2. Infer Why might a scientist need to repeat a step in scientific methods?

_____

_____

_____

3. Identify What are two ways that scientists can share the results of their experiments?

_____

_____

4. Define What is a prediction?

_____

_____

5. Explain Why might a scientist repeat an experiment?

_____

_____

_____

6. Describe What can scientists do if the results of an experiment do not support a hypothesis?

_____

_____

_____

Copyright © by Holt, Rinehart and Winston; a Division of Houghton Mifflin Harcourt Publishing Company. All rights reserved.

Name _____ Class _____ Date _____

TN Tennessee Science Standards
GLE 0707.Inq.2
GLE 0707.Inq.3
GLE 0707.Inq.4
GLE 0707.Inq.5

BEFORE YOU READ

After you read this section, you should be able to answer these questions:

• What are the three types of scientific models?

• How do scientists use models to help them understand scientific information?

What Is a Scientific Model?

How can you see the parts of a cell? Unless you had superhuman eyesight, you couldn't see inside most cells without using a microscope. What would you do if you didn't have a microscope? Looking at a model of a cell would help. A model of a cell can help you understand what the parts of a cell look like.

A **model** is something that scientists use to represent an object or process. For example, models of human body systems help you learn how the body works. Models can also help you learn about the past or predict the future. However, a model cannot tell you everything about the thing it represents. This is because the model always has a least a few differences from what it represents.

There are three common kinds of scientific models: physical models, mathematical models, and conceptual models.

STUDY TIP

Compare After you read this section, make a chart comparing scientific theories and scientific laws.

Critical Thinking

1. Compare What are some differences you might find between a model and the thing it represents?

_____

_____

_____

_____

This physical model of a skyscraper is not the same as the real building in every way. This is both an advantage and a drawback of the model.

TAKE A LOOK

2. Explain What is one advantage to using a model of a skyscraper? What is one drawback?

_____

_____

_____

_____

_____

Copyright © by Holt, Rinehart and Winston; a Division of Houghton Mifflin Harcourt Publishing Company. All rights reserved.

SECTION 3 Scientific Models *continued*

PHYSICAL MODELS

A *physical model* is a model you can see and touch. Some physical models can help you study things that are too small to see. For example, a ball-and-stick model can show the arrangement of atoms in a molecule.

Other physical models can help you study things that are too large to see all at once. For example, a model of of the solar system helps you picture what it looks like. Models can also represent things that do not exist. For example, scientists have built models of dinosaurs based on information from fossils and other observations.

A physical model can also help you understand a concept. Launching a model of a space shuttle can help you understand how a real space shuttle is launched.

MATHEMATICAL MODELS

A *mathematical model* is made of mathematical equations and data. You can't see a mathematical model the way you can see a physical model. However, you can use it to understand systems and make predictions.

Some mathematical models are simple. Others are very complicated and are operated by computers. For example, computer-operated models are often used to help predict the weather.

CONCEPTUAL MODELS

Conceptual models are used to help explain ideas. Conceptual models may be based on systems of ideas or data from experiments. Some conceptual models help explain natural events or processes that cannot be observed.

For example, scientists use conceptual models to help understand animal behavior. Scientists can use a model to predict how an animal might respond to a certain action.

📣 Say It

Brainstorm In a small group, come up with a list of five kinds of physical models. Talk about how they are similar to the things they represent and how they are different.

TAKE A LOOK

3. Identify Fill in the blanks in the table to tell whether the example is a physical, mathematical, or conceptual model.

Model	Type of Model
Computer simulation of the path of a tornado	
Ball-and-stick model of a molecule	
Animal behavior classification	

Copyright © by Holt, Rinehart and Winston; a Division of Houghton Mifflin Harcourt Publishing Company. All rights reserved.

SECTION 3 Scientific Models *continued*

Even though dinosaurs no longer exist, this computer-generated model can be used to show what they may have looked like.

LIMITS OF MODELS

Models are useful, but they are not perfect. For example, the figure above shows a model that gives scientists an idea of how a dinosaur looked. However, it cannot tell scientists how strong the dinosaur's jaws were. Scientists can build other models to conduct tests of dinosaur jaw strength. But without a live dinosaur, they can never know if their results are correct.

How Do Models Help Build Scientific Knowledge?

Models are often used to help explain scientific theories. In science, a **theory** is an explanation for many hypotheses and observations. A theory can explain why something happens and can also predict what will happen in the future. ☑

Scientists use models to help them look for new scientific information. The new information can support a theory or show that it is wrong. As scientists make new observations, new theories are developed to replace old theories that are shown to be wrong.

SCIENTIFIC LAWS

When a theory and its model correctly predict the results of many different experiments, a scientific law can be made. In science, a **law** is a summary of many experimental results and observations. It tells you how things work. ☑

A law is different from a theory. A law tells you only what happens, not why it happens. An example of a scientific law is the *law of gravity*. It states that objects will always fall toward the center of Earth. The law does not explain why objects will always fall toward the center of Earth. The law just states that they do.

Critical Thinking

4. Apply Concepts What kind of model could scientists use to find out a dinosaur's jaw strength?

_____

_____

_____

_____

_____

✓ **READING CHECK**

5. Define What is a scientific theory?

_____

_____

_____

✓ **READING CHECK**

6. Define What is a scientific law?

_____

_____

Copyright © by Holt, Rinehart and Winston; a Division of Houghton Mifflin Harcourt Publishing Company. All rights reserved.

Section 3 Review

GLE 0707.Inq.2, GLE 0707.Inq.3, GLE 0707.Inq.4, GLE 0707.Inq.5 TN

SECTION VOCABULARY

law a summary of many experimental results and observations; a law tells how things work	**theory** an explanation for some phenomenon that is based on observation, experimentation, and reasoning
model a pattern, plan, representation, or description designed to show the structure or workings of an object, system, or concept	

1. Identify What are three types of models used by scientists?

_____

2. Explain What are some ways that scientists use models?

_____

_____

_____

3. Identify A model of a molecule can help you imagine what a molecule looks like. What are two ways that this model is different from the object it represents?

_____

_____

4. Compare What is the difference between a scientific theory and a scientific law?

_____

_____

_____

5. Explain How can theories change over time?

_____

_____

_____

6. Describe Which kind of model would you use to represent a human heart? Explain the reason for your choice.

_____

_____

Copyright © by Holt, Rinehart and Winston; a Division of Houghton Mifflin Harcourt Publishing Company. All rights reserved.

CHAPTER 1 Science in Our World

SECTION 4 Science and Engineering

BEFORE YOU READ

After you read this section, you should be able to answer these questions:

• What is engineering?

• How does engineering help meet society's needs?

• How do engineers design new products?

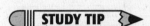
TN **Tennessee Science Standards**

GLE 0707.T/E.1
GLE 0707.T/E.2
GLE 0707.T/E.3
GLE 0707.T/E.4

What Is Technology?

Imagine that you have an assignment that you have to write. At one time, you would have had to type the assignment on a typewriter or write it by hand. Now you can use a computer to write the paper. These improvements are the result of scientists' and engineers' problem-solving efforts. These scientists and engineers created new technology to serve a need.

Technology refers to the products and processes that are designed to serve our needs. This is only part of the definition. Technology also refers to the tools and methods for creating these products and processes. A computer is an example of technology. The tools and processes used to make computers are also examples of technology. Technology applies to any product, process, or knowledge that is created to meet a need.

STUDY TIP

Clarify Concepts Take turns reading this section out loud with a partner. Stop to discuss ideas that seem confusing.

Computers have many different pieces of hardware that we can use in our everyday lives.

**Say It**

Discuss In a group, talk about some other possible hypotheses that the students could have come up with.

Copyright © by Holt, Rinehart and Winston; a Division of Houghton Mifflin Harcourt Publishing Company. All rights reserved.

How Does Science Relate to Technology?

Science is knowledge of the natural world. Engineering is closely related to science, but it is not the same. Engineers use science and mathematics to create new technologies that serve human needs. ☑

THE INTERNATIONAL SYSTEM OF UNITS

Engineering is the process of creating technology. When you think of an engineer, you might think of a person who designs bridges or skyscrapers. This is just one kind of engineer. There are also many other types of engineers who develop many different products.

Hybrid cars, computers, and disease-resistant corn were all developed by engineers. Engineers also designed the tools and processes needed to make these new products. For example, engineers not only created computers but also designed the machines used to make computers. The software used to run programs on a computer was also designed by engineers. ☑

Professional engineers have produced a lot of technology. However, you don't have to be an engineer to participate in engineering. Scientists, inventors, artists, and even students have also engineered new technologies. Anyone can follow the engineering design process to solve a problem or address a need.

> **READING CHECK**
> **1. Explain** How is engineering related to science and math?
>
> *Used computer*
> _____
> _____

> **READING CHECK**
> **2. Identify** What are three categories of technologies that engineers create?
>
> *Hybrid cars*
> *Computer*
> _____

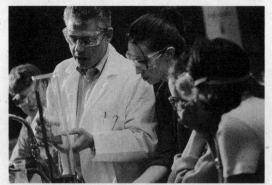

Professional engineers spend years learning the math and science they need to use to create new products and processes.

What Is the Engineering Design Process?

The **engineering design process** is similar to the scientific process. Like the scientific process, some steps may require repeating or modifying to fit different needs. Learning the process will help you understand how new technology is created.

Copyright © by Holt, Rinehart and Winston; a Division of Houghton Mifflin Harcourt Publishing Company. All rights reserved.

STEP 1: IDENTIFYING AND RESEARCHING A NEED

The first step in the engineering design process is finding a need or problem that engineers want to solve. Engineers define and write down the need or problem they are trying to solve.

For example, some people may be missing arms or legs because of a disease or an accident. To solve this problem, an engineer might attempt to create an artificial limb that would allow them to perform everyday activities. Engineers might research a person's needs, current prosthetic devices, and how human limbs work.

STEP 2: DEVELOPING POSSIBLE SOLUTIONS

Once the need has been identified and researched, the second step is to think about possible solutions. This can include brainstorming. When people *brainstorm*, they get together in a group and share ideas. Brainstorming usually leads to ideas that no one would have thought of on his or her own. In other cases, it takes more time and thought. Sometimes people can get an idea from a product that already exists. ☑

STEP 3: MAKING A PROTOTYPE

After the best idea is chosen, the third step is building a **prototype**. A prototype is a model of the product. Engineers test prototypes to see if their design works the way they expect it to.

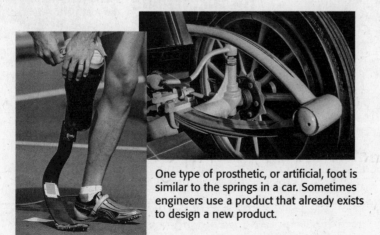

One type of prosthetic, or artificial, foot is similar to the springs in a car. Sometimes engineers use a product that already exists to design a new product.

Critical Thinking
3. Apply Concepts Suppose you were asked to design a new type of coat for researchers in Antarctica. What are two topics you might research to get started on your project?

brain storm

 **READING CHECK**

4. Define What is brainstorming?

when people brain storming they ge together

Copyright © by Holt, Rinehart and Winston; a Division of Houghton Mifflin Harcourt Publishing Company. All rights reserved.

STEP 4: TESTING AND EVALUATING

Then engineers try to find out whether the technology does the job it was designed to do. They test the prototype and evaluate how well it works.

Engineers make sure the cost of designing and producing the new product is worth its benefit. This is called a **cost-benefit analysis**. Sometimes engineers come up with a design that works very well but is too expensive to produce.

STEP 5: MODIFYING AND RETESTING THE SOLUTION

When a prototype is not successful, engineers follow the fifth step in the engineering design process. They either modify their prototype or try a new solution. It is important that engineers consider what they learned from the first prototype. They begin the design process again with their new knowledge and continue working on the problem.

Scientists and engineers also look for other possible uses for the new product. For example, computed tomography (CAT) scans have been used to see the internal structures of a body. Scientists are now using CAT scans to help model limbs to improve the fit of artificial limbs.

Math Focus

5. Calculate Engineers are working on a cost-benefit analysis for a new light bulb. The new bulb uses 40 watts of energy in an hour. Electricity costs $0.01 per watt. How much does it cost to run the bulb for a day?

Cat _____

_____

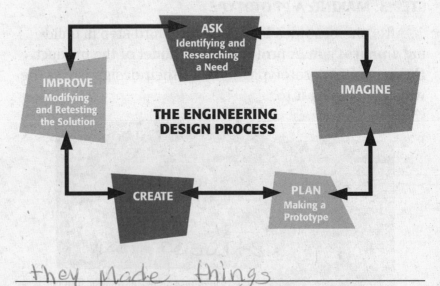

THE ENGINEERING DESIGN PROCESS

ASK — Identifying and Researching a Need

IMAGINE

IMPROVE — Modifying and Retesting the Solution

CREATE

PLAN — Making a Prototype

TAKE A LOOK

6. Identify Fill in the blanks in the engineering design process diagram.

they made things _____

_____

Copyright © by Holt, Rinehart and Winston; a Division of Houghton Mifflin Harcourt Publishing Company. All rights reserved.

SECTION 4 Science and Engineering *continued*

COMMUNICATION

Engineers need to share their successes, failures, and reasoning with others. They may explain and promote the technology to customers, or they may communicate with the public through news releases or advertisements. Engineers may also publish details of the design process in journals so other engineers can build on their work. ☑

HOW DOES TECHNOLOGY AFFECT SOCIETY?

Technology provides solutions for many types of social, political, and economic needs. For example, telephone companies have an economic need for less expensive telephone and radio towers. To solve this problem, electrical engineers have developed new materials for building more durable. ☑

City governments have a political need for information to improve police, firefighting, and medical services. To fulfill this need, computer engineers write software that makes collecting information from emergency calls easier.

INTENDED BENEFITS

Think about how refrigeration has affected the way we live. With the invention of the refrigerator, people are able to store and preserve food easily. Before the refrigerator, people had to rely on salt and canning to preserve food. Ice and cold food were luxuries during the summer.

Refrigerators fulfill their *intended benefit*—to provide fresh food to people throughout the year. An intended benefit is the purpose for which a technology is designed. An intended benefit is always positive.

> **READING CHECK**
>
> **7. Identify** List three ways in which engineers might share their results.
>
> _News_ _____
>
> _____
>
> _____

> **READING CHECK**
>
> **8. Identify** What are three types of needs that new technologies might serve?
>
> _radio_ _____
>
> _____
>
> _____

Critical Thinking

9. Apply Concepts What is one of the intended benefits of cell phones?

 # min _____

_____

Refrigeration allows people to eat a variety of foods throughout the year. This is the intended benefit of refrigeration.

Copyright © by Holt, Rinehart and Winston; a Division of Houghton Mifflin Harcourt Publishing Company. All rights reserved.

SECTION 4 Science and Engineering *continued*

TN **TENNESSEE STANDARDS CHECK**

GLE 0607.T/E.3 Compare the intended benefits with the unintended consequences of a new technology.

Word Help: technology tools, including electronic products

10. Apply Concepts Name one positive unintended consequence of automobiles. Name one negative unintended consequence.

unintended

consequences

_____

_____

_____

✓ *READING CHECK*

11. Identify How did engineers solve the problem of toxic materials in refrigerators?

Society

_____

_____

UNINTENDED CONSEQUENCES

Refrigeration provides the intended benefit of preserving food, but it has had both positive and negative unintended consequences. *Unintended consequences* are uses or results that engineers do not purposely include in the design of products. An unintended consequence can be positive. For example, the ability for people to have food from distant areas is a positive unintended consequence of refrigeration.

Some unintended consequences have a negative impact on society. For example, early refrigerators used toxic gases to help cool the air inside the refrigerator. Later, chlorofluorocarbons (CFCs) were introduced to eliminate the problems that earlier toxic gases caused. These CFCs hurt the environment. Old, discarded refrigerators become the unintended consequence of garbage in junkyards. ☑

These are just some of the effects of refrigeration and refrigerators on our society. Some effects are positive, while some are negative. Not all technologies have had such an impact on society, but many have. Think about how society has been changed by electricity, plastics, and automobiles.

Technologies can have unintended consequences, such as these refrigerators that contain toxic materials. Frequently, engineers must use the engineering design process to fix problems that come from unintended consequences.

What Is Bioengineering?

The engineering design process can even be applied to living things. **Bioengineering** is the use of engineering to help living things, such as humans and plants. Bioengineers and scientists study problems that occur in living organisms and their environments. They use their skills, knowledge, and technology to develop solutions to these problems. ☑

✓ *READING CHECK*

12. Identify What is bioengineering?

is the use of

engineer

_____

Copyright © by Holt, Rinehart and Winston; a Division of Houghton Mifflin Harcourt Publishing Company. All rights reserved.

SECTION 4 Science and Engineering *continued*

ASSISTIVE BIOENGINEERING

Some bioengineered technologies are classified as either **assistive** or **adaptive**. Assistive technologies are developed to help organisms without changing them. For example, contact lenses are an assistive technology that helps people see more clearly. Crutches and canes are other examples of the many assistive bioengineered products that can improve our lives.

ADAPTIVE BIOENGINEERING

Other bioengineered technologies are classified as adaptive. Unlike assistive technologies, adaptive technologies actually change the living organism. Adaptive bioengineering has been used for many exciting new technologies. Some of these technologies give scientists and engineers the ability to examine and change the material inside a cell (DNA). ☑

Scientists are looking at this technology to help prevent and cure some diseases. A similar adaptive biotechnology is the replacement of damaged cells with healthy cells. Bioengineers have even developed artificial skin that can help wounds heal faster.

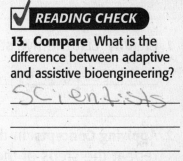

✓ **READING CHECK**

13. Compare What is the difference between adaptive and assistive bioengineering?

Scientists _____

_____

_____

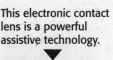

This electronic contact lens is a powerful assistive technology.
▼

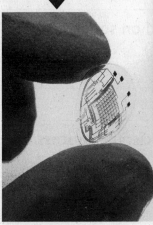

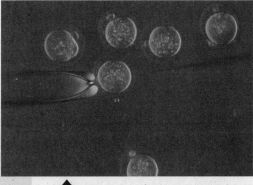

▲
Injecting a cell with genetic material is an example of adaptive bioengineering that helps people with certain diseases or conditions. One day this adaptive technology may prevent the need for the assistive technology of contact lenses.

Copyright © by Holt, Rinehart and Winston; a Division of Houghton Mifflin Harcourt Publishing Company. All rights reserved.

Section 4 Review

GLE 0707.T/E.1, GLE 0707.T/E.2, GLE 0707.T/E.3, GLE 0707.T/E.4 TN

SECTION VOCABULARY

adaptive bioengineering engineering that results in a product or process that changes living organisms

assistive bioengineering engineering that results in a product or process that helps living organisms but does not change them permanently

bioengineering the application of engineering to living things, such as humans and plants

cost-benefit analysis the process of determining whether the cost of doing something is worth the benefit provided

engineering the process of creating technology

engineering design process the process engineers use to develop a new technology

prototype a test model of a product

prosthetic an artificial device that replaces a missing part of the body

technology the products and processes that are designed to serve our needs

1. Identify List the five steps in the engineering design process.

the process the process enginners

2. Applying Concepts In the engineering design process, why would you need to repeat the steps of the process?

the products and processes

3. Consumer Focus Many people have poor vision. Name two bioengineered products that have been developed to deal with this need.

the process of determine

4. Apply Concepts What is one benefit that computers have had on society? What is one unintended negative consequence?

the application of energ

5. Compare How are the engineering design process and the scientific process alike?

engineering that results.

Copyright © by Holt, Rinehart and Winston; a Division of Houghton Mifflin Harcourt Publishing Company. All rights reserved.

CHAPTER 1 Science in Our World

SECTION 5 Tools, Measurement, and Safety

BEFORE YOU READ

After you read this section, you should be able to answer these questions:

• What tools do scientists use to make measurements?

• What is the International System of Units?

• How can you stay safe in science class?

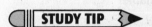

TN Tennessee Science Standards

GLE 0707.Inq.2
GLE 0707.Inq.3
GLE 0707.Inq.5

What Kinds of Tools Do Scientists Use?

Scientists use many tools in their experiments. A *tool* is anything that helps you do something. To do an experiment correctly, you need the right tools.

Some tools are used to take measurements. These include stopwatches, meter sticks, thermometers, balances, and spring scales. Making measurements is one way that scientists gather data.

STUDY TIP

Outline As you read, underline the main ideas in each paragraph. When you finish reading, make an outline of the section using the ideas you underlined.

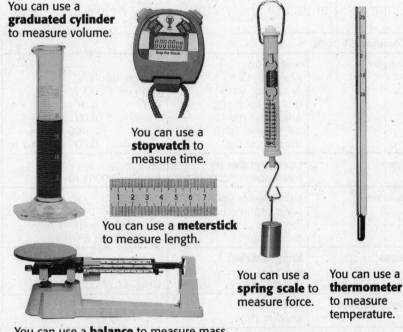

You can use a **graduated cylinder** to measure volume.

You can use a **stopwatch** to measure time.

You can use a **meterstick** to measure length.

You can use a **spring scale** to measure force.

You can use a **thermometer** to measure temperature.

You can use a **balance** to measure mass.

TAKE A LOOK
1. **Identify** What tool can you use to measure mass?

_____

Some tools help you analyze your data. A pencil and graph paper are simple tools you can use to graph data. A calculator can help you do calculations quickly. A graphing calculator can show your data in a graph. You can use a computer to display your data many different ways.

Copyright © by Holt, Rinehart and Winston; a Division of Houghton Mifflin Harcourt Publishing Company. All rights reserved.

How Do Scientists Take Measurements?

Many systems of measurement are used throughout the world. At one time in England, the standard for an inch was three grains of barley lined up. Other units of measurement were originally based on parts of the body, such as the foot. These systems did not work well because not every grain of barley or human foot is the same size. ☑

THE INTERNATIONAL SYSTEM OF UNITS

In the late 1700s, the French Academy of Sciences created a system of measurements that would be used by scientists throughout the world. Today, this system is called the *International System of Units*, which is abbreviated as SI. You might know it as the metric system. Using the same system of measurements makes it easier for scientists to share their observations and discoveries.

All the SI units are written in multiples of 10. This makes it easy to change from one unit to another. Some common SI units are shown in the table below. ☑

✔ READING CHECK

2. Explain Why were measurements based on grains or parts of the body unreliable?

_____

_____

_____

✔ READING CHECK

3. Describe Why is it easy to change from one SI unit to another?

_____

_____

Common SI Units		
Length	meter (m)	
	kilometer (km)	1 km = 1,000 m
	decimeter (dm)	1 dm = 0.1 m
	centimeter (cm)	1 cm = 0.01 m
	millimeter (mm)	1 mm = 0.001 m
	micrometer (μm)	1 μm = 0.000 001 m
	nanometer (nm)	1 nm = 0.000 000 001 m
Area	square meter (m²)	
	square centimeter (cm²)	$1 \ cm^2 = 0.0001 \ m^2$
Volume	cubic meter (m³)	
	cubic centimeter (cm³)	$1 \ cm^3 = 0.000 \ 001 \ m^3$
	liter (L)	$1 \ L = 1 \ dm^3 = 0.001 \ m^3$
	milliliter (mL)	$1 \ mL = 0.001 \ L = 1 \ cm^3$
Mass	kilogram (kg)	
	gram (g)	1 g = 0.001 kg
	milligram (mg)	1 mg = 0.000001 kg
Temperature	Kelvin (K)	0 K = −273 °C
	Celsius (°C)	0°C = 273 K

Math Focus

4. Calculate How many meters are in 50 kilometers?

_____

Copyright © by Holt, Rinehart and Winston; a Division of Houghton Mifflin Harcourt Publishing Company. All rights reserved.

SECTION 5 Tools, Measurement, and Safety *continued*

LENGTH

The **meter** (m) is the basic SI unit for length. An Olympic-sized swimming pool, for example, is 50 m long. Recall that all SI units are based on the number 10. Therefore, 1 millimeter (mm) is one 1/1,000th of a meter (m). That means that if 1 m is divided into 1,000 equal parts, each part is 1 mm.

AREA

Knowing the length of an object can help you calculate its area. **Area** is a measure of the surface of an object. Area is based on two measurements: length and width. Therefore, differently shaped objects can have the same surface area.

The following equation shows how the area of a square or rectangle is calculated:

$$area = length \times width$$

The units for area are called square units, such as square kilometers (km^2) and square meters (m^2).

MASS

Mass is a measure of how much matter is in an object. The *kilogram* (kg) is the basic SI unit for mass. The kilogram is used to describe the mass of large objects, such as cars. *Grams* (g) are used to describe the mass of smaller objects, such as apples. A medium-sized apple, for example, has a mass of about 100 g. One thousand grams equals one kilogram. ☑

VOLUME

Volume is the amount of space that something takes up. Volume is based on three measurements: length, width, and height. The following equation shows how the volume of a box is calculated:

$$volume = length \times width \times height$$

The volume of a large solid object is measured in cubic meters (m^3). A smaller object is measured in cubic centimeters (cm^3). The volume of a liquid, however, is measured in *liters* (L). One cubic meter equals 1,000 L. One cubic centimeter equals one milliliter (mL).

Critical Thinking

5. Explain How can two differently shaped objects have the same area?

_____

_____

_____

_____

READING CHECK

6. Define What is mass?

_____

_____

Math Focus

7. Calculate What is the volume of an object that is 3 cm long, 4 cm wide, and 6 cm high? Show your work.

_____

_____

Copyright © by Holt, Rinehart and Winston; a Division of Houghton Mifflin Harcourt Publishing Company. All rights reserved.

SECTION 5 Tools, Measurement, and Safety *continued*

Critical Thinking

8. Explain Why is it necessary to use a different method to measure the volume of an irregularly shaped object?

_____

_____

_____

_____

_____

Many objects have an irregular shape. The figure above shows how to measure the volume of an irregularly shaped object, such as a rock. Notice in the left figure that the volume of the water in the graduated cylinder is 70 mL. Notice in the right figure that adding the rock to the water increases the volume to 80 mL. This increase of 10 mL represents the volume of the rock. Because 1 mL = 1 cm³, the volume of the rock is 10 cm³.

DENSITY

If you know the mass and volume of an object, you can calculate its density. **Density** is a measure of how much matter is in a given volume. You can find the density of an object using the following equation: ☑

$$density = \frac{mass}{volume}$$

If mass is measured in grams, and volume is measured in milliliters, then density is expressed as grams per milliliter (g/mL). Because 1 mL = 1 cm³, density can also be expressed as g/cm³.

The following table lists the densities of some common substances.

✓ READING CHECK

9. Define What is density?

_____

_____

Substance	Density (g/cm³)
Water	0.997
Table salt	2.16
Iron	7.86
Silver	10.5
Gold	19.3

Copyright © by Holt, Rinehart and Winston; a Division of Houghton Mifflin Harcourt Publishing Company. All rights reserved.

TEMPERATURE

The **temperature** of a substance is the measurement of how hot or cold it is. You are probably most familiar with degrees Fahrenheit (°F), because that is the unit of temperature we use in the United States. However, the kelvin (K, without a degree sign) is the SI unit for temperature. Another unit for temperature, degrees Celsius (°C), is the one you will probably use in the laboratory. ☑

The figure below shows the relationship between the Celsius scale and the Fahrenheit scale.

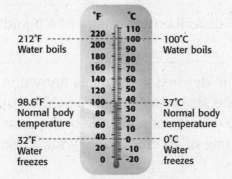

212°F Water boils — 100°C Water boils
98.6°F Normal body temperature — 37°C Normal body temperature
32°F Water freezes — 0°C Water freezes

READING CHECK

10. Define What is the SI unit for temperature?

_____

TAKE A LOOK

11. Calculate The equation to convert from degrees Celsius to Kelvins is $K = °C + 273$. What temperature does water boil on the Kelvin scale?

_____

How Can You Stay Safe in Science Class?

Science is exciting and fun, but it can be dangerous. There are many safety rules that you must follow whenever you do an experiment.
• Always listen to your teacher's instructions.
• Don't take shortcuts.
• Read lab directions carefully.
• Pay attention to safety information. ☑
The safety symbols in the figure below are important. Learn them so that you and others will be safe in the lab.

 Eye Protection

 Clothing Protection

 Hand Safety

 Heating Safety

 Electric Safety

 Sharp Object

 Chemical Safety

 Animal Safety

Plant Safety

READING CHECK

12. Identify List four general safety rules.

_____
_____
_____
_____
_____

Copyright © by Holt, Rinehart and Winston; a Division of Houghton Mifflin Harcourt Publishing Company. All rights reserved.

Section 5 Review

GLE 0707.Inq.2, GLE 0707.Inq.3, GLE 0707.Inq.5 🔲TN

SECTION VOCABULARY

area a measure of the size of a surface or a region	**meter** the basic unit of length in the SI (symbol, m)
density the ratio of the mass of a substance to the volume of a substance	**temperature** a measure of how hot (or cold) something is
mass a measure of the amount of matter in an object	**volume** a measure of the size of a body or region in three-dimensional space

1. List What two quantities are needed to calculate area?

_____

2. Identify What unit is usually used to describe the volume of a liquid?

_____

3. Calculate Osmium and iridium are the densest substances known. A piece of either osmium or iridium that was the size of a football would be too heavy to lift because of their density. What is the density of a sample of osmium or iridium with a mass of 226 grams and a volume of 10 cm³? Show your work.

4. Identify Which of the safety symbols shown below would you see in a lab procedure that asks you to pour acid into a beaker? Explain your choice(s).

_____

_____

_____

Copyright © by Holt, Rinehart and Winston; a Division of Houghton Mifflin Harcourt Publishing Company. All rights reserved.

SECTION
1
The Diversity of Cells

BEFORE YOU READ

After you read this section, you should be able to answer these questions:

• What is a cell?

• What do all cells have in common?

• What are the two kinds of cells?

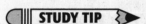

TN Tennessee Science Standards

GLE 0707.Inq.2
GLE 0707.Inq.5
GLE 0707.1.1

What Is a Cell?

Most cells are so small that they cannot be seen by the naked eye. So how did scientists find cells? By accident! The first person to see cells wasn't even looking for them.

A **cell** is the smallest unit that can perform all the functions necessary for life. All living things are made of cells. Some living things are made of only one cell. Others are made of millions of cells. ☑

Robert Hooke was the first person to describe cells. In 1665, he built a microscope to look at tiny objects. One day he looked at a piece of cork. Cork is found in the bark of cork trees. Hooke thought the cork looked like it was made of little boxes. He named these boxes *cells*, which means "little rooms" in Latin.

STUDY TIP

Organize As you read this section, make lists of things that are found in prokaryotic cells, things that are found in eukaryotic cells, and things that are found in both kinds of cells.

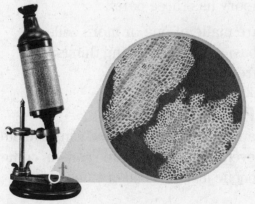

Hooke discovered cells using this microscope. Hooke's drawing of cork cells is shown to the right of his microscope.

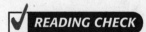

READING CHECK

1. Identify What is the basic unit of all living things?

_____

_____

In the late 1600s, a Dutch merchant named Anton van Leeuwenhoek studied many different kinds of cells. He made his own microscopes. With them, he looked at tiny pond organisms called protists. He also looked at blood cells, yeasts, and bacteria.

Copyright © by Holt, Rinehart and Winston; a Division of Houghton Mifflin Harcourt Publishing Company. All rights reserved.

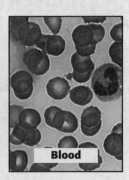

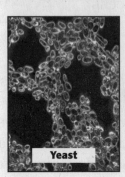

Leeuwenhoek looked at many different kinds of cells with his microscope. He was the first person to see bacteria. Bacterial cells are usually much smaller than most other types of cells.

TAKE A LOOK
2. Identify Which of these cells is probably the smallest? Explain your answer.

_____

_____

_____

What Is the Cell Theory?

Since Hooke first saw cork cells, many discoveries have been made about cells. Cells from different organisms can be very different from one another. Even cells from different parts of the same organism can be very different. However, all cells have several important things in common. These observations are known as the *cell theory*. The cell theory has three parts:

1. All organisms are made of one or more cells.
2. The cell is the basic unit of all living things.
3. All cells come from existing cells.

What Are the Parts of a Cell?

Cells come in many shapes and sizes and can have different functions. However, all cells have three parts in common: a cell membrane, genetic material, and organelles. ☑

CELL MEMBRANE

✓ **READING CHECK**

3. List What three parts do all cells have in common?

_____

_____

_____

All cells are surrounded by a cell membrane. The **cell membrane** is a layer that covers and protects the cell. The membrane separates the cell from its surroundings. The cell membrane also controls all material going in and out of the cell. Inside the cell is a fluid called *cytoplasm*.

Copyright © by Holt, Rinehart and Winston; a Division of Houghton Mifflin Harcourt Publishing Company. All rights reserved.

GENETIC MATERIAL

All cells contain DNA (deoxyribonucleic acid) at some point in their lives. *DNA* is the genetic material that carries information needed to make proteins, new cells, and new organisms. DNA is passed from parent cells to new cells and it controls the activities of the cell.

The DNA in some cells is found inside a structure called the **nucleus**. Most of your cells have a nucleus.

ORGANELLES

Cells have structures called **organelles** that do different jobs for the cell. Most organelles have a membrane covering them. Different types of cells can have different organelles.

Parts of a Cell

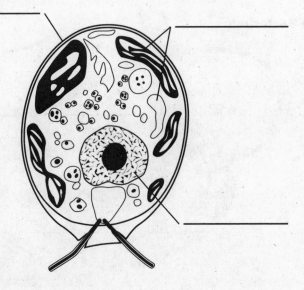

TN TENNESSEE STANDARDS CHECK

GLE 0707.1.1 Make observations and describe the <u>structure</u> and <u>function</u> of organelles found in plant and animal cells.

Word Help: <u>structure</u> the arrangement of the parts of a whole

Word Help: <u>function</u> use or purpose

4. Identify Which organelle contains DNA in many types of cells?

_____

_____

_____

_____

_____

TAKE A LOOK
5. Identify Use the following words to fill in the blank labels on the figure: DNA, cell membrane, organelles.

What Are the Two Kinds of Cells?

There are two basic kinds of cells—cells with a nucleus and cells without a nucleus. Those without a nucleus are called *prokaryotic cells*. Those with a nucleus are called *eukaryotic cells*. ☑

What Are Prokaryotes?

A **prokaryote** is an organism made of one cell that does not have a nucleus or other organelles covered by a membrane. Prokaryotes are made of prokaryotic cells. There are two types of prokaryotes: bacteria and archaea.

✓ READING CHECK

6. Compare What is one way prokaryotic and eukaryotic cells differ?

_____

_____

_____

Copyright © by Holt, Rinehart and Winston; a Division of Houghton Mifflin Harcourt Publishing Company. All rights reserved.

SECTION 1 The Diversity of Cells *continued*

BACTERIA

The most common prokaryotes are bacteria (singular, *bacterium*). Bacteria are the smallest known cells. These tiny organisms live almost everywhere. Some bacteria live in the soil and water. Others live on or inside other organisms. You have bacteria living on your skin and teeth and in your digestive system. The following are some characteristics of bacteria:

- no nucleus
- circular DNA shaped like a twisted rubber band
- no membrane-covered (or *membrane-bound*) organelles
- a cell wall outside the cell membrane
- a *flagellum* (plural, *flagella*), a tail-like structure that some bacteria use to help them move

Critical Thinking

7. Make Inferences Why do you think bacteria can live in your digestive system without making you sick?

_____

_____

_____

A Bacterium

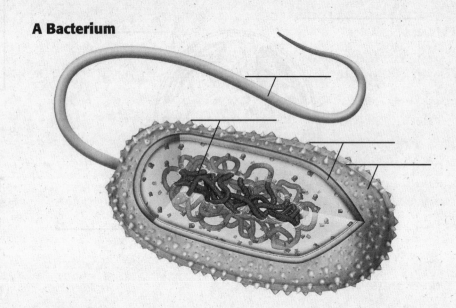

TAKE A LOOK
8. Identify Label the parts of the bacterium using the following terms: DNA, flagellum, cell membrane, cell wall.

ARCHAEA

Archaea (singular, *archaeon*) and bacteria share the following characteristics:

- no nucleus
- no membrane-bound organelles
- circular DNA
- a cell wall

Copyright © by Holt, Rinehart and Winston; a Division of Houghton Mifflin Harcourt Publishing Company. All rights reserved.

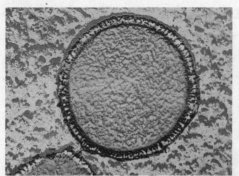

This photograph was taken with an electron microscope. This archaeon lives in volcanic vents deep in the ocean. Temperatures at these vents are very high. Most other living things could not survive there.

Archaea have some other features that no other cells have. For example, the cell wall and cell membrane of archaea are made of different substances from those of bacteria. Some archaea live in places where no other organisms could live. For example, some can live in the boiling water of hot springs. Others can live in toxic places such as volcanic vents filled with sulfur. Still others can live in very salty water in places such as the Dead Sea. ☑

What Are Eukaryotes?

Eukaryotic cells are the largest cells. They are about 10 times larger than bacteria cells. However, you still need a microscope to see most eukaryotic cells.

Eukaryotes are organisms made of eukaryotic cells. These organisms can have one cell or many cells. Yeast, which makes bread rise, is an example of a eukaryote with one cell. Multicellular organisms, or those made of many cells, include plants and animals.

Unlike prokaryotic cells, eukaryotic cells have a nucleus that holds their DNA. Eukaryotic cells also have membrane-bound organelles. ☑

☑ READING CHECK

9. Compare Name two ways that archaea differ from bacteria.

_____

_____

_____

_____

☑ READING CHECK

10. Identify Name two things eukaryotic cells have that prokaryotic cells do not.

_____

_____

Eukaryotic Cell

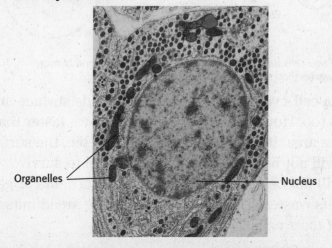

Organelles ———— | ———— Nucleus

Copyright © by Holt, Rinehart and Winston; a Division of Houghton Mifflin Harcourt Publishing Company. All rights reserved.

Organelles in a Typical Eukaryotic Cell

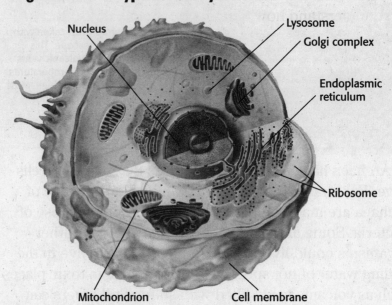

TAKE A LOOK

11. Identify Where is the genetic material found in this cell?

_____

_____

Critical Thinking

12. Apply Concepts The yolk of a chicken egg is a very large cell. Unlike most cells, egg yolks do not have to take in any nutrients. Why does this allow the cell to be so big?

_____

_____

_____

_____

_____

_____

Why Are Cells So Small?

Your body is made of trillions of cells. Most cells are so small you need a microscope to see them. More than 50 human cells can fit on the dot of this letter *i*. However, some cells are big. For example, the yolk of a chicken egg is one big cell! Why, then, are most cells small?

Cells take in food and get rid of waste through their outer surfaces. As a cell gets larger, it needs more food to survive. It also produces more waste. This means that more materials have to pass through the surface of a large cell than a small cell.

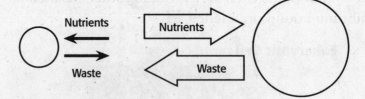

Large cells have to take in more nutrients and get rid of more wastes than small cells.

As a cell's volume increases, its outside surface area grows too. However, volume always grows faster than surface area. If the cell volume gets too big, the surface area will not be large enough for the cell to survive. The cell will not be able to take in enough nutrients or get rid of all its wastes. This means that surface area limits the size of most cells.

Copyright © by Holt, Rinehart and Winston; a Division of Houghton Mifflin Harcourt Publishing Company. All rights reserved.

SECTION 1 The Diversity of Cells *continued*

SURFACE AREA AND VOLUME OF CELLS

To understand how surface area limits the size of a cell, study the figures below. Imagine that the cubes are cells. You can calculate the surface areas and volumes of the cells using these equations:

volume of cube = side × side × side

surface area of cube = number of sides × area of side

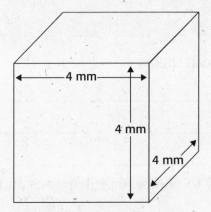

The volume of this cell is 64 mm³. Its surface area is 96 mm².

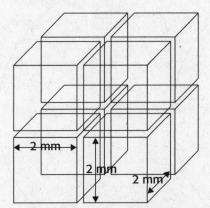

When the large cell is broken up into 8 smaller cells, the total volume stays the same. However, all of the small cells together have more surface area than the large cell. The total surface area of the small cells is 192 mm².

Math Focus
13. Calculate Ratios
Scientists say that most cells are small because of the surface area-to-volume ratio. What is this ratio for the large cell?

_____
_____

TAKE A LOOK
14. Compare Which cell has a greater surface area compared to its volume—the large cell or one of the smaller cells?

_____
_____

The large cell takes in and gets rid of the same amount of material as all of the smaller cells. However, the large cell does not have as much surface area as the smaller cells. Therefore, it cannot take in nutrients or get rid of wastes as easily as each of the smaller cells.

Copyright © by Holt, Rinehart and Winston; a Division of Houghton Mifflin Harcourt Publishing Company. All rights reserved.

Name _____ Class _____ Date _____

Section 1 Review

GLE 0707.Inq.2, GLE 0707.Inq.5, GLE 0707.1.1   **TN**

SECTION VOCABULARY

cell in biology, the smallest unit that can perform all life processes; cells are covered by a membrane and have DNA and cytoplasm

cell membrane a phospholipid layer that covers a cell's surface; acts as a barrier between the inside of a cell and the cell's environment

eukaryote an organism made up of cells that have a nucleus enclosed by a membrane; eukaryotes include animals, plants, and fungi, but not archaea or bacteria

nucleus in a eukaryotic cell, a membrane-bound organelle that contains the cell's DNA and that has a role in processes such as growth, metabolism, and reproduction

organelle one of the small bodies in a cell's cytoplasm that are specialized to perform a specific function

prokaryote an organism that consists of a single cell that does not have a nucleus

1. Identify What are the three parts of the cell theory?

_____

_____

_____

2. Compare Fill in the Venn Diagram below to compare prokaryotes and eukaryotes. Be sure to label the circles.

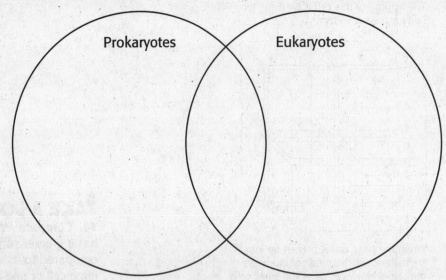

Prokaryotes   Eukaryotes

3. Apply Concepts You have just discovered a new organism. It has only one cell and was found on the ocean floor, at a vent of boiling hot water. The organism has a cell wall but no nucleus. Explain how you would classify this organism.

_____

_____

_____

Copyright © by Holt, Rinehart and Winston; a Division of Houghton Mifflin Harcourt Publishing Company. All rights reserved.

CHAPTER 2 Cells: The Basic Units of Life
SECTION 2 Eukaryotic Cells

TN Tennessee Science Standards
GLE 0707.Inq.5
GLE 0707.1.1
GLE 0707.3.1

BEFORE YOU READ

After you read this section, you should be able to answer these questions:

• What are the parts of a eukaryotic cell?

• What is the function of each part of a eukaryotic cell?

What Are the Parts of a Eukaryotic Cell?

Plant cells and animal cells are two types of eukaryotic cells. A eukaryotic cell has many parts that help the cell stay alive.

STUDY TIP

Organize As you read this section, make a chart comparing plant cells and animal cells.

CELL WALL

All plant cells have a cell wall. The **cell wall** is a stiff structure that supports the cell and surrounds the cell membrane. The cell wall of a plant cell is made of a type of sugar called cellulose.

Fungi (singular *fungus*), such as yeasts and mushrooms, also have cell walls. The cell walls of fungi are made of a sugar called *chitin*. Prokaryotic cells such as bacteria and archaea also have cell walls. ☑

READING CHECK

1. Identify Name two kinds of eukaryotes that have a cell wall.

_____

_____

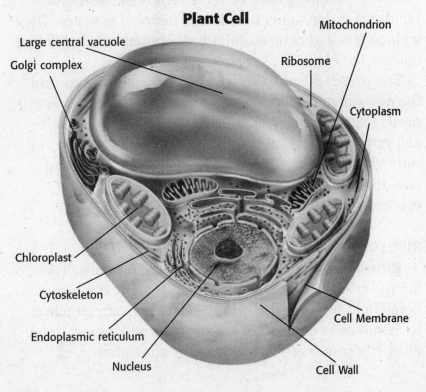

Plant Cell

Large central vacuole
Golgi complex
Mitochondrion
Ribosome
Cytoplasm
Chloroplast
Cytoskeleton
Endoplasmic reticulum
Nucleus
Cell Membrane
Cell Wall

TAKE A LOOK

2. Identify Describe where the cell wall is located.

_____

_____

Copyright © by Holt, Rinehart and Winston; a Division of Houghton Mifflin Harcourt Publishing Company. All rights reserved.

SECTION 2 Eukaryotic Cells *continued*

TAKE A LOOK

3. Compare Compare the pictures of an animal cell and a plant cell. Name three structures found in both.

_____

_____

_____

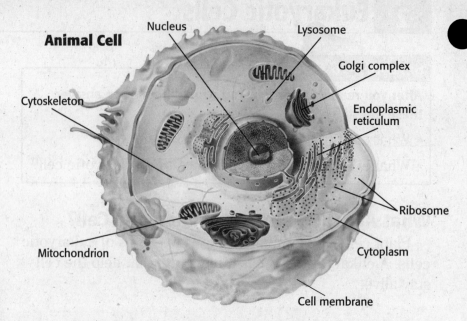

Animal Cell

Nucleus
Lysosome
Golgi complex
Cytoskeleton
Endoplasmic reticulum
Ribosome
Mitochondrion
Cytoplasm
Cell membrane

TN TENNESSEE STANDARDS CHECK

GLE 0707.1.1 Make observations and describe the <u>structure</u> and <u>function</u> of organelles found in plant and animal cells.

Word Help: <u>structure</u> the arrangement of the parts of a whole

Word Help: <u>function</u> use or purpose

4. Explain What is the main function of the cell membrane?

_____

_____

_____

_____

_____

_____

5. Compare How are ribosomes different from other organelles?

_____

_____

CELL MEMBRANE

All cells have a cell membrane. The cell membrane is a protective barrier that surrounds the cell. It separates the cell from the outside environment. In cells that have a cell wall, the cell membrane is found just inside the cell wall.

The cell membrane is made of different materials. It contains proteins, lipids, and phospholipids. Proteins are molecules made by the cell for a variety of functions. Lipids are compounds that do not dissolve in water. They include fats and cholesterol. Phospholipids are lipids that contain the element phosphorous.

The proteins and lipids in the cell membrane control the movement of materials into and out of the cell. A cell needs materials such as nutrients and water to survive and grow. Nutrients and wastes go in and out of the cell through the proteins in the cell membrane. Water can pass through the cell membrane without the help of proteins.

RIBOSOMES

Ribosomes are organelles that make proteins. They are the smallest organelles. A cell has many ribosomes. Some float freely in the cytoplasm. Others are attached to membranes or to other organelles. Unlike most organelles, ribosomes are not covered by a membrane. ☑

Copyright © by Holt, Rinehart and Winston; a Division of Houghton Mifflin Harcourt Publishing Company. All rights reserved.

SECTION 2 Eukaryotic Cells *continued*

Ribosome
This organelle is where amino acids are hooked together to make proteins.

NUCLEUS

The nucleus is a large organelle in a eukaryotic cell. It contains the cell's genetic material, or DNA. DNA has the instructions that tell a cell how to make proteins.

The nucleus is covered by two membranes. Materials pass through pores in the double membrane. The nucleus of many cells has a dark area called the *nucleolus*.

The Nucleus

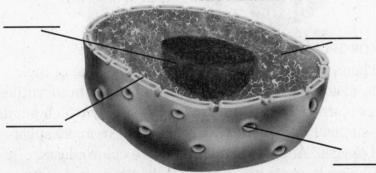

TAKE A LOOK
6. Identify Label the diagram of a nucleus using these terms: pore, DNA, nucleolus, double membrane.

ENDOPLASMIC RETICULUM

Many chemical reactions take place in the cell. Many of these reactions happen on or inside the endoplasmic reticulum. The **endoplasmic reticulum** (ER) is a system of membranes with many folds in which proteins, lipids, and other materials are made.

The ER is also part of the cell's delivery system. Its folds have many tubes and passageways. Materials move through the ER to other parts of the cell.

There are two types of ER: rough and smooth. Smooth ER makes lipids and helps break down materials that could damage the cell. Rough ER has ribosomes attached to it. The ribosomes make proteins. The proteins are then delivered to other parts of the cell by the ER. ☑

READING CHECK
7. Compare What is the difference between smooth ER and rough ER?

_____

_____

_____

_____

_____

Endoplasmic reticulum
This organelle makes lipids, breaks down drugs and other substances, and packages proteins for the Golgi complex.

Copyright © by Holt, Rinehart and Winston; a Division of Houghton Mifflin Harcourt Publishing Company. All rights reserved.

MITOCHONDRIA

A **mitochondrion** (plural, *mitochondria*) is the organelle in which sugar is broken down to make energy. It is the main power source for a cell.

A mitochondrion is covered by two membranes. Most of a cell's energy is made in the inside membrane. Energy released by mitochondria is stored in a molecule called ATP. The cell uses ATP to do work.

Mitochondria are about the same size as some bacteria. Like bacteria, mitochondria have their own DNA. The DNA in mitochondria is different from the cell's DNA. ☑

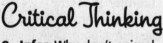

8. Compare How are mitochondria like bacteria?

_____

_____

Mitochondrion
This organelle breaks down food molecules to make ATP.

CHLOROPLASTS

Plants and algae have chloroplasts in some of their cells. *Chloroplasts* are organelles in which photosynthesis takes place. *Photosynthesis* is a process by which plants use sunlight, carbon dioxide, and water to make sugar and oxygen. Animal cells do not have chloroplasts.

Chloroplasts are green because they contain a green molecule called *chlorophyll*. Chlorophyll traps the energy of sunlight. Mitochondria then use the sugar made in photosynthesis to make ATP.

Critical Thinking

9. Infer Why don't animal cells need chloroplasts?

_____

_____

_____

Chloroplast
This organelle uses the energy of sunlight to make food.

CYTOSKELETON

The cytoskeleton is a web of proteins inside the cell. It acts as both a skeleton and a muscle. The cytoskeleton helps the cell keep its shape. It also helps some cells, such as bacteria, to move.

VESICLES

A **vesicle** is a small sac that surrounds material to be moved. The vesicle moves material to other areas of the cell or into or out of the cell. All eukaryotic cells have vesicles.

Copyright © by Holt, Rinehart and Winston; a Division of Houghton Mifflin Harcourt Publishing Company. All rights reserved.

GOLGI COMPLEX

The **Golgi complex** is the organelle that packages and distributes proteins. It is the "post office" of the cell. The Golgi complex looks like the smooth ER.

The ER delivers lipids and proteins to the Golgi complex. The Golgi complex can change the lipids and proteins to do different jobs. The final products are then enclosed in a piece of the Golgi complex's membrane. This membrane pinches off to form a vesicle. The vesicle transports the materials to other parts of the cell or out of the cell. ☑

Golgi complex
This organelle processes and transports proteins and other materials out of cell.

✓ **READING CHECK**

10. Define What is the function of the Golgi complex?

_____

_____

_____

LYSOSOMES

Lysosomes are organelles that contain digestive enzymes. The enzymes destroy worn-out or damaged organelles, wastes, and invading particles.

Lysosomes are found mainly in animal cells. The cell wraps itself around a particle and encloses it in a vesicle. Lysosomes bump into the vesicle and pour enzymes into it. The enzymes break down the particles inside the vesicle. Without lysosomes, old or dangerous materials could build up and damage or kill the cell.

Lysosome
This organelle digests food particles, wastes, cell parts, and foreign invaders.

VACUOLES

A vacuole is a vesicle. In plant and fungal cells, some vacuoles act like lysosomes. They contain enzymes that help a cell digest particles. The large central vacuole in plant cells stores water and other liquids. Large vacuoles full of water help support the cell. Some plants wilt when their vacuoles lose water. ☑

✓ **READING CHECK**

11. Identify Vacuoles are found in what types of eukaryotic cells?

_____

_____

_____

Large central vacuole
This organelle stores water and other materials.

Copyright © by Holt, Rinehart and Winston; a Division of Houghton Mifflin Harcourt Publishing Company. All rights reserved.

Section 2 Review

GLE 0707.Inq.5, GLE 0707.1.1, GLE 0707.3.1 **TN**

SECTION VOCABULARY

cell wall a rigid structure that surrounds the cell membrane and provides support to the cell	**lysosome** a cell organelle that contains digestive enzymes
endoplasmic reticulum a system of membranes that is found in a cell's cytoplasm and that assists in the production, processing, and transport of proteins and in the production of lipids	**mitochondrion** in eukaryotic cells, the cell organelle that is surrounded by two membranes and that is the site of cellular respiration
Golgi complex cell organelle that helps make and package materials to be transported out of the cell	**ribosome** cell organelle composed of RNA and protein; the site of protein synthesis
	vesicle a small cavity or sac that contains materials in a eukaryotic cell

1. Compare Name three parts of a plant cell that are not found in an animal cell.

_____

_____

2. Explain How does a cell get water and nutrients?

_____

_____

_____

3. Explain What would happen to an animal cell if it had no lysosomes?

_____

_____

_____

4. Apply Concepts Which kind of cell in the human body do you think would have more mitochondria—a muscle cell or a skin cell? Explain.

_____

_____

_____

5. List What are two functions of the cytoskeleton?

_____

_____

Copyright © by Holt, Rinehart and Winston; a Division of Houghton Mifflin Harcourt Publishing Company. All rights reserved.

CHAPTER 2 Cells: The Basic Units of Life

SECTION
3 **The Organization of Living Things**

TN Tennessee Science Standards
GLE 0707.Inq.5
GLE 0707.1.2

BEFORE YOU READ

After you read this section, you should be able to answer these questions:

• What are the advantages of being multicellular?

• What are the four levels of organization in living things?

• How are structure and function related in an organism?

What Is an Organism?

Anything that can perform life processes by itself is an **organism**. An organism made of a single cell is called a *unicellular organism*. An organism made of many cells is a multicellular organism. The cells in a multicellular organism depend on each other for the organism to survive. ☑

What Are the Benefits of Having Many Cells?

Some organisms exist as one cell. Others can be made of trillions of cells. A *multicellular organism* is an organism made of many cells.

There are three benefits of being multicellular: larger size, longer life, and specialization of cells.

LARGER SIZE

Most multicellular organisms are bigger than one-celled organisms. In general, a larger organism, such as an elephant, has few predators. ☑

LONGER LIFE

A multicellular organism usually lives longer than a one-celled organism. A one-celled organism is limited to the life span of its one cell. The life span of a multicellular organism, however, is not limited to the life span of any one of its cells.

SPECIALIZATION

In a multicellular organism, each type of cell has a particular job. Each cell does not have to do everything the organism needs. Specialization makes the organism more efficient.

STUDY TIP

Outline As you read, make an outline of this section. Use the heading questions from the section in your outline.

✓ READING CHECK

1. Define What is an organism?

_____

_____

_____

✓ READING CHECK

2. Identify Name one way that being large can benefit an organism.

_____

_____

_____

Copyright © by Holt, Rinehart and Winston; a Division of Houghton Mifflin Harcourt Publishing Company. All rights reserved.

SECTION 3 The Organization of Living Things *continued*

TN TENNESSEE STANDARDS CHECK

GLE 0707.1.2 Summarize how the different levels of organization are integrated within living systems.

3. List What are the four levels of organization for an organism?

_____

_____

_____

_____

What Are the Four Levels of Organization of Living Things?

Multicellular organisms have four levels of organization:

Cell

Tissue

Cells form tissues.

Organ

Tissues form organs.

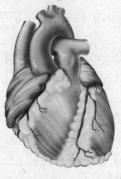

Organ system

Organs form organ systems.

Organ systems form organisms such as you.

TAKE A LOOK
4. Explain Are the cells that make up heart tissue prokaryotic or eukaryotic? How do you know?

_____

_____

_____

_____

Copyright © by Holt, Rinehart and Winston; a Division of Houghton Mifflin Harcourt Publishing Company. All rights reserved.

CELLS WORK TOGETHER AS TISSUES

A **tissue** is a group of cells that work together to perform a specific job. Heart muscle tissue, for example, is made of many heart muscle cells.

TISSUES WORK TOGETHER AS ORGANS

A structure made of two or more tissues that work together to do a certain job is called an **organ**. Your heart, for example, is an organ made of different tissues. The heart has muscle tissues and nerve tissues that work together.

ORGANS WORK TOGETHER AS ORGAN SYSTEMS

A group of organs working together to do a job is called an **organ system**. An example of an organ system is your digestive system. Organ systems depend on each other to help the organism function. For example, the digestive system depends on the cardiovascular and respiratory systems for oxygen.

HOW DOES STRUCTURE RELATE TO FUNCTION?

In an organism, the structure and function of part are related. **Function** is the job the part does. **Structure** is the arrangement of parts in an organism. It includes the shape of a part or the material the part is made of.

Critical Thinking

5. Apply Concepts Do prokaryotes have tissues? Explain.

_____

_____

_____

_____

_____

Say It

Name With a partner, name as many of the organs in the human body as you can.

Oxygen-poor blood

The function of the lungs is to bring oxygen to the body and get rid of carbon dioxide. The structure of the lungs helps them to perform their function.

Oxygen-rich blood

The lungs contain tiny, spongy sacs that blood can flow through. Carbon dioxide moves out of the blood and into the sacs. Oxygen flows from the sacs into the blood. If the lungs didn't have this structure, it would be hard for them to perform their function.

Blood vessels

Copyright © by Holt, Rinehart and Winston; a Division of Houghton Mifflin Harcourt Publishing Company. All rights reserved.

Section 3 Review

GLE 0707.Inq.5, GLE 0707.1.2 ◢TN◣

SECTION VOCABULARY

function the special, normal, or proper activity of an organ or part	**organism** a living thing; anything that can carry out life processes independently
organ a collection of tissues that carry out a specialized function of the body	**structure** the arrangement of parts in an organism
organ system a group of organs that work together to perform body functions	**tissue** a group of similar cells that perform a common function

1. List What are three benefits of being multicellular?

_____

_____

_____

2. Apply Concepts Could an organism have organs but no tissues? Explain.

_____

_____

3. Compare How are structure and function different?

_____

_____

_____

4. Explain What does "specialization of cells" mean?

_____

_____

5. Apply Concepts Why couldn't your heart have only cardiac tissue?

_____

_____

6. Explain Why do multicellular organisms generally live longer than unicellular organisms?

_____

_____

Copyright © by Holt, Rinehart and Winston; a Division of Houghton Mifflin Harcourt Publishing Company. All rights reserved.

SECTION
1 Exchange with the Environment

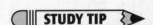
TN Tennessee Science Standards
GLE 0707.Inq.5
GLE 0707.1.5

BEFORE YOU READ

After you read this section, you should be able to answer these questions:

• How do cells take in food and get rid of wastes?

• What is diffusion?

Where Do Cells Get the Materials They Need?

What would happen to a factory if its power were shut off or its supply of materials never arrived? What would happen if the factory couldn't get rid of its garbage? Like a factory, an organism must be able to get energy and raw materials and get rid of wastes. These jobs are done by an organism's cells. Materials move in and out of the cell across the cell membrane. Many materials, such as water and oxygen, can cross the membrane by diffusion.

What Is Diffusion?

The figure below shows what happens when dye is placed on top of a layer of gelatin. Over time, the dye mixes with the gelatin. Why does this happen?

Everything, including the gelatin and the dye, is made of tiny moving particles. Particles tend to move from places where they are crowded to places where they are less crowded. When there are many of one type of particle, this is a high concentration. When there are fewer of one kind of particle, this is a low concentration. The movement from areas of high concentration to areas of low concentration is called **diffusion**. ☑

STUDY TIP

Compare As you read, make a chart comparing diffusion and osmosis. In your chart, show how they are similar and how they are different.

✓ **READING CHECK**

1. Define What is diffusion?

_____

_____

_____

_____

At first, the dye and the gelatin are separate from each other.

Dye —

Gelatin —

After a while, the particles in the dye move into the gelatin. This process is called diffusion.

TAKE A LOOK

2. Identify How do dye particles move through the water?

_____

_____

Copyright © by Holt, Rinehart and Winston; a Division of Houghton Mifflin Harcourt Publishing Company. All rights reserved.

SECTION 1 Exchange with the Environment *continued*

DIFFUSION OF WATER

Substances, such as water, are made up of particles called *molecules*. Pure water has the highest concentration of water molecules. This means that 100% of the molecules are water molecules. If you mix another substance, such as food coloring, into the water, you lower the concentration of water molecules. This means that water molecules no longer make up 100% of the total molecules.

The figure below shows a container that has been divided by a membrane. The membrane is *semipermeable*—that is, only some substances can pass through it. The membrane lets smaller molecules, such as water, pass through. Larger molecules, such as food coloring, cannot pass through. Water molecules will move across the membrane. The diffusion of water through a membrane is called **osmosis**.

Critical Thinking

3. Apply Concepts Which of the following has a higher concentration of water molecules—200 molecules of water, or a mixture of 300 molecules of water and 100 molecules of food coloring? Explain your answer.

_____
_____
_____
_____
_____
_____

Osmosis

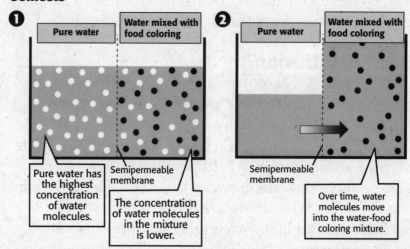

TAKE A LOOK
4. Explain Why does the volume of liquid in the right-hand side of the container increase with time?

_____
_____
_____

A cell membrane is a type of semipermeable membrane. This means that water can pass through the cell membrane, but most other substances cannot. The cells of organisms are surrounded by and filled with fluids. These fluids are made mostly of water. Water moves in and out of a cell by osmosis. ☑

 **READING CHECK**

5. Identify How does water move into and out of cells?

_____
_____

Copyright © by Holt, Rinehart and Winston; a Division of Houghton Mifflin Harcourt Publishing Company. All rights reserved.

How Do Small Particles Enter and Leave a Cell?

Small particles, such as sugars, can cross the cell membrane through passageways called *channels*. These channels in the cell membrane are made of proteins. Particles can travel through these channels by passive transport or by active transport.

During **passive transport**, particles move through the cell membrane without using energy from the cell. During passive transport, particles move from areas of high concentration to areas of lower concentration. Diffusion and osmosis are examples of passive transport.

During **active transport**, the cell has to use energy to move particles through channels. During active transport, particles usually move from areas of low concentration to areas of high concentration. ☑

How Do Large Particles Enter and Leave a Cell?

Large particles cannot move across a cell membrane in the same ways as small particles. Larger particles must move in and out of the cell by endocytosis and exocytosis. Both processes require energy from the cell.

Endocytosis happens when a cell surrounds a large particle and encloses it in a vesicle. A *vesicle* is a sac formed from a piece of cell membrane.

Endocytosis

❶ The cell comes into contact with a particle.

❷ The cell membrane begins to wrap around the particle.

❸ Once the particle is completely surrounded, a vesicle pinches off.

Exocytosis happens when a cell uses a vesicle to move a particle from within the cell to outside the cell. Exocytosis is how cells get rid of large waste particles.

Exocytosis

❶ Large particles that must leave the cell are packaged in vesicles.

❷ The vesicle travels to the cell membrane and fuses with it.

❸ The cell releases the particle to the outside of the cell.

☑ **READING CHECK**

6. Identify What is needed to move particles from areas of low concentration to areas of high concentration?

_____

_____

TAKE A LOOK
7. Identify Label the vesicle in the figure.

Copyright © by Holt, Rinehart and Winston; a Division of Houghton Mifflin Harcourt Publishing Company. All rights reserved.

Section 1 Review

GLE 0707.Inq.5, GLE 0707.1.5 **TN**

SECTION VOCABULARY

active transport the movement of substances across the cell membrane that requires the cell to use energy

diffusion the movement of particles from regions of higher density to regions of lower density

endocytosis the process by which a cell membrane surrounds a particle and encloses the particle in a vesicle to bring the particle into the cell

exocytosis the process in which a cell releases a particle by enclosing the particle in a vesicle that then moves to the cell surface and fuses with the cell membrane

osmosis the diffusion of water through a semipermeable membrane

passive transport the movement of substances across a cell membrane without the use of energy by the cell

1. Compare How is endocytosis different from exocytosis? How are they similar?

_____

_____

_____

2. Explain How is osmosis related to diffusion?

_____

_____

_____

3. Compare What are the differences between active and passive transport?

_____

_____

_____

_____

_____

4. Identify What structures allow small particles to cross cell membranes?

_____

5. Apply Concepts Draw an arrow in the figure below to show the direction that water molecules will move in.

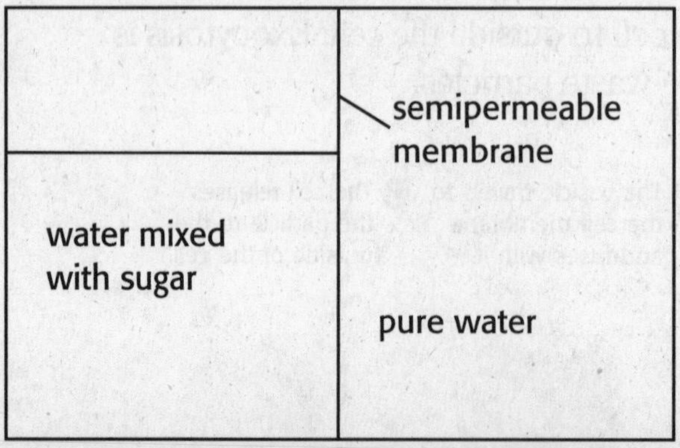

semipermeable membrane

water mixed with sugar

pure water

Copyright © by Holt, Rinehart and Winston; a Division of Houghton Mifflin Harcourt Publishing Company. All rights reserved.

CHAPTER 3 | The Cell in Action
SECTION 2 | **Cell Energy**

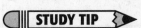 **Tennessee Science Standards**

GLE 0707.Inq.5
GLE 0707.1.1
GLE 0707.3.1
GLE 0707.3.2

BEFORE YOU READ

After you read this section, you should be able to answer these questions:

• How do plant cells make food?

• How do plant and animal cells get energy from food?

How Does a Plant Make Food?

The sun is the major source of energy for life on Earth. Plants use carbon dioxide, water, and the sun's energy to make food in a process called **photosynthesis**. The food that plants make gives them energy. When animals eat plants, the plants become sources of energy for the animals.

Plant cells have molecules called *pigments* that absorb light energy. Chlorophyll is the main pigment used in photosynthesis. Chlorophyll is found in chloroplasts. The food plants make is a simple sugar called *glucose*. Photosynthesis also produces oxygen. ☑

STUDY TIP

Compare As you read this section, make a Venn Diagram to compare cellular respiration and fermentation.

READING CHECK

1. Identify In which cell structures does photosynthesis take place?

_____

_____

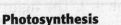

Photosynthesis

$$6CO_2 + 6H_2O + \text{light energy} \longrightarrow C_6H_{12}O_6 + 6O_2$$

Carbon dioxide   Water   Glucose   Oxygen

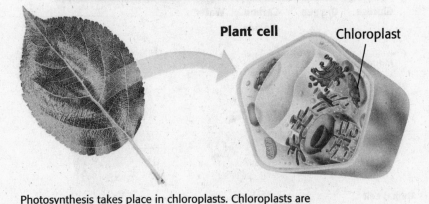

Plant cell   Chloroplast

Photosynthesis takes place in chloroplasts. Chloroplasts are found inside plant cells.

TAKE A LOOK

2. Identify What two materials are produced during photosynthesis?

_____

_____

Copyright © by Holt, Rinehart and Winston; a Division of Houghton Mifflin Harcourt Publishing Company. All rights reserved.

SECTION 2 Cell Energy *continued*

TN TENNESSEE
STANDARDS CHECK
GLE 0707.1.1 Distinguish
between the basic features of
photosynthesis and respiration.
3. Define What is cellular
respiration?

_____

_____

How Do Organisms Get Energy from Food?

Both plant and animal cells must break down food molecules to get energy from them. There are two ways cells get energy: cellular respiration and fermentation.

During **cellular respiration**, cells use oxygen to break down food. During **fermentation**, food is broken down without oxygen. Cellular respiration releases more energy from food than fermentation. Most eukaryotes, such as plants and animals, use cellular respiration.

What Happens During Cellular Respiration?

When you hear the word *respiration*, you might think of breathing. However, cellular respiration is different from breathing. Cellular respiration is a chemical process that happens in cells. In eukaryotic cells, such as plant and animal cells, cellular respiration takes place in structures called *mitochondria*.

Recall that to get energy, cells must break down glucose. During cellular respiration, glucose is broken down into carbon dioxide (CO_2) and water (H_2O), and energy is released. This energy is stored in a molecule called *ATP* (adenosine triphosphate). The figure below shows how energy is released when a cow eats grass.

TAKE A LOOK

4. Identify What two materials are needed for cellular respiration?

_____

_____

5. List What three things are produced during cellular respiration?

_____

_____

_____

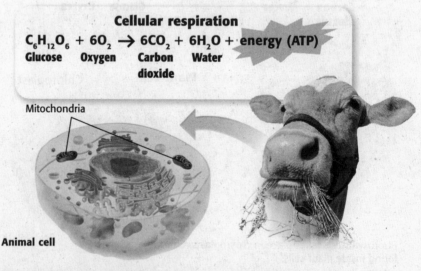

Cellular respiration

$$C_6H_{12}O_6 + 6O_2 \rightarrow 6CO_2 + 6H_2O + \text{energy (ATP)}$$

Glucose   Oxygen   Carbon   Water
                dioxide

Mitochondria

Animal cell

The mitochondria in the cells of this cow will use cellular respiration to release the energy stored in the grass.

Copyright © by Holt, Rinehart and Winston; a Division of Houghton Mifflin Harcourt Publishing Company. All rights reserved.

| SECTION 2 | Cell Energy *continued* |

The Connection Between Photosynthesis and Cellular Respiration

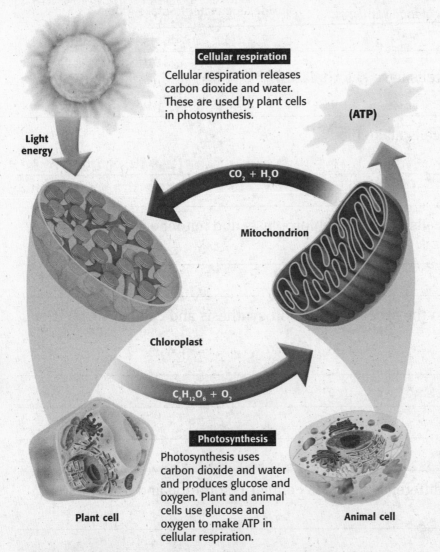

Cellular respiration
Cellular respiration releases carbon dioxide and water. These are used by plant cells in photosynthesis.

(ATP)

Light energy

$CO_2 + H_2O$

Mitochondrion

Chloroplast

$C_6H_{12}O_6 + O_2$

Photosynthesis
Photosynthesis uses carbon dioxide and water and produces glucose and oxygen. Plant and animal cells use glucose and oxygen to make ATP in cellular respiration.

Plant cell

Animal cell

Critical Thinking

6. Apply Concepts What would happen if oxygen were not produced during photosynthesis?

_____
_____
_____
_____
_____
_____

TAKE A LOOK
7. Complete Plant and animal cells use glucose and oxygen to make

_____.

How Is Fermentation Different from Cellular Respiration?

During fermentation, cells break down glucose without oxygen. Some bacteria and fungi rely only on fermentation to release energy from food. However, cells in other organisms may use fermentation when there is not enough oxygen for cellular respiration.

When you exercise, your muscles use up oxygen very quickly. When cells don't have enough oxygen, they must use fermentation to get energy. Fermentation creates a byproduct called *lactic acid*. This is what makes your muscles ache if you exercise too hard or too long.

**Say It**

Research Use the school library or the Internet to research an organism that uses fermentation. What kind of organism is it? Where is it found? Is this organism useful to humans? Present your findings to the class.

Copyright © by Holt, Rinehart and Winston; a Division of Houghton Mifflin Harcourt Publishing Company. All rights reserved.

Section 2 Review

GLE 0707.Inq.5, GLE 0707.1.1, GLE 0707.3.1, GLE 0707.3.2 **TN**

SECTION VOCABULARY

cellular respiration the process by which cells use oxygen to produce energy from food **fermentation** the breakdown of food without the use of oxygen	**photosynthesis** the process by which plants, algae, and some bacteria use sunlight, carbon dioxide, and water to make food.

1. Identify What kind of cells have chloroplasts?

_____

2. Explain How do plant cells make food?

_____

_____

3. Explain Why do plant cells need both chloroplasts and mitochondria?

_____

_____

4. Apply Concepts How do the processes of photosynthesis and cellular respiration work together?

_____

_____

_____

_____

5. Compare What is one difference between cellular respiration and fermentation?

_____

_____

_____

6. Explain Do your body cells always use cellular respiration to break down glucose? Explain your answer.

_____

_____

_____

Copyright © by Holt, Rinehart and Winston; a Division of Houghton Mifflin Harcourt Publishing Company. All rights reserved.

CHAPTER 3 The Cell in Action

SECTION 3 The Cell Cycle

BEFORE YOU READ

After you read this section, you should be able to answer these questions:

- How are new cells made?
- What is mitosis?
- How is cell division different in animals and plants?

TN Tennessee Science Standards
GLE 0707.Inq.5
GLE 0707.1.4

How Are New Cells Made?

As you grow, you pass through different stages in your life. Cells also pass through different stages in their life cycles. These stages are called the **cell cycle**. The cell cycle starts when a cell is made, and ends when the cell divides to make new cells.

Before a cell divides, it makes a copy of its DNA (deoxyribonucleic acid). *DNA* is a molecule that contains all the instructions for making new cells. The DNA is stored in structures called **chromosomes**. Cells make copies of their chromosomes so that new cells have the same chromosomes as the parent cells. Although all cells pass through a cell cycle, the process differs in prokaryotic and eukaryotic cells. ☑

STUDY TIP

Summarize As you read this section, make a diagram showing the stages of the eukaryotic cell cycle.

How Do Prokaryotic Cells Divide?

Prokaryotes have only one cell. Prokaryotic cells have no nucleus. They also have no organelles that are surrounded by membranes. The DNA for prokaryotic cells, such as bacteria, is found on one circular chromosome. The cell divides by a process called *binary fission*. During binary fission, the cell splits into two parts. Each part has one copy of the cell's DNA.

READING CHECK

1. Explain What must happen before a cell can divide?

_____

_____

_____

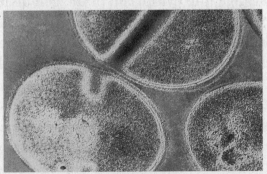

Bacteria reproduce by binary fission.

TAKE A LOOK
2. Complete Prokaryotic cells divide by _____

_____.

Copyright © by Holt, Rinehart and Winston; a Division of Houghton Mifflin Harcourt Publishing Company. All rights reserved.

SECTION 3 The Cell Cycle *continued*

How Do Eukaryotic Cells Divide?

Different kinds of eukaryotes have different numbers of chromosomes. However, complex eukaryotes do not always have more chromosomes than simpler eukaryotes. For example, potatoes have 48 chromosomes, but humans have 46. Many eukaryotes, including humans, have pairs of similar chromosomes. These pairs are called **homologous chromosomes**. One chromosome in a pair comes from each parent.

Cell division in eukaryotic cells is more complex than in prokaryotic cells. The cell cycle of a eukaryotic cell has three stages: interphase, mitosis, and cytokinesis.

The first stage of the cell cycle is called *interphase*. During interphase, the cell grows and makes copies of its chromosomes and organelles. The two copies of a chromosome are called *chromatids*. The two chromatids are held together at the *centromere*.

Critical Thinking

3. Compare What is the difference between a chromosome and a chromatid?

_____

_____

_____

_____

_____

This duplicated chromosome consists of two chromatids. The chromatids are joined at the centromere.

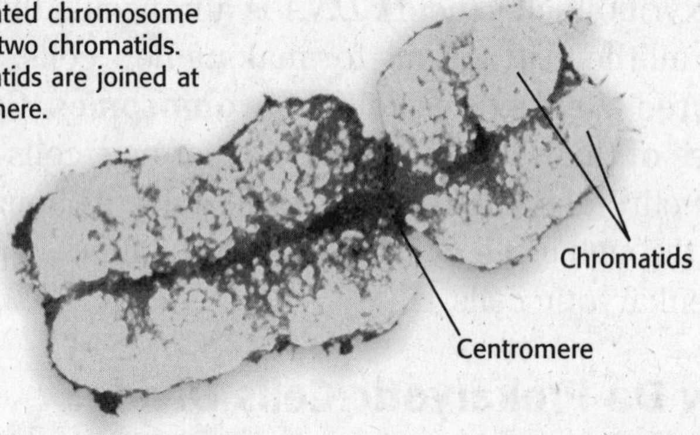

Chromatids

Centromere

The second stage of the cell cycle is called **mitosis**. During this stage, the chromatids separate. This allows each new cell to get a copy of each chromosome. Mitosis happens in four phases, as shown in the figure on the next page: prophase, metaphase, anaphase, and telophase.

The third stage of the cell cycle is called **cytokinesis**. During this stage, the cytoplasm of the cell divides to form two cells. These two cells are called *daughter cells*. The new daughter cells are exactly the same as each other. They are also exactly the same as the original cell. ☑

THE CELL CYCLE

The figure on the following page shows the cell cycle. In this example, the stages of the cell cycle are shown in a eukaryotic cell that has only four chromosomes.

READING CHECK

4. Identify What are the three stages of the eukaryotic cell cycle?

_____

_____

_____

Copyright © by Holt, Rinehart and Winston; a Division of Houghton Mifflin Harcourt Publishing Company. All rights reserved.

Interphase Before mitosis begins, chromosomes are copied. Each chromosome is then made of two chromatids.

Mitosis Phase 1 (Prophase) Mitosis begins. Chromatids condense from long strands to thick rods.

Mitosis Phase 2 (Metaphase) The nuclear membrane dissolves. Chromosome pairs line up around the equator of the cell.

Mitosis Phase 3 (Anaphase) Chromatids separate and move to opposite sides of the cell.

Mitosis Phase 4 (Telophase) A nuclear membrane forms around each set of chromosomes. The chromosomes unwind. Mitosis is complete.

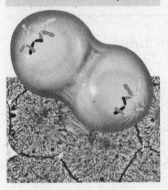

Cytokinesis In cells with no cell wall, the cell pinches in two.

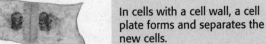

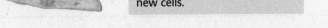

In cells with a cell wall, a cell plate forms and separates the new cells.

Math Focus

5. Calculate Cell A takes 6 h to complete division. Cell B takes 8 h to complete division. After 24 h, how many more copies of cell A than cell B will there be?

_____

TAKE A LOOK

6. List What are the four phases of mitosis?

_____

_____

_____

7. Identify What structure do plant cells have during cytokinesis that animal cells do not have?

_____

_____

Copyright © by Holt, Rinehart and Winston; a Division of Houghton Mifflin Harcourt Publishing Company. All rights reserved.

Section 3 Review

GLE 0707.Inq.5, GLE 0707.1.4 ▰TN▰

SECTION VOCABULARY

cell cycle the life cycle of a cell	**homologous chromosomes** chromosomes that have the same sequence of genes and the same structure
chromosome in a eukaryotic cell, one of the structures in the nucleus that are made up of DNA and protein; in a prokaryotic cell, the main ring of DNA	**mitosis** in eukaryotic cells, a process of cell division that forms two new nuclei, each of which has the same number of chromosomes
cytokinesis the division of cytoplasm of a cell	

1. Compare How does the DNA of prokaryotic and eukaryotic cells differ?

_____

_____

2. Summarize Complete the Process Chart to explain the three stages of the cell cycle. Include the four phases of mitosis.

```
┌─────────────────────────────────────────────┐
│                                             │
│                                             │
└─────────────────────────────────────────────┘
                      │
                      ▼
┌─────────────────────────────────────────────┐
│ Mitosis begins with prophase. The chromosomes condense. │
└─────────────────────────────────────────────┘
                      │
                      ▼
┌─────────────────────────────────────────────┐
│                                             │
│                                             │
└─────────────────────────────────────────────┘
                      │
                      ▼
┌─────────────────────────────────────────────┐
│ During telophase the nuclear membrane forms. The chromosomes lengthen and mitosis ends. │
└─────────────────────────────────────────────┘
                      │
                      ▼
┌─────────────────────────────────────────────┐
│                                             │
└─────────────────────────────────────────────┘
```

3. Explain Why does a cell make a copy of its DNA before it divides?

_____

_____

_____

4. Infer Why is cell division in eukaryotic cells more complex than in prokaryotic cells?

_____

_____

Copyright © by Holt, Rinehart and Winston; a Division of Houghton Mifflin Harcourt Publishing Company. All rights reserved.

CHAPTER 4 Heredity

SECTION 1 # Mendel and His Peas

BEFORE YOU READ

After you read this section, you should be able to answer these questions:

• What is heredity?

• How did Gregor Mendel study heredity?

TN **Tennessee Science Standards**
GLE 0707.Inq.5
GLE 0707.4.2
GLE 0707.4.4

What Is Heredity?

Why don't you look like a rhinoceros? The answer to that question seems simple. Neither of your parents is a rhinoceros. Only a human can pass on its traits to make another human. Your parents passed some of their traits on to you. The passing of traits from parents to offspring is called **heredity**.

About 150 years ago, a monk named Gregor Mendel performed experiments on heredity. His discoveries helped establish the field of genetics. *Genetics* is the study of how traits are passed on, or inherited. ☑

STUDY TIP

Define As you read this section, make a list of all of the underlined and italicized words. Write a definition for each of the words.

READING CHECK

1. **Define** What is genetics?

_____

_____

_____

Who Was Gregor Mendel?

Gregor Mendel was born in Austria in 1822. He grew up on a farm where he learned a lot about flowers and fruit trees. When he was 21 years old, Mendel entered a monastery. A monastery is a place where monks study and practice religion. The monks at Mendel's monastery also taught science and performed scientific experiments.

Mendel studied pea plants in the monastery garden to learn how traits are passed from parents to offspring. He used garden peas because they grow quickly. They also have many traits, such as height and seed color, that are easy to see. His results changed the way people think about how traits are passed on. ☑

READING CHECK

2. **Explain** Why did Mendel choose to study pea plants?

_____

_____

_____

Gregor Mendel discovered the principles of heredity while studying pea plants.

Copyright © by Holt, Rinehart and Winston; a Division of Houghton Mifflin Harcourt Publishing Company. All rights reserved.

REPRODUCTION IN PEAS

Like many flowering plants, pea plants have both male and female reproductive parts. Many flowering plants can reproduce by cross-pollination. In most plants, sperm are carried in structures called pollen. In *cross-pollination*, sperm in the pollen of one plant fertilize eggs in the flower of another plant. Pollen can be carried by organisms, such as insects. It may also be carried by the wind from one flower to another.

Some flowering plants must use cross-pollination. They need another plant to reproduce. However, some plants, including pea plants, can also reproduce by self-pollination. In *self-pollination*, sperm from one plant fertilize the eggs of the same plant.

Mendel used self-pollination in pea plants to grow true-breeding plants for his experiments. When a *true-breeding* plant self-pollinates, its offspring all have the same traits as the parent. For example, a true-breeding plant with purple flowers always has offspring with purple flowers.

Critical Thinking

3. Compare What is the difference between cross-pollination and self-pollination?

_____

_____

_____

_____

_____

TAKE A LOOK

4. Identify What are two ways pollen can travel from one plant to another during cross-pollination?

_____

_____

_____

_____

Self-pollination

Cross-pollination by animals

Stigma

Cross-pollination by wind

Pollen

Anther

Ovary

Petal

Ovule

During pollination, pollen from the anther (male) is carried to the stigma (female). Fertilization happens when a sperm from the pollen moves through the stigma and enters an egg in an ovule.

Copyright © by Holt, Rinehart and Winston; a Division of Houghton Mifflin Harcourt Publishing Company. All rights reserved.

SECTION 1 Mendel and His Peas *continued*

CHARACTERISTICS

A *characteristic* is a feature that has different forms. For example, hair color is a characteristic of humans. The different forms or colors, such as brown or red hair, are *traits*. ☑

Mendel studied one characteristic of peas at a time. He used plants that had different traits for each characteristic he studied. One characteristic he studied was flower color. He chose plants that had purple flowers and plants that had white flowers. He also studied other characteristics, such as seed shape, pod color, and plant height.

CROSSING PEA PLANTS

Mendel was careful to use true-breeding plants in his experiments. By choosing these plants, he would know what to expect if his plants self-pollinated. He decided to find out what would happen if he bred, or crossed, two plants that had different traits.

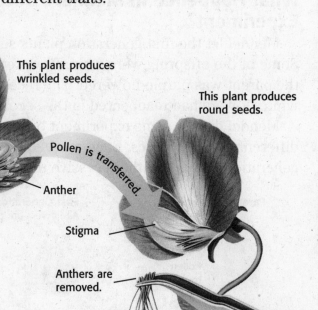

This plant produces wrinkled seeds.

This plant produces round seeds.

Pollen is transferred.

Anther

Stigma

Anthers are removed.

Mendel removed the anthers from a plant that made round seeds. Then, he used pollen from a plant that made wrinkled seeds to fertilize the plant that made round seeds.

☑ **READING CHECK**

5. Explain How are characteristics and traits related?

_____

_____

_____

_____

Say It

Describe How would you describe yourself? Make a list of your physical traits, such as height, hair color, and eye color. List other traits you have that you weren't born with. Share this list with your classmates. Which of these traits did you inherit?

TAKE A LOOK

6. Describe How did Mendel make sure that the plant with round seeds did not self-pollinate?

_____

_____

_____

_____

Copyright © by Holt, Rinehart and Winston; a Division of Houghton Mifflin Harcourt Publishing Company. All rights reserved.

What Happened in Mendel's First Experiments?

Mendel studied seven different characteristics in his first experiments with peas. He crossed plants that were true-breeding for different traits. For example, he crossed plants that had purple flowers with plants that had white flowers. The offspring from such a cross are called *first-generation plants*. All of the first-generation plants in this cross had purple flowers. What happened to the trait for white flowers?

Mendel got similar results for each cross. One trait was always present in the first generation and the other trait seemed to disappear. Mendel called the trait that appeared the **dominant trait**. He called the other trait the **recessive trait**. To *recede* means "to go away or back off." To find out what happened to the recessive trait, Mendel did another set of experiments. ☑

What Happened in Mendel's Second Experiment?

Mendel let the first-generation plants self-pollinate. Some of the offspring were white-flowered, even though the parent was purple-flowered. The recessive trait for white flowers had reappeared in the second generation.

Mendel did the same experiment on plants with seven different characteristics. Each time, some of the second-generation plants had the recessive trait.

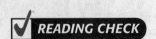

READING CHECK

7. Identify What kind of trait appeared in the first generation?

_____

_____

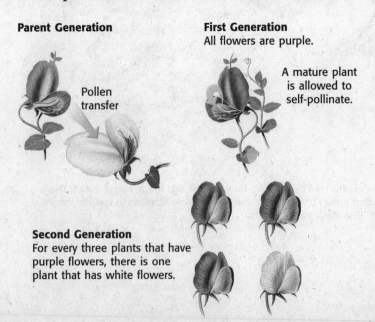

Parent Generation

Pollen transfer

First Generation
All flowers are purple.

A mature plant is allowed to self-pollinate.

Second Generation
For every three plants that have purple flowers, there is one plant that has white flowers.

TAKE A LOOK

8. Identify What type of traits appeared in the second generation?

_____

_____

Copyright © by Holt, Rinehart and Winston; a Division of Houghton Mifflin Harcourt Publishing Company. All rights reserved.

RATIOS IN MENDEL'S EXPERIMENTS

Mendel counted the number of plants that had each trait in the second generation. He hoped that this might help him explain his results.

As you can see from the table below, the recessive trait did not show up as often as the dominant trait. Mendel decided to figure out the ratio of dominant traits to recessive traits. A *ratio* is a relationship between two numbers. It is often written as a fraction. For example, the second generation produced 705 plants with purple flowers and 224 plants with white flowers. Mendel used this formula to calculate the ratios:

$$\frac{705}{224} = \frac{3.15}{1} \text{ or } 3.15:1$$

Characteristic	Dominant trait	Recessive trait	Ratio
Flower color	705 purple	224 white	3.15:1
Seed color	6,002 yellow	2,001 green	
Seed shape	5,474 round	1,850 wrinkled	
Pod color	428 green	152 yellow	
Pod shape	882 smooth	299 bumpy	
Flower position	651 along stem	207 at tip	
Plant height	787 tall	277 short	

Math Focus

9. Find Ratios Calculate the ratios of the other pea plant characteristics in the table.

Math Focus

10. Round Round off all numbers in the ratios to whole numbers. What ratio do you get?

_____

What Did Mendel Conclude?

Mendel knew that his results could be explained only if each plant had two sets of instructions for each characteristic. He concluded that each parent gives one set of instructions to the offspring. The dominant set of instructions determines the offspring's traits.

Copyright © by Holt, Rinehart and Winston; a Division of Houghton Mifflin Harcourt Publishing Company. All rights reserved.

Section 1 Review

GLE 0707.Inq.5, GLE 0707.4.2, GLE 0707.4.4 TN

SECTION VOCABULARY

dominant trait the trait observed in the first generation when parents that have different traits are bred **heredity** the passing of genetic traits from parent to offspring	**recessive trait** a trait that is apparent only when two recessive alleles for the same characteristic are inherited

1. Define What is a true-breeding plant?

_____

_____

_____

2. Apply Concepts Cats may have straight or curly ears. A curly-eared cat mated with a straight-eared cat. All the kittens had curly ears. Are curly ears a dominant or recessive trait? Explain your answer.

_____

_____

3. Summarize Complete the cause and effect map to summarize Mendel's experiments on flower color in pea plants.

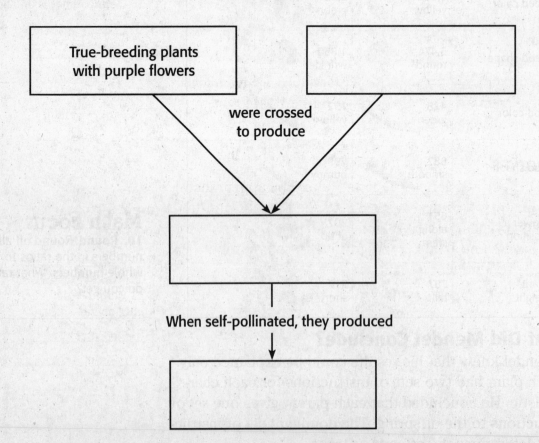

True-breeding plants with purple flowers

were crossed to produce

When self-pollinated, they produced

Copyright © by Holt, Rinehart and Winston; a Division of Houghton Mifflin Harcourt Publishing Company. All rights reserved.

CHAPTER 4 | Heredity

SECTION 2 Traits and Inheritance

TN Tennessee Science Standards
GLE 0707.4.3
GLE 0707.4.4

> **BEFORE YOU READ**
>
> After you read this section, you should be able to answer these questions:
> - What did Mendel's experiments tell him about heredity?
> - Are there exceptions to Medel's laws of heredity?

What Did Mendel Learn About Heredity?

Mendel knew from his pea plant experiments that there must be two sets of instructions for each characteristic. All of the first-generation plants showed the dominant trait. However, they could give the recessive trait to their offspring. Instructions for an inherited trait are called **genes**. Offspring have two sets of genes—one from each parent.

The two sets of genes that parents give to offspring are never exactly the same. The same gene might have more than one version. The different versions of a gene are called **alleles**. ☑

Alleles may be dominant or recessive. A trait for an organism is usually identified with two letters, one for each allele. Dominant alleles are given capital letters (*A*). Recessive alleles are given lowercase letters (*a*). If a dominant allele is present, it will hide a recessive allele. An organism can have a recessive trait only if it gets a recessive allele for that trait from both parents.

PHENOTYPE

An organism's genes affect its traits. The appearance of an organism, or how it looks, is called its **phenotype**. The phenotypes for flower color in Mendel's pea plants were purple and white. The figure below shows one example of a human phenotype. ☑

Albinism is an inherited disorder that affects a person's phenotype in many ways.

<table>
<tr><td>

🖊 **STUDY TIP**

Organize As you read, make a Concept Map using the vocabulary words highlighted in the section.

</td></tr>
</table>

☑ **READING CHECK**

1. Define What is an allele?

_____

_____

_____

☑ **READING CHECK**

2. Define What is a phenotype?

_____

_____

Copyright © by Holt, Rinehart and Winston; a Division of Houghton Mifflin Harcourt Publishing Company. All rights reserved.

SECTION 2 Traits and Inheritance *continued*

GENOTYPE

A **genotype** is the combination of alleles that an organism gets from its parents. A plant with two dominant or two recessive alleles (*PP, pp*) is *homozygous. Homo* means "the same." A plant with one dominant allele and one recessive allele (*Pp*) is *heterozygous. Hetero* means "different." The allele for purple flowers (*P*) in pea plants is dominant. The plant will have purple flowers even if it has only one *P* allele. ☑

READING CHECK

3. Identify What kind of alleles does a heterozygous individual have?

_____

_____

PUNNETT SQUARES

A **Punnett square** is used to predict the possible genotypes of offspring from certain parents. It can be used to show the alleles for any trait. In a Punnett square, the alleles for one parent are written along the top of the square. The alleles for the other parent are written along the side of the square. The possible genotypes of offspring are found by combining the letters at the top and side of each square.

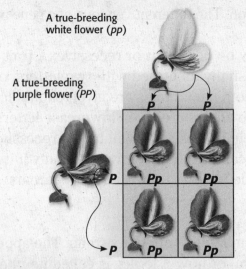

A true-breeding white flower (*pp*)

A true-breeding purple flower (*PP*)

TAKE A LOOK
4. Identify Is the plant with white flowers homozygous or heterozygous? How can you tell?

_____

_____

_____

All of the offspring for this cross have the same genotype—*Pp*.

The figure shows a Punnett square for a cross of two true-breeding plants. One has purple flowers and the other has white flowers. The alleles for a true-breeding purple-flowered plant are written as *PP*. The alleles for a true-breeding white flowered plant are written as *pp*. Offspring get one of their two alleles from each parent. All of the offspring from this cross will have the same genotype: *Pp*. Because they have a dominant allele, all of the offspring will have purple flowers.

Copyright © by Holt, Rinehart and Winston; a Division of Houghton Mifflin Harcourt Publishing Company. All rights reserved.

SECTION 2 Traits and Inheritance *continued*

MORE EVIDENCE FOR INHERITANCE

In his second experiments, Mendel let the first-generation plants self-pollinate. He did this by covering the flowers of the plant. This way, no pollen from another plant could fertilize its eggs. The Punnett square below shows a cross of a plant that has the genotype *Pp*.

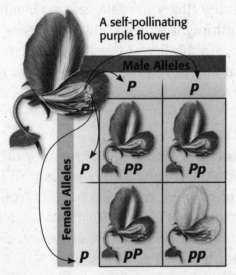

A self-pollinating purple flower

Male Alleles

This Punnett square shows the possible results from the cross *Pp × Pp*.

Female Alleles

TAKE A LOOK
5. List What are the possible genotypes of the offspring in this cross?

_____

_____

Notice that one square shows the genotype *Pp* and another shows *pP*. These are exactly the same genotype. They both have one *p* allele and one *P* allele. The combinations *PP*, *Pp*, and *pP* have the same phenotype—purple flowers. This is because they all have at least one dominant allele, *P*. ☑

Only one combination, *pp*, produces plants that have white flowers. The ratio of dominant phenotypes to recessive phenotypes is 3:1. This means that three out of four offspring from that cross will have purple flowers. This is the same ratio Mendel found.

☑ **READING CHECK**

6. Explain Why do the genotypes *PP*, *Pp*, and *pP* all have the same phenotype?

_____

_____

_____

What Is the Chance That Offspring Will Receive a Certain Allele?

Each parent has two alleles for each gene. When an individual reproduces, it passes one of its two alleles to its offspring. When a parent has two different alleles for a gene, such as *Pp*, offspring may receive either of the alleles. Both alleles have an equal chance to be passed from the parent to the offspring.

Think of a coin toss. When you toss the coin, there is a 50% chance you will get heads, and a 50% chance you will get tails. The chance of the offspring receiving one allele or another from a parent is as random as a coin toss.

Copyright © by Holt, Rinehart and Winston; a Division of Houghton Mifflin Harcourt Publishing Company. All rights reserved.

PROBABILITY

The mathematical chance that something will happen is known as **probability**. Probability is usually written as a fraction or percentage. If you toss a coin, the probability of tossing tails is 1/2, or 50%. In other words, you will get tails half of the time.

What is the probability that you will toss two heads in a row? To find out, multiply the probability of tossing the first head (1/2) by the probability of tossing the second head (1/2). The probability of tossing two heads in a row is 1/4.

GENOTYPE PROBABILITY

Finding the probability of certain genotypes for offspring is like predicting the results of a coin toss. To have white flowers, a pea plant must receive a *p* allele from each parent. Each offspring of a *Pp* × *Pp* cross has a 50% chance of receiving either allele from either parent. So, the probability of inheriting two *p* alleles is 1/2 × 1/2. This equals 1/4, or 25%.

Math Focus

7. Complete Complete the Punnett square to show the cross between two heterozygous parents. What percentage of the offspring are homozygous?

_____

_____

	P	**P**
P		
P		

Are There Exceptions to Mendel's Principles?

Mendel's experiments helped show the basic principles of how genes are passed from one generation to the next. Mendel studied sets of traits such as flower color and seed shape. The traits he studied in pea plants are easy to predict because there are only two choices for each trait. ☑

Traits in other organisms are often harder to predict. Some traits are affected by more than one gene. A single gene may affect more than one trait. As scientists learned more about heredity, they found exceptions to Mendel's principles.

READING CHECK

8. Explain Why were color and seed shape in pea plants good traits for Mendel to study?

_____

_____

_____

Copyright © by Holt, Rinehart and Winston; a Division of Houghton Mifflin Harcourt Publishing Company. All rights reserved.

SECTION 2 **Traits and Inheritance** *continued*

INCOMPLETE DOMINANCE

Sometimes, one trait isn't completely dominant over another. These traits do not blend together, but each allele has an influence on the traits of offspring. This is called *incomplete dominance*. For example, the offspring of a true-breeding red snapdragon and a true-breeding white snapdragon are all pink. This is because both alleles for the gene influence color.

The offspring of two true-breeding show incomplete dominance.

ONE GENE, MANY TRAITS

In Mendel's studies, one gene controlled one trait. However, some genes affect more than one trait. For example, some tigers have white fur instead of orange. These white tigers also have blue eyes. This is because the gene that controls fur color also affects eye color.

MANY GENES, ONE TRAIT

Some traits, such as the color of your skin, hair, and eyes, are the result of several genes acting together. In humans, different combinations of many alleles can result in a variety of heights. ☑

THE IMPORTANCE OF ENVIRONMENT

Genes are not the only things that can affect an organism's traits. Traits are also affected by factors in the environment. For example, human height is affected not only by genes. Height is also influenced by nutrition. An individual who has plenty of food to eat may be taller than one who does not.

Critical Thinking

9. Infer If snapdragons showed complete dominance like pea plants, what would the offspring look like?

_____

_____

_____

Critical Thinking

10. Compare How is the allele for fur color in tigers different from the allele for flower color in pea plants?

_____

_____

_____

_____

☑ **READING CHECK**

11. Identify Give an example of a single trait that is affected by more than one gene.

_____

Copyright © by Holt, Rinehart and Winston; a Division of Houghton Mifflin Harcourt Publishing Company. All rights reserved.

Section 2 Review

GLE 0707.4.3, GLE 0707.4.4 TN

SECTION VOCABULARY

allele one of the alternative forms of a gene that governs a characteristic, such as hair color

gene one set of instructions for an inherited trait

genotype the entire genetic makeup of an organism; also the combination of genes for one or more specific traits

phenotype an organism's appearance or other detectable characteristic

probability the likelihood that a possible future event will occur in any given instance of the event

Punnett square a diagram used to organize all the possible combinations of offspring from particular parents

1. Identify Relationships How are genes and alleles related?

_____

_____

2. Explain How is it possible for two individuals to have the same phenotype but different genotype for a trait?

_____

_____

_____

3. Punnett Square Mendel allowed a pea plant that was heterozygous for yellow seeds (*Y*) to self-pollinate. Fill in the Punnett square below for this cross. What percentage of the offspring will have green (*y*) seeds?

_____

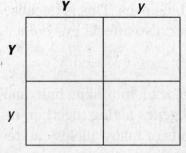

4. Discuss How is human height an exception to Mendel's principles of heredity?

_____

_____

_____

Copyright © by Holt, Rinehart and Winston; a Division of Houghton Mifflin Harcourt Publishing Company. All rights reserved.

CHAPTER 4 Heredity

SECTION 3 Meiosis

TN Tennessee Science Standards
GLE 0707.1.4
GLE 0707.4.1
GLE 0707.4.3
GLE 0707.4.4

BEFORE YOU READ

After you read this section, you should be able to answer these questions:

• What are sex cells?

• How does meiosis help explain Mendel's results?

How Do Organisms Reproduce?

When organisms reproduce, their genetic information is passed on to their offspring. There are two kinds of reproduction: asexual and sexual.

ASEXUAL REPRODUCTION

In asexual reproduction, only one parent is needed to produce offspring. Asexual reproduction produces offspring with exact copies of the parent's genotype.

SEXUAL REPRODUCTION

In sexual reproduction, cells from two parents join to form offspring. Sexual reproduction produces offspring that share traits with both parents. However, the offspring are not exactly like either parent.

What Are Homologous Chromosomes?

Recall that genes are the instructions for inherited traits. Genes are located on chromosomes. Each human body cell has a total of 46 chromosomes, or 23 pairs. A pair of chromosomes that carry the same sets of genes are called **homologous chromosomes**. One chromosome from a pair comes from each parent. ☑

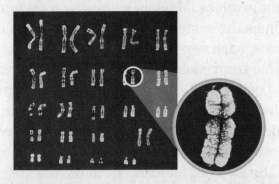

Human body cells have 23 pairs of chromosomes. One member of a pair of homologous chomosomes has been magnified.

STUDY TIP

Summarize Make flashcards that show the steps of meiosis. On the front of the cards, write the steps of meiosis. On the back of the cards, write what happens at each step. Practice arranging the steps in the correct order.

✓ READING CHECK

1. Define What are homologous chromosomes?

_____

_____

_____

TAKE A LOOK

2. Identify How many total chromosomes are in each human body cell?

_____

_____

Copyright © by Holt, Rinehart and Winston; a Division of Houghton Mifflin Harcourt Publishing Company. All rights reserved.

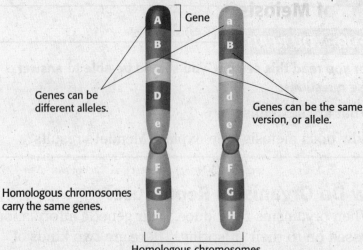

Genes can be different alleles.

Gene

Genes can be the same version, or allele.

Homologous chromosomes carry the same genes.

Homologous chromosomes

What Are Sex Cells?

In sexual reproduction, cells from two parents join to make offspring. However, only certain cells can join. Cells that can join to make offspring are called *sex cells*. An egg is a female sex cell. A sperm is a male sex cell. Unlike ordinary body cells, sex cells do not have homologous chromosomes. ☑

Imagine a pair of shoes. Each shoe is like a chromosome and the pair represents a homologous pair of chromosomes. Recall that your body cells have a total of 23 pairs of "shoes," or homologous chromosomes. Each sex cell, however, has only one of the chromosomes from each homologous pair. Sex cells have only one "shoe" from each pair. How do sex cells end up with only one chromosome from each pair?

How Are Sex Cells Made?

Sex cells are made during meiosis. **Meiosis** is a copying process that produces cells with half the usual number of chromosomes. Meiosis keeps the total number of chromosomes the same from one generation to the next.

In meiosis, each sex cell that is made gets only one chromosome from each homologous pair. For example, a human egg cell has 23 chromosomes and a sperm cell has 23 chromosomes. When these sex cells later join together during reproduction, they form pairs. The new cell has 46 chromosomes, or 23 pairs. The figure on the next page describes the steps of meiosis. To make the steps easy to see, only four chromosomes are shown.

READING CHECK

3. Explain How are sex cells different from ordinary body cells?

_____

_____

_____

TN⟋ TENNESSEE STANDARDS CHECK

GLE 0707.1.4 Illustrate how cell division <u>occurs</u> in sequential stages to maintain the chromosome number of a species.

Word Help: <u>illustrate</u> to make clear or show

Word Help: <u>occur</u> to happen

4. Define What is the function of meiosis?

_____

_____

_____

Copyright © by Holt, Rinehart and Winston; a Division of Houghton Mifflin Harcourt Publishing Company. All rights reserved.

SECTION 3 Meiosis *continued*

Steps of Meiosis
First cell division

1 The chromosomes are copied before meiosis begins. The identical copies, or chromatids, are joined together.

2 The nuclear membrane disappears. Pairs of homologous chromosomes line up at the equator of the cell.

3 The chromosomes separate from their homologous partners. Then they move to the opposite ends of the cell.

4 The nuclear membrane re-forms, and the cell divides. The paired chromatids are still joined.

Second cell division

5 Each cell contains one member of the homologous chromosome pair. The chromosomes are not copied again between the two cell divisions.

6 The nuclear membrane disappears. The chromosomes line up along the equator of each cell.

7 The chromatids pull apart and move to opposite ends of the cell. The nuclear membranes re-form, and the cells divide.

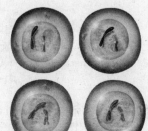

8 Four new cells have formed from the original cell. Each new cell has half the number of chromosomes as the original cell.

Critical Thinking

5. Predict What would happen if meiosis did not occur?

_____

_____

_____

_____

TAKE A LOOK

6. Identify How many times does the cell nucleus divide during meiosis?

_____

7. Identify At the end of meiosis, how many sex cells have been produced from one cell?

_____

Copyright © by Holt, Rinehart and Winston; a Division of Houghton Mifflin Harcourt Publishing Company. All rights reserved.

How Does Meiosis Explain Mendel's Results?

Mendel knew that eggs and sperm give the same amount of information to offspring. However, he did not know how traits were actually carried in the cell. Many years later, a scientist named Walter Sutton was studying grasshopper sperm cells. He knew about Mendel's work. When he saw chromosomes separating during meiosis, he made an important conclusion: genes are located on chromosomes.

The figure below shows what happens to chromosomes during meiosis and fertilization in pea plants. The cross shown is between two true-breeding plants. One produces round seeds and the other produces wrinkled seeds.

Meiosis and Dominance

Male Parent In the plant cell nucleus below, each homologous chromosome has an allele for seed shape. Each allele carries the same instructions: to make wrinkled seeds.

Female Parent In the plant cell nucleus below, each homologous chromosome has an allele for seed shape. Each allele carries the same instructions: to make round seeds.

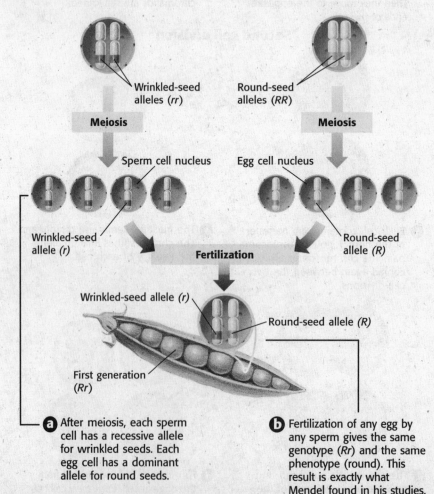

Wrinkled-seed alleles (*rr*)

Round-seed alleles (*RR*)

Meiosis

Meiosis

Sperm cell nucleus

Egg cell nucleus

Wrinkled-seed allele (*r*)

Round-seed allele (*R*)

Fertilization

Wrinkled-seed allele (*r*)

Round-seed allele (*R*)

First generation (*Rr*)

a After meiosis, each sperm cell has a recessive allele for wrinkled seeds. Each egg cell has a dominant allele for round seeds.

b Fertilization of any egg by any sperm gives the same genotype (*Rr*) and the same phenotype (round). This result is exactly what Mendel found in his studies.

Critical Thinking

8. Identify Relationships How did Sutton's work build on Mendel's work?

_____

_____

_____

_____

_____

_____

TAKE A LOOK

9. Explain In this figure, how many genotypes are possible for the offspring? Explain your answer.

_____

_____

_____

_____

Copyright © by Holt, Rinehart and Winston; a Division of Houghton Mifflin Harcourt Publishing Company. All rights reserved.

What Are Sex Chromosomes?

Information contained on chromosomes determines many of our traits. **Sex chromosomes** carry genes that determine sex. In humans, females have two X chromosomes. Human males have one X chromosome and one Y chromosome. ☑

During meiosis, one of each of the chromosome pairs ends up in a sex cell. Females have two X chromosomes in each body cell. When meiosis produces egg cells, each egg gets one X chromosome. Males have both an X chromosome and a Y chromosome in each body cell. Meiosis produces sperm with either an X or a Y chromosome.

An egg fertilized by a sperm with an X chromosome will produce a female. If the sperm contains a Y chromosome, the offspring will be male.

READING CHECK

10. Identify What combination of sex chromosomes makes a human male?

_____

_____

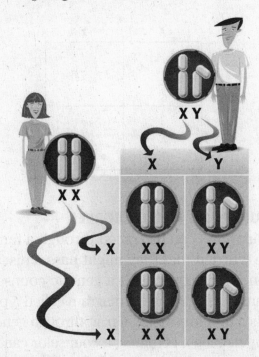

Egg and sperm join to form either the XX or XY combination.

SEX-LINKED DISORDERS

Hemophilia is a disorder that prevents blood from clotting. People with hemophilia bleed for a long time after small cuts. This disorder can be fatal. Hemophilia is an example of a sex-linked disorder. The genes for *sex-linked disorders* are carried on the X chromosome. Colorblindness is another example of a sex-linked disorder. Men are more likely than women to have sex-linked disorders. Why is this? ☑

TAKE A LOOK
11. Identify Circle the offspring in the figure that will be female.

READING CHECK
12. Define What is a sex-linked disorder?

_____

_____

_____

Copyright © by Holt, Rinehart and Winston; a Division of Houghton Mifflin Harcourt Publishing Company. All rights reserved.

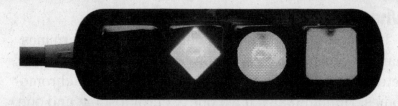

This stoplight in Canada was made to help the colorblind see signals easily.

The Y chromosome does not carry all of the genes that an X chromosome does. Females have two X chromosomes, so they carry two copies of each gene found on the X chromosome. This makes a backup gene available if one becomes damaged. Males have only one copy of each gene on their one X chromosome. If a male gets an allele for a sex-linked disorder, he will have the disorder, even if the allele is recessive.

TAKE A LOOK

13. Complete A particular sex-linked disorder is recessive. Fill in the Punnett Square to show how the disorder is passed from a carrier to its offspring. The chromosome carrying the trait for the disorder is underlined.

14. Identify Which individual will have the disorder?

_____

	X	Y
<u>X</u>		
X		

GENETIC COUNSELING AND PEDIGREES

Genetic disorders can be traced through a family tree. If people are worried that they might pass a disease to their children, they may consult a genetic counselor.

These counselors often use a diagram called a pedigree. A **pedigree** is a tool for tracing a trait through generations of a family. By making a pedigree, a counselor can often predict whether a person is a carrier of a hereditary disease.

The pedigree on the next page traces a disease called *cystic fibrosis*. Cystic fibrosis causes serious lung problems. People with this disease have inherited two recessive alleles. Both parents need to be carriers of the gene for the disease to show up in their children.

Copyright © by Holt, Rinehart and Winston; a Division of Houghton Mifflin Harcourt Publishing Company. All rights reserved.

SECTION 3 Meiosis *continued*

Pedigree for a Recessive Disease

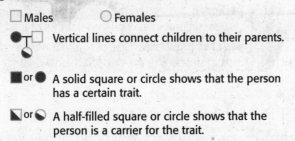

☐ Males   ◯ Females

● ─┬─ ☐ Vertical lines connect children to their parents.

◖

■ or ● A solid square or circle shows that the person has a certain trait.

◪ or ◐ A half-filled square or circle shows that the person is a carrier for the trait.

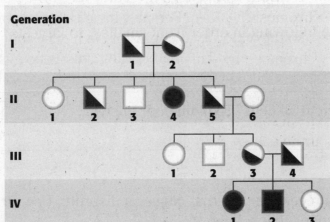

Generation

I

II

III

IV

TAKE A LOOK
15. Identify Circle all of the individuals in the pedigree who have the disorder. Draw a line under the individuals that carry the trait, but do not have the disorder.

You could draw a pedigree to trace almost any trait through a group of people who are biologically related. For example, a pedigree can show how you inherited your hair color. Many different pedigrees could be drawn for related individuals.

What Is Selective Breeding?

For thousands of years, humans have bred plants and animals to produce individuals with traits that they liked. This is known as *selective breeding*. Breeders may choose a plant or animal with traits they would like to see in the offspring. They breed that individual with another that also has those traits. For example, farmers might breed fruit trees that bear larger fruits.

You may see example of selective breeding every day. Different breeds of dogs, such as chihuahuas and German sheperds, were produced by selective breeding. Many flowers, such as roses, have been bred to produce large flowers. Wild roses are usually much smaller than roses you would buy at a flower store or plant nursery.

**Say It**

Discuss In a small group, come up with other examples of organisms that humans have changed through selective breeding. What traits do you think people wanted the organism to have? How is this trait helpful to humans?

Copyright © by Holt, Rinehart and Winston; a Division of Houghton Mifflin Harcourt Publishing Company. All rights reserved.

Section 3 Review

GLE 0707.1.4, GLE 0707.4.1, GLE 0707.4.3, GLE 0707.4.4 ⬛ TN

SECTION VOCABULARY

homologous chromosomes chromosomes that have the same sequence of genes and the same structure

meiosis a process in cell division during which the number of chromosomes decreases to half the original number by two divisions of the nucleus, which results in the production of sex cells

pedigree a diagram that shows the occurrence of a genetic trait in several generations of a family

sex chromosomes one of the pair of chromosomes that determine the sex of an individual

1. Identify Relationships Put the following in order from smallest to largest: chromosome, gene, cell.

_____

2. Explain Does meiosis happen in all cells? Explain your answer.

_____

The pedigree below shows a recessive trait that causes a disorder. Use the pedigree to answer the questions that follow.

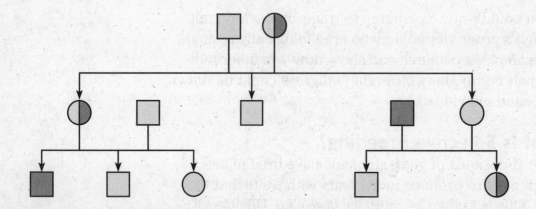

3. Identify Circle all individuals on the pedigree that are heterozygous for the trait. Are these individuals male or female?

_____

4. Identify Put a square around all individuals that have the disorder. Are these individuals male or female?

_____

5. Interpret Is the trait sex-linked? Explain your answer.

_____

Copyright © by Holt, Rinehart and Winston; a Division of Houghton Mifflin Harcourt Publishing Company. All rights reserved.

CHAPTER 5 Genes and DNA

SECTION 1 # What Does DNA Look Like?

TN Tennessee Science
Standards
GLE 0707.Inq.5

BEFORE YOU READ

After you read this section, you should be able to answer these questions:

- What units make up DNA?
- What does DNA look like?
- How does DNA copy itself?

What Is DNA?

Remember that *inherited traits* are traits that are passed from generation to generation. To understand how inherited traits are passed on, you must understand the structure of DNA. **DNA** (*deoxyribonucleic acid*) is the molecule that carries the instructions for inherited traits. In cells, DNA is wrapped around proteins to form *chromosomes*. Stretches of DNA that carry the information for inherited traits are called *genes*.

What Is DNA Made Of?

DNA is made up of smaller units called nucleotides. A *nucleotide* is made of three parts: a sugar, a phosphate, and a base. The sugar and the phosphate are the same for each nucleotide. However, different nucleotides may have different bases.

There are four different bases found in DNA nucleotides. They are *adenine, thymine, guanine*, and *cytosine*. Scientists often refer to a base by its first letter: *A* for adenine, *T* for thymine, *G* for guanine, and *C* for cytosine. Each base has a different shape.

STUDY TIP

Clarify Concepts As you read the text, make a list of ideas that are confusing. Discuss these with a small group. Ask your teacher to explain things that your group is unsure about.

The Four Nucleotides of DNA

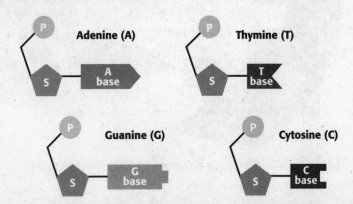

Adenine (A)
P
S A base

Thymine (T)
P
S T base

Guanine (G)
P
S G base

Cytosine (C)
P
S C base

TAKE A LOOK
1. Identify What are two things that are the same in all nucleotides?

_____

_____

Copyright © by Holt, Rinehart and Winston; a Division of Houghton Mifflin Harcourt Publishing Company. All rights reserved.

SECTION 1 What Does DNA Look Like? *continued*

What Does DNA Look Like?

As you can see in the figure below, a strand of DNA looks like a twisted ladder. This spiral shape is called a *double helix*. The two sides of the ladder are made of the sugar and phosphate parts of nucleotides. The sugars and phosphates alternate along each side of the ladder. The rungs of the DNA ladder are made of pairs of bases. ☑

The bases in DNA can only fit together in certain ways, like puzzle pieces. Adenine on one side of a DNA strand always pairs with thymine on the other side. Guanine always pairs with cytosine. This means that adenine is *complementary* to thymine, and guanine is complementary to cytosine. Because the pairs of bases in DNA are complementary, the two sides of a strand of DNA are also complementary.

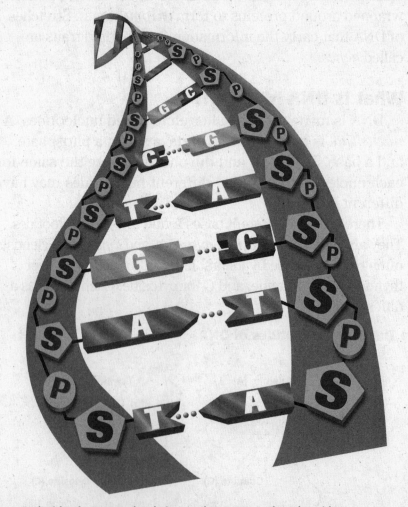

Each side of a DNA molecule is complementary to the other side.

☑ **READING CHECK**

2. Identify What are the sides of the DNA "ladder" made of?

_____

_____

Critical Thinking

3. Apply Concepts Imagine that you are a scientist studying DNA. You measure the number of cytosines and thymines in a small strand of DNA. There are 45 cytosines and 55 thymines. How many guanines are there in the strand? How many adenines are there?

_____

_____

TAKE A LOOK

4. Identify Give the ways that DNA bases can pair up.

_____

_____

Copyright © by Holt, Rinehart and Winston; a Division of Houghton Mifflin Harcourt Publishing Company. All rights reserved.

How Does DNA Copy Itself?

Before a cell divides, it makes a copy of its genetic information for the new cell. The pairing of bases allows the cell to *replicate*, or make copies of, DNA. Remember that bases are complementary and can only fit together in certain ways. Therefore, the order of bases on one side of the DNA strand controls the order of bases on the other side of the strand. For example, the base order CGAC can only fit with the order GCTG. ☑

When DNA replicates, the pairs of bases separate and the DNA splits into two strands. The bases on each side of the original strand are used as a pattern to build a new strand. As the bases on the original strands are exposed, the cell adds nucleotides to form a new strand.

Finally, two DNA strands are formed. Half of each of the two DNA strands comes from the original strand. The other half is built from new nucleotides.

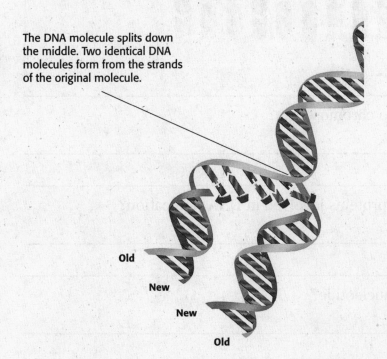

The DNA molecule splits down the middle. Two identical DNA molecules form from the strands of the original molecule.

Old
New
New
Old

READING CHECK

5. Explain What happens to DNA before a cell divides?

_____

_____

TAKE A LOOK

6. Compare What is the difference between an "old" and a "new" strand of DNA?

_____

_____

_____

DNA is copied every time a cell divides. Each new cell gets a complete copy of the entire DNA strand. Proteins in the cell unwind, copy, and rewind the DNA.

Copyright © by Holt, Rinehart and Winston; a Division of Houghton Mifflin Harcourt Publishing Company. All rights reserved.

Section 1 Review

GLE 0707.Inq.5

SECTION VOCABULARY

DNA deoxyribonucleic acid, a molecule that is present in all living cells and that contains the information that determines the traits that a living thing inherits and needs to live	**nucleotide** in a nucleic-acid chain, a subunit that consists of a sugar, a phosphate, and a nitrogenous base

1. Identify Where are genes located? What do they do?

_____

_____

2. Compare How are the four kinds of DNA nucleotides different from each other?

_____

3. Apply Concepts The diagram shows part of a strand of DNA. Using the order of bases given in the top of the strand, write the letters of the bases that belong on the bottom strand.

_____

4. Describe How is DNA related to chromosomes?

_____

_____

5. Identify Relationships How are proteins involved in DNA replication?

_____

_____

6. List What are three parts of a nucleotide?

_____

_____

_____

Copyright © by Holt, Rinehart and Winston; a Division of Houghton Mifflin Harcourt Publishing Company. All rights reserved.

CHAPTER 5 Genes and DNA

SECTION 2 How DNA Works

TN Tennessee Science Standards
GLE 0707.T/E.3
GLE 0707.T/E.4
GLE 0707.4.3

BEFORE YOU READ

After you read this section, you should be able to answer these questions:

• What does DNA look like in different cells?

• How does DNA help make proteins?

• What happens if a gene changes?

What Does DNA in Cells Look Like?

The human body contains trillions of cells, which carry out many different functions. Most cells are very small and can only be seen with a microscope. A typical skin cell, for example, has a diameter of about 0.0025 cm. However, almost every cell contains about 2 m of DNA. How can so much DNA fit into the nucleus of such a small cell? The DNA is bundled.

STUDY TIP

Compare After you read this section, make a table comparing chromatin, chromatids, and chromosomes.

Math Focus

1. Convert About how long is the DNA in a cell in inches?
1 in. = 2.54 cm

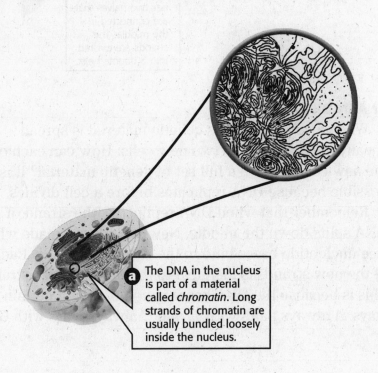

(a) The DNA in the nucleus is part of a material called *chromatin*. Long strands of chromatin are usually bundled loosely inside the nucleus.

TAKE A LOOK

2. Identify In what form is the DNA in the nucleus?

Copyright © by Holt, Rinehart and Winston; a Division of Houghton Mifflin Harcourt Publishing Company. All rights reserved.

SECTION 2 How DNA Works *continued*

FITTING DNA INTO THE CELL

Large amounts of DNA can fit inside a cell because the DNA is tightly bundled by proteins. The proteins found with DNA help support the structure and function of DNA. Together, the DNA and the proteins it winds around make up a chromosome. ☑

DNA's structure allows it to hold a lot of information. Remember that a gene is made of a string of nucleotides. That is, it is part of the 2 m of DNA in a cell. Because there is an enormous amount of DNA, there can be a large variety of genes.

✔ **READING CHECK**

3. Identify What are two things that are found in a chromosome?

_____

_____

TAKE A LOOK
4. Describe What is chromatin made of?

_____

_____

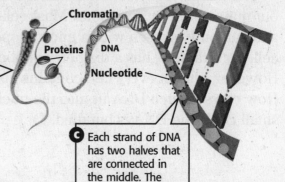

b A single strand of chromatin is made up of a long strand of DNA that is coiled around proteins.

Chromatin

Proteins   DNA

Nucleotide

c Each strand of DNA has two halves that are connected in the middle. The strands are twisted into a double helix.

Critical Thinking

5. Predict Consequences Imagine that DNA did not replicate before cell division. What would happen to the amount of DNA in each of the new cells formed during cell division?

_____

_____

_____

_____

_____

DNA IN DIVIDING CELLS

When a cell divides, its genetic material is spread equally into each of the two new cells. How can each of the new cells receive a full set of genetic material? It is possible because DNA replicates before a cell divides.

Remember that when DNA replicates, the strand of DNA splits down the middle. New strands are made when free nucleotide bases bind to the exposed strands. Each of the new strands is identical to the original DNA strand. This is because the DNA bases can join only in certain ways. *A* always pairs with *T*, and *C* always pairs with *G*.

Copyright © by Holt, Rinehart and Winston; a Division of Houghton Mifflin Harcourt Publishing Company. All rights reserved.

When a cell is ready to divide, it has already copied its DNA. The copies initially are attached as two chromatids.

TN TENNESSEE STANDARDS CHECK

GLE 0707.4.3 Explain the relationship among genes, chromosomes, and inherited traits.

6. Analyze Relationships What is the relationship between DNA and proteins?

_____

_____

_____

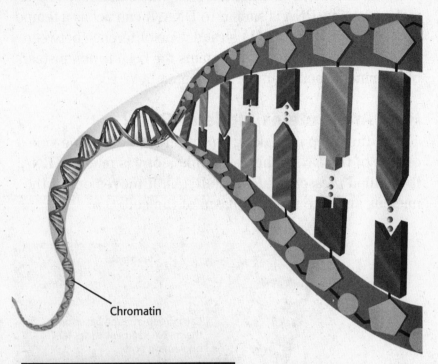

Chromatin

Chromatids

How Does DNA Help Make Proteins?

Proteins are found throughout cells. They cause most of the differences that you can see among organisms. A single organism can have thousands of different proteins.

Proteins act as chemical messengers for many of the activities in cells, helping the cells to work together. They also affect traits, such as the color of your eyes and how tall you will grow.

Proteins are made from many subunits called *amino acids*. A long string of amino acids forms a protein.

The order of bases in DNA is a code. The code tells how to make proteins. A group of three DNA bases acts as a code for one amino acid. For example, the group of DNA bases CAA *codes for*, or stands for, the amino acid valine. A gene usually contains instructions for making one specific protein.

Math Focus
7. Calculate How many DNA bases are needed to code for five amino acids?

_____

Copyright © by Holt, Rinehart and Winston; a Division of Houghton Mifflin Harcourt Publishing Company. All rights reserved.

HELP FROM RNA

RNA, or *ribonucleic acid*, is a chemical that helps DNA make proteins. RNA is similar to DNA. It can act as a temporary copy of part of a DNA strand. One difference between DNA and RNA is that RNA contains the base *uracil* instead of thymine. Uracil is often represented by *U*. ☑

How Are Proteins Made in Cells?

The first step in making a protein is to copy one side of part of the DNA. This mirrorlike copy is made of RNA. It is called *messenger RNA* (mRNA). It moves out of the nucleus and into the cytoplasm of the cell.

✓ **READING CHECK**

8. Identify What is one difference between RNA and DNA?

_____

_____

_____

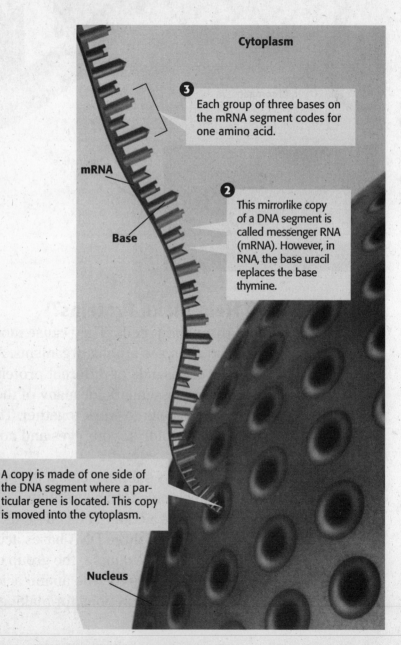

Cytoplasm

3 Each group of three bases on the mRNA segment codes for one amino acid.

mRNA

2 This mirrorlike copy of a DNA segment is called messenger RNA (mRNA). However, in RNA, the base uracil replaces the base thymine.

Base

1 A copy is made of one side of the DNA segment where a particular gene is located. This copy is moved into the cytoplasm.

Nucleus

TAKE A LOOK

9. Compare How does the shape of RNA differ from the shape of DNA?

_____

_____

_____

Copyright © by Holt, Rinehart and Winston; a Division of Houghton Mifflin Harcourt Publishing Company. All rights reserved.

SECTION 2 How DNA Works *continued*

RIBOSOMES

In the cytoplasm, the messenger RNA enters a protein assembly line. The "factory" that runs this assembly line is a ribosome. A **ribosome** is a cell organelle composed of RNA and protein. The mRNA moves through a ribosome as a protein is made.

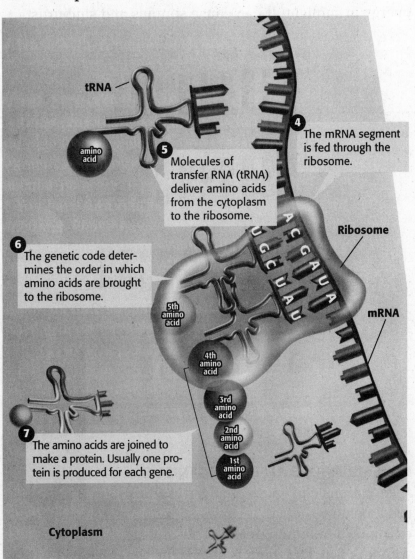

tRNA

amino acid

5 Molecules of transfer RNA (tRNA) deliver amino acids from the cytoplasm to the ribosome.

4 The mRNA segment is fed through the ribosome.

Ribosome

6 The genetic code determines the order in which amino acids are brought to the ribosome.

5th amino acid

4th amino acid

3rd amino acid

2nd amino acid

1st amino acid

mRNA

7 The amino acids are joined to make a protein. Usually one protein is produced for each gene.

Cytoplasm

What Happens If Genes Change?

Read this sentence: "Put the book on the desk." Does it make sense? What about this sentence: "Rut the zook in the tesk."? Changing only a few letters in a sentence can change what the sentence means. It can even keep the sentence from making any sense at all! In a similar way, even small changes in a DNA sequence can affect the protein that the DNA codes for. A change in the nucleotide-base sequence of DNA is called a **mutation**. ☑

Copyright © by Holt, Rinehart and Winston; a Division of Houghton Mifflin Harcourt Publishing Company. All rights reserved.

Critical Thinking

10. Explain Proteins are made in the cytoplasm, but DNA never leaves the nucleus of a cell. How does DNA control how proteins are made?

_____

_____

_____

_____

_____

_____

TAKE A LOOK

11. Identify What does tRNA do?

_____

_____

_____

☑ READING CHECK

12. Define What is a mutation?

_____

_____

_____

SECTION 2 How DNA Works *continued*

HOW MUTATIONS HAPPEN

Some mutations happen because of mistakes when DNA is copied. Other mutations happen when DNA is damaged. Things that can cause mutations are called *mutagens*. Examples of mutagens include X rays and ultraviolet radiation. Ultraviolet radiation is one type of energy in sunlight. It can cause suntans and sunburns.

Original sequence

Base pair replaced

Base pair added

Base pair removed

Mutations can happen in different ways. A nucleotide may be replaced, added, or removed.

TAKE A LOOK
13. Compare What happens to one strand of DNA when there is a change in a base on the other strand?

_____

_____

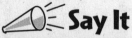 **Say It**

Brainstorm Whether a mutation is helpful or harmful to an organism often depends on the organism's environment. In a group, discuss how the same mutation could be helpful in one environment but harmful in another.

HOW MUTATIONS AFFECT ORGANISMS

Mutations can cause changes in traits. Some mutations produce new traits that can help an organism survive. For example, a mutation might allow an organism to survive with less water. If there is a drought, the organism will be more likely to survive.

Many mutations produce traits that make an organism less likely to survive. For example, a mutation might make an animal a brighter color. This might make the animal easier for predators to find.

Some mutations are neither helpful nor harmful. If a mutation does not cause a change in a protein, then the mutation will not help or hurt the organism.

Copyright © by Holt, Rinehart and Winston; a Division of Houghton Mifflin Harcourt Publishing Company. All rights reserved.

SECTION 2 How DNA Works *continued*

PASSING ON MUTATIONS

Cells make proteins that can find and fix many mutations. However, not all mutations can be fixed.

If a mutation happens in egg or sperm cells, the changed gene can be passed from one generation to the next. For example, sickle cell disease is caused by a genetic mutation that can be passed to future generations.

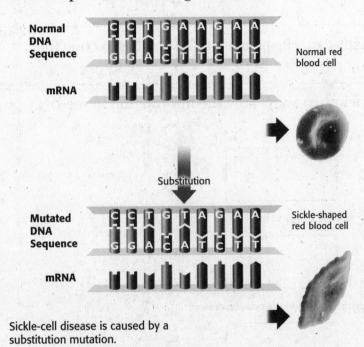

Sickle-cell disease is caused by a substitution mutation.

TAKE A LOOK
14. Identify What kind of mutation causes sickle cell disease: deletion, insertion, or substitution?

_____

How Can We Use Genetic Knowledge?

Scientists use their knowledge of genetics in many ways. Most of these ways are helpful to people. However, other ways can cause ethical and scientific concerns.

GENETIC ENGINEERING

Scientists have learned how to change individual genes within organisms. This is called *genetic engineering*. In some cases, scientists transfer genes from one organism to another. For example, scientists can transfer genes from people into bacteria. The bacteria can then make proteins for people who are sick. ☑

GENETIC IDENTIFICATION

Your DNA is unique, so it can be used like a fingerprint to identify you. *DNA fingerprinting* identifies the unique patterns in a person's DNA. Scientists can use these genetic fingerprints as evidence in criminal cases. They can also use genetic information to determine whether people are related.

✓ READING CHECK
15. **Define** What is genetic engineering?

_____

_____

Copyright © by Holt, Rinehart and Winston; a Division of Houghton Mifflin Harcourt Publishing Company. All rights reserved.

Section 2 Review

GLE 0707.T/E.3, GLE 0707.T/E.4, GLE 0707.4.3 **TN**

SECTION VOCABULARY

mutation a change in the nucleotide-base sequence of a gene or DNA molecule **ribosome** a cell organelle composed of RNA and protein; the site of protein synthesis	**RNA** ribonucleic acid, a molecule that is present in all living cells and that plays a role in protein production

1. Identify What structures in cells contain DNA and proteins?

_____

2. Calculate How many amino acids can a sequence of 24 DNA bases code for?

_____

3. Explain Fill in the flow chart below to show how the information in the DNA code becomes a protein.

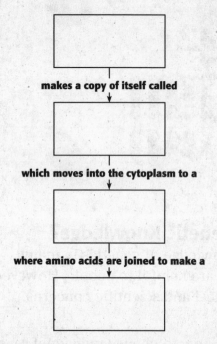

makes a copy of itself called

which moves into the cytoplasm to a

where amino acids are joined to make a

4. Draw Conclusions How can a mutation in a DNA base sequence cause a change in a gene and a trait? What determines whether the mutation is passed on to offspring?

_____

_____

_____

5. Identify Give two ways that genetic fingerprinting can be used.

_____

_____

Copyright © by Holt, Rinehart and Winston; a Division of Houghton Mifflin Harcourt Publishing Company. All rights reserved.

CHAPTER 6 Introduction to Plants
SECTION 1 # What Is a Plant?

BEFORE YOU READ

After you read this section, you should be able to answer these questions:

• What characteristics do all plants share?

• What are two differences between plant cells and animal cells?

TN Tennessee Science Standards
GLE 0707.Inq.3
GLE 0707.1.1

What Are the Characteristics of Plants?

A plant is an organism that uses sunlight to make food. Trees, grasses, ferns, cactuses, and dandelions are all types of plants. Plants can look very different, but they all share four characteristics.

STUDY TIP

Organize As you read, make a diagram to show the major groups of plants. Be sure to include the characteristics of each group.

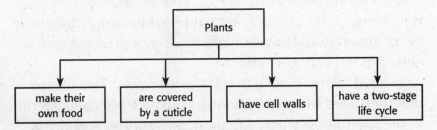

PHOTOSYNTHESIS

Plants make their own food from carbon dioxide, water, and energy from sunlight. This process is called *photosynthesis*. Photosynthesis takes place in special organelles called *chloroplasts*. Inside the chloroplasts, a green pigment called *chlorophyll* collects energy from the sun for photosynthesis. Chlorophyll is what makes most plants look green. Animal cells do not have chloroplasts. ☑

CUTICLES

Every plant has a cuticle that covers and protects it. A *cuticle* is a waxy layer that coats a plant's leaves and stem. The cuticle keeps plants from drying out by keeping water inside the plant.

CELL WALLS

How do plants stay upright? They do not have skeletons, as many animals do. Instead, each plant cell is surrounded by a stiff cell wall. The cell wall is outside the cell membrane. Cell walls support and protect the plant cell. Animal cells do not have cell walls.

READING CHECK

1. Define What is chlorophyll?

_____

_____

_____

Copyright © by Holt, Rinehart and Winston; a Division of Houghton Mifflin Harcourt Publishing Company. All rights reserved.

Structures in a Plant Cell

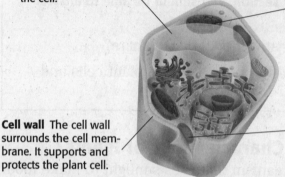

Large central vacuole
A vacuole stores water and helps support the cell.

Chloroplast Chloroplasts contain chlorophyll. Chlorophyll captures energy from the sun. Plants use this energy to make food.

Cell wall The cell wall surrounds the cell membrane. It supports and protects the plant cell.

Cell membrane The cell membrane surrounds a plant cell and lies under the cell wall.

TAKE A LOOK
2. Identify What structure in a plant cell stores water?

_____

_____

3. Identify Where is chlorophyll found?

_____

_____

TWO-STAGE LIFE CYCLE

Many organisms, including plants, produce offspring when a sperm joins with an egg. This is called *sexual reproduction*. In animals, sexual reproduction happens in every generation. However, plants do not produce sperm and eggs in every generation.

Instead, plants have a two-stage life cycle. This means that they need two generations to produce eggs and sperm. In the *sporophyte* stage, a plant makes spores. A spore is a cell that can divide and grow into a new plant. This new plant is called a *gametophyte*. In the gametophyte stage, the plants produce sperm and eggs. The sperm and eggs must join to produce a new sporophyte. ☑

✓ READING CHECK
4. List What are the two stages of the plant life cycle?

_____

_____

What Are the Main Groups of Plants?

There are two main groups of plants: vascular and nonvascular. A **vascular plant** has specialized vascular tissues. *Vascular tissues* move water and nutrients from one part of a plant to another. A **nonvascular plant** does not have vascular tissues to move water and nutrients.

The Main Groups of Plants

Nonvascular Plants	Vascular Plants		
	Seedless plants	Seed plants	
Mosses, liverworts, and hornworts	Ferns, horsetails, and club mosses	Nonflowering	Flowering
		Gymnosperms	Angiosperms

Copyright © by Holt, Rinehart and Winston; a Division of Houghton Mifflin Harcourt Publishing Company. All rights reserved.

SECTION 1 What Is a Plant? *continued*

NONVASCULAR PLANTS

Nonvascular plants depend on diffusion to move water and nutrients through the plant. In *diffusion*, water and nutrients move through a cell membrane and into a cell. Each cell must get water and nutrients from the environment or a cell that is close by.

Nonvascular plants can rely on diffusion because they are small. If a nonvascular plant were large, not all of its cells would get enough water and nutrients. Most nonvascular plants live in damp areas, so each of their cells is close to water.

VASCULAR PLANTS

Many of the plants we see in gardens and forests are vascular plants. Vascular plants are divided into two groups: seedless plants and seed plants. Seed plants are divided into two more groups—flowing and nonflowering. Nonflowering seed plants, such as pine trees, are called **gymnosperms**. Flowering seed plants, such as magnolias, are called **angiosperms**.

What Are the Ancestors of Plants?

What would you see if you traveled back in time about 440 million years? The Earth would be a strange, bare place. There would be no plants on land. Where did plants come from?

The green alga in the figure below may look like a plant, but it is not. However, it does share some characteristics with plants. Both algae and plants have the same kind of chlorophyll and make their food by photosynthesis. Like plants, algae also have a two-stage life cycle. Scientists think these similarities show that plants and green algae share a common ancestor.

Critical Thinking

5. Apply Concepts Do you think a sunflower is a gymnosperm or an angiosperm? Explain your answer.

_____

_____

_____

TN TENNESSEE STANDARDS CHECK

GLE 0707.1.1 Make observations and describe the <u>structure</u> and <u>function</u> of organelles found in plant and animal cells.

Word Help: <u>structure</u> the arrangement of parts of a whole

Word Help: <u>function</u> use or purpose

6. Apply Concepts Green algae and plants both make their food by photosynthesis. What organelle do they have in common? Explain.

_____

_____

_____

_____

A modern green alga and plants, such as ferns, share several characteristics. Because of this, scientists think that both types of organisms shared an ancient ancestor.

Copyright © by Holt, Rinehart and Winston; a Division of Houghton Mifflin Harcourt Publishing Company. All rights reserved.

Section 1 Review

GLE 0707.Inq.3, GLE 0707.1.1 **TN**

SECTION VOCABULARY

angiosperm a flowering plant that produces seeds within a fruit **gymnosperm** a woody, vascular seed plant whose seeds are not enclosed by an ovary or fruit	**nonvascular plant** the three groups of plants (liverworts, hornworts, and mosses) that lack specialized conducting tissues and true roots, stems, and leaves **vascular plant** a plant that has specialized tissues that conduct materials from one part of the plant to another

1. Explain What are the two main differences between a plant cell and an animal cell?

_____

_____

_____

_____

2. Organize Fill in each box in the figure below with one of the main characteristics of plants.

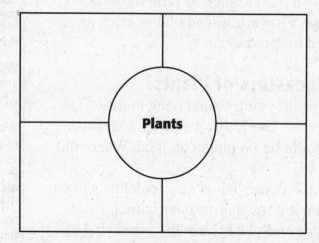

3. Predict What would happen to a plant if its chloroplasts stopped working? Explain your answer.

_____

_____

_____

4. Compare What is the main difference between vascular and nonvascular plants?

_____

_____

_____

Copyright © by Holt, Rinehart and Winston; a Division of Houghton Mifflin Harcourt Publishing Company. All rights reserved.

CHAPTER 6 Introduction to Plants
SECTION 2 Seedless Plants

TN Tennessee Science Standards
GLE 0707.Inq.2
GLE 0707.Inq.3
GLE 0707.Inq.5

BEFORE YOU READ

After you read this section, you should be able to answer these questions:

• What are the differences between seedless vascular plants and nonvascular plants?

• How can plants reproduce without seeds?

What Are Seedless Plants?

When you think of plants, you probably think of plants that make seeds, such as flowers and trees. However, there are many plants that don't make seeds.

Remember that plants are divided into two main groups: nonvascular plants and vascular plants. All nonvascular plants are seedless, and some vascular plants are seedless, as well.

What Are the Features of Nonvascular Plants?

Mosses, liverworts, and hornworts are types of nonvascular plants. Remember that nonvascular plants do not have vascular tissue to deliver water and nutrients. Instead, each plant cell gets water and nutrients directly from the environment or from a nearby cell. Therefore, nonvascular plants usually live in places that are damp.

Nonvascular plants do not have true stems, roots, or leaves. However, they do have features that help them to get water and stay in place. For example, a **rhizoid** is a rootlike structure that holds some nonvascular plants in place. Rhizoids also help plants get water and nutrients. ☑

Nonvascular plants
• have no vascular tissue
• have no true roots, stems, leaves, or seeds
• are usually small
• live in damp places

REPRODUCTION IN NONVASCULAR PLANTS

Like all plants, nonvascular plants have a two-stage life cycle. They have a sporophyte generation, which produces spores, and a gametophyte generation, which produces eggs and sperm. Sperm from these plants need water so they can swim to the eggs. Nonvascular plants can also reproduce asexually, that is, without eggs and sperm.

STUDY TIP

Organize As you read this section, make a chart that compares vascular plants and nonvascular plants.

Critical Thinking

1. Apply Concepts Why wouldn't you expect to see nonvascular plants in the desert?

_____

_____

_____

_____

READING CHECK

2. List What are two functions of the rhizoid?

_____

_____

_____

Copyright © by Holt, Rinehart and Winston; a Division of Houghton Mifflin Harcourt Publishing Company. All rights reserved.

Moss Life Cycle

ⓑ The sporophyte releases spores into the air.

ⓒ Spores land in a moist place, crack open, and grow into leafy gametophytes.

Spores

Gametophyte

Sporophyte

Male

ⓐ The fertilized egg grows into a sporophyte. The sporophyte grows from the top of the gametophyte.

Female

Gametophyte

Egg

Sporophyte

Fertilized egg

Sperm

ⓓ Sperm swim through water from the male gametophyte to fertilize the egg at the top of the female gametophyte.

TAKE A LOOK

3. Identify Are the male and female gametophytes separate plants or part of the same plant?

_____

_____

Critical Thinking

4. Apply Concepts Why do you think nonvascular plants can be the first plants to grow in a new environment?

_____

_____

_____

_____

☑ READING CHECK

5. Explain How do the cells of a seedless vascular plant get water?

_____

_____

_____

IMPORTANCE OF NONVASCULAR PLANTS

Nonvascular plants are usually the first plants to live in a new environment, such as newly exposed rock. When these plants die, they break down and help form a thin layer of soil. Then plants that need soil in order to grow can move into these areas.

Some nonvascular plants are important as food or nesting material for animals. A nonvascular plant called peat moss is important to humans. When it turns to peat, it can be burned as a fuel.

What Are the Features of Seedless Vascular Plants?

Vascular plants have specialized tissues that carry water and nutrients to all their cells. These tissues generally make seedless vascular plants larger than nonvascular plants. Because they have tissues to move water, vascular plants do not have to live in places that are damp. ☑

Many seedless vascular plants, such as ferns, have a structure called a rhizome. The **rhizome** is an underground stem that produces new leaves and roots.

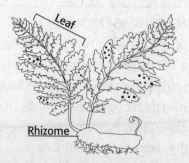

Leaf

Rhizome

Copyright © by Holt, Rinehart and Winston; a Division of Houghton Mifflin Harcourt Publishing Company. All rights reserved.

REPRODUCTION IN SEEDLESS VASCULAR PLANTS

The life cycles of vascular plants and nonnvascular plants are similar. Sperm and eggs are produced in gametophytes. They join to form a sporophyte. The sporophyte produces spores. The spores are released. They grow into new gametophytes. ☑

Seedless vascular plants can also reproduce asexually in two ways. New plants can branch off from older plants. Pieces of a plant can fall off and begin to grow as new plants.

Fern Life Cycle

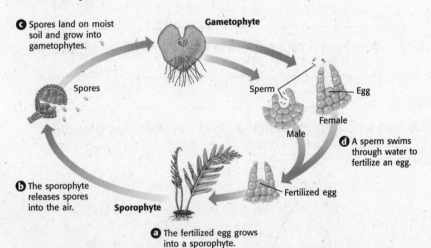

ⓒ Spores land on moist soil and grow into gametophytes.

Gametophyte

Spores

Sperm

Egg

Female

Male

ⓓ A sperm swims through water to fertilize an egg.

Fertilized egg

ⓑ The sporophyte releases spores into the air.

Sporophyte

ⓐ The fertilized egg grows into a sporophyte.

READING CHECK

6. **Identify** What do spores grow into?

_____
_____

TAKE A LOOK
7. **Apply Concepts** Does this figure show sexual or asexual reproduction? Explain your answer.

_____
_____
_____
_____

IMPORTANCE OF SEEDLESS VASCULAR PLANTS

Did you know that seedless vascular plants that lived 300 million years ago are important to people today? After these ancient ferns, horsetails, and club mosses died, they formed coal and oil. Coal and oil are fossil fuels that people remove from Earth's crust to use for energy. They are called *fossil fuels* because they formed from plants (or animals) that lived long ago. ☑

Another way seedless vascular plants are important is they help make and preserve soil. Seedless vascular plants help form new soil when they die and break down. Their roots can make the soil deeper, which allows other plants to grow. Their roots also help prevent soil from washing away.

Many seedless vascular plants are used by humans. Ferns and some club mosses are popular houseplants. Horsetails are used in some shampoos and skincare products.

READING CHECK

8. **Explain** Where does coal come from?

_____
_____
_____
_____

Copyright © by Holt, Rinehart and Winston; a Division of Houghton Mifflin Harcourt Publishing Company. All rights reserved.

Name _____ Class _____ Date _____

Section 2 Review

GLE 0707.Inq.2, GLE 0707.Inq.3, GLE 0707.Inq.5 TN

SECTION VOCABULARY

rhizoid a rootlike structure in nonvascular plants that holds the plants in place and helps plants get water and nutrients	**rhizome** a horizontal underground stem that produces new leaves, shoots, and roots

1. Compare What are two differences between a rhizoid and a rhizome?

_____

_____

_____

_____

2. Explain In which generation does sexual reproduction occur? Explain your answer.

_____

_____

3. Compare Use a Venn Diagram to compare vascular and nonvascular plants.

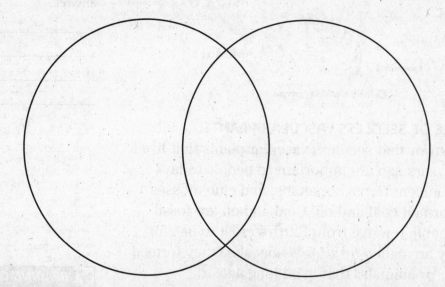

4. Describe What are two ways in which seedless nonvascular plants reproduce asexually?

_____

_____

5. Apply Concepts Nonvascular plants are usually very small. How does their structure limit their size?

_____

_____

Copyright © by Holt, Rinehart and Winston; a Division of Houghton Mifflin Harcourt Publishing Company. All rights reserved.

CHAPTER 6 Introduction to Plants
SECTION 3 Seed Plants

TN **Tennessee Science Standards**
GLE 0707.Inq.2
GLE 0707.Inq.3
GLE 0707.Inq.5
GLE 0707.4.2

BEFORE YOU READ

After you read this section, you should be able to answer these questions:

• How are seed plants different from seedless plants?

• What are the parts of a seed?

• How do gymnosperms and angiosperms reproduce?

What Are Seed Plants?

Think about the seed plants that you use during the day. You probably use dozens of seed plants, including the food you eat and the paper you write on. Seed plants include trees, such as oaks and pine trees, as well as flowers, such as roses and dandelions. Seed plants are one of the two main groups of vascular plants.

Like all plants, seed plants have a two-stage life cycle. However, seed plants differ from seedless plants, as shown below.

Seedless plants	Seed plants
They do not produce seeds.	They produce seeds.
The gametophyte grows as an independent plant.	The gametophyte lives inside the sporophyte.
Sperm need water to swim to the eggs.	Sperm are carried to the eggs by pollen.

Seed plants do not depend on moist habitats for reproduction, the way seedless plants do. Because of this, seed plants can live in many more places than seedless plants can. Seed plants are the most common plants on Earth today.

What Is a Seed?

A *seed* is a structure that feeds and protects a young plant. It forms after fertilization, when a sperm and an egg join. A seed has the following three main parts:

• a young plant, or sporophyte

• *cotyledons*, early leaves that provide food for the young plant

• a seed coat that covers and protects the young plant ☑

STUDY TIP

Organize As you read this section, make cards showing the parts of the life cycle of seed plants. Practice arranging the cards in the correct sequence.

TAKE A LOOK

1. Compare How do the gametophytes of seedless plants differ from those of seed plants?

_____

_____

_____

_____

READING CHECK

2. Identify What process must occur before a seed can develop?

_____

Copyright © by Holt, Rinehart and Winston; a Division of Houghton Mifflin Harcourt Publishing Company. All rights reserved.

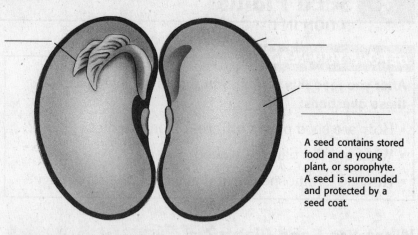

A seed contains stored food and a young plant, or sporophyte. A seed is surrounded and protected by a seed coat.

TAKE A LOOK

3. Label Label the parts of a seed with these terms: young plant, seed coat, cotyledon.

TN TENNESSEE STANDARDS CHECK

GLE 0707.4.2 Demonstrate an understanding of <u>sexual</u> reproduction in flowering plants.

Word Help: <u>sexual</u> having to do with sex

4. Identify What are two advantages seeds have over spores?

_____

_____

_____

_____

_____

_____

ADVANTAGES OF HAVING SEEDS

Seeds have some advantages over spores. For example, when the young plant inside a seed begins to grow, it uses the food stored in the seed. In contrast, the spores of seedless plants don't have stored food to help a new plant grow. Therefore, they will live only if they start growing when and where there are enough resources available.

Another advantage is that seeds can be spread by animals. The spores of seedless plants are usually spread by wind. Animals often spread seeds more efficiently than the wind spreads spores. Therefore, seeds that are spread by animals are more likely to find a good place to grow.

What Kinds of Plants Have Seeds?

Seed plants are divided into two main groups: gymnosperms and angiosperms. *Gymnosperms* are non-flowering plants, and *angiosperms* are flowering plants.

GYMNOSPERMS

Gymnosperms are seed plants that do not have flowers or fruits. They include plants such as pine trees and redwood trees. Many gymnosperms are evergreen, which means that they keep their leaves all year. Gymnosperm seeds usually develop in a cone, such as a pine cone.

Pine cone

Seeds

SECTION 3 Seed Plants *continued*

REPRODUCTION IN GYMNOSPERMS

The most well-known gymnosperms are the conifers. Conifers are evergreen trees and shrubs, such as pines, spruces, and firs, that make cones to reproduce. They have male cones and female cones. Spores in male cones develop into male gametophytes, and spores in female cones develop into female gametophytes. The gameto-phytes produce sperm and eggs.

A **pollen** grain contains the tiny male gametophyte. The wind carries pollen from the male cones to the female cones. This movement of pollen to the female cones is called **pollination**. Pollination is part of sexual reproduction in plants. ☑

After pollination, sperm fertilize the eggs in the female cones. A fertilized egg develops into a new sporophyte inside a seed. Eventually, the seeds fall from the cone. If the conditions are right, the seeds will grow into plants.

☑ **READING CHECK**

5. Explain How is gymnosperm pollen carried from one plant to another?

_____

_____

The Life Cycle of a Pine Tree

Gametophyte

c Sperm and eggs are produced in the cones.

Egg

Fertilized egg

Female cone (cutaway view)

Male cone (cutaway view)

Pollen

d Wind carries pollen to the egg. A sperm from a pollen grain fertilizes the egg.

b Spores are produced. They grow into gametophytes.

Male cones

Seed

Female cones

Sporophyte

a The seed contains a young sporophyte, which grows into an adult sporophyte.

TAKE A LOOK

6. Explain Does this picture show an example of sexual or asexual reproduction? Explain.

_____

_____

_____

_____

IMPORTANCE OF GYMNOSPERMS

Gymnosperms are used to make many products, such as medicines, building materials, and household products. Some conifers produce a drug used to fight cancer. Many trees are cut so that their wood can be used to build homes and furniture. Pine trees make a sticky substance called resin. Resin can be used to make soap, paint, and ink.

Copyright © by Holt, Rinehart and Winston; a Division of Houghton Mifflin Harcourt Publishing Company. All rights reserved.

What Are Angiosperms?

Angiosperms are seed plants that produce flowers and fruit. Maple trees, daisies, and blackberries are all examples of angiosperms. There are more angiosperms on Earth than any other kind of plant. They can be found in almost every land ecosystem, including grasslands, deserts, and forests.

TWO KINDS OF ANGIOSPERMS

There are two kinds of angiosperms: monocots and dicots. These plants are grouped based on how many cotyledons, or seed leaves, the seeds have. *Monocots* have seeds with one cotyledon. Grasses, orchids, palms, and lilies are all monocots.

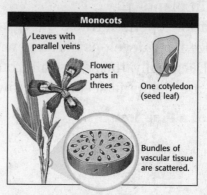

Dicots have seeds with two cotyledons. Roses, sunflowers, peanuts, and peas are all dicots.

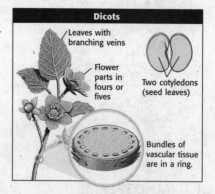

Math Focus

7. Calculate Percentages
More than 265,000 species of plants have been discovered. About 235,000 of those species are angiosperms. What percentage of plants are angiosperms?

_____

TAKE A LOOK
8. Complete Fill in the table to show the differences between monocots and dicots.

Monocots	Dicots
	flower parts in fours or fives
leaves with parallel veins	
	two cotyledons
bundles of vascular tissue scattered	

Copyright © by Holt, Rinehart and Winston; a Division of Houghton Mifflin Harcourt Publishing Company. All rights reserved.

SECTION 3 Seed Plants *continued*

REPRODUCTION IN ANGIOSPERMS

In angiosperms, pollination takes place in flowers. Some angiosperms depend on the wind for pollination. Others rely on animals such as bees and birds to carry pollen from flower to flower.

Angiosperm seeds develop inside fruits. Some fruits and seeds, like those of a dandelion, are made to help the wind carry them. Other fruits, such as blackberries, attract animals that eat them. The animals drop the seeds in new places, where they can grow into plants. Some fruits, such as burrs, travel by sticking to animal fur. ☑

✓ READING CHECK
9. Identify Where do angiosperm seeds develop?

_____

_____

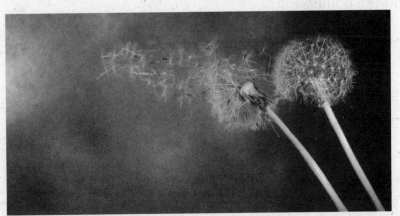

Each of the fluffy structures on this dandelion is actually a fruit. Each of the fruits contains a seed.

TAKE A LOOK
10. Identify How are the fruits of this dandelion spread?

_____

_____

IMPORTANCE OF ANGIOSPERMS

Like many other plants, flowering plants provide food for animals. A mouse that eats seeds and berries uses flowering plants directly as food. An owl that eats a field mouse uses flowering plants indirectly as food. Flowering plants can also provide food for the animals that pollinate them.

People use flowering plants, too. Major food crops, such as corn, wheat, and rice, come from flowering plants. Many flowering trees, such as oak trees, can be used for building materials. Plants such as cotton and flax are used to make clothing and rope. Flowering plants are also used to make medicines, rubber, and perfume oils.

📣 Say It
Describe Think of all the products you used today that came from angiosperms. Describe to the class five items you used in some way and what kind of angiosperm they came from.

Copyright © by Holt, Rinehart and Winston; a Division of Houghton Mifflin Harcourt Publishing Company. All rights reserved.

Name _____ Class _____ Date _____

Section 3 Review

GLE 0707.Inq.2, GLE 0707.Inq.3, GLE 0707.Inq.5, GLE 0707.4.2 TN

SECTION VOCABULARY

pollen the tiny granules that contain the male gametophyte of seed plants	**pollination** the transfer of pollen from the male reproductive structures to the female reproductive structures of seed plants

1. Compare How are the gametophytes of seed plants different from the gameto-phytes of seedless plants?

_____

_____

_____

2. Describe What happens during pollination?

_____

_____

3. Compare Use a Venn Diagram to compare gymnosperms and angiosperms.

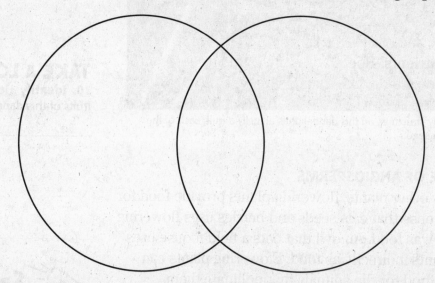

4. List What are the three main parts of a seed? What does each part do?

_____

_____

_____

5. Explain Does a plant benefit if its fruit is eaten by a bird? Why or why not?

_____

6. Identify What two structures are unique to angiosperms?

_____

Copyright © by Holt, Rinehart and Winston; a Division of Houghton Mifflin Harcourt Publishing Company. All rights reserved.

CHAPTER 6 Introduction to Plants
SECTION 4 Structures of Seed Plants

TN Tennessee Science Standards
GLE 0707.Inq.3
GLE 0707.1.2
GLE 0707.4.2

BEFORE YOU READ

After you read this section, you should be able to answer these questions:

• What are the functions of roots and stems?

• What is the function of leaves?

• What is the function of a flower?

What Structures Are Found in a Seed Plant?

Remember that seed plants include trees, such as oaks and pine trees, as well as flowers, such as roses and dandelions. Seed plants are one of the two main groups of vascular plants.

You have different body systems that carry out many functions. Plants have systems too. Vascular plants have a root system, a shoot system, and a reproductive system. A plant's root and shoot systems help the plant to get water and nutrients. Roots are often found underground. Shoots include stems and leaves. They are usually found above ground. ☑

STUDY TIP

List As you read this section, make a chart listing the structures of seed plants and their functions.

READING CHECK

1. Identify What are the three main parts of a seed plant?

_____

_____

Onion

Dandelion

Carrots

The roots of plants absorb and store water and nutrients.

Copyright © by Holt, Rinehart and Winston; a Division of Houghton Mifflin Harcourt Publishing Company. All rights reserved.

SECTION 4 Structures of Seed Plants *continued*

VASCULAR TISSUE

Like all vascular plants, seed plants have specialized tissues that move water and nutrients through the plant. There are two kinds of vascular tissue: xylem and phloem. **Xylem** moves water and minerals from the roots to the shoots. **Phloem** moves food molecules to all parts of the plant. The vascular tissues in the roots and shoots are connected. ☑

What Are Roots?

Roots are organs that have three main functions:

• to absorb water and nutrients from the soil

• to hold plants in the soil

• to store extra food made in the leaves

Roots have several structures that help them do these jobs. The *epidermis* is a layer of cells that covers the outside of the root, like skin. Some cells of the epidermis, called *root hairs*, stick out from the root. These hairs expose more cells to water and minerals in the soil. This helps the root absorb more of these materials.

Roots grow longer at their tips. A *root cap* is a group of cells found at the tip of a root. These cells produce a slimy substance. This helps the root push through the soil as it grows.

READING CHECK

2. Describe What are the functions of xylem and phloem?

_____

_____

_____

_____

Critical Thinking

3. Apply Concepts What do you think happens to water and minerals right after they are absorbed by roots?

_____

_____

_____

TAKE A LOOK

4. Identify Where is the vascular tissue located in this root?

_____

_____

The Parts of a Root

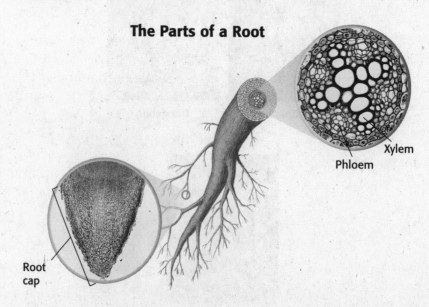

Root cap

Xylem

Phloem

Copyright © by Holt, Rinehart and Winston; a Division of Houghton Mifflin Harcourt Publishing Company. All rights reserved.

SECTION 4 Structures of Seed Plants *continued*

TYPES OF ROOT SYSTEMS

There are two kinds of root systems: taproot systems and fibrous root systems. A *taproot system* has one main root, or taproot, that grows downward. Many smaller roots branch from the taproot. Taproots can reach water deep underground. Carrots are plants that have taproot systems.

A *fibrous root system* has several roots that spread out from the base of a plant's stem. The roots are usually the same size. Fibrous roots usually get water from near the soil surface. Many grasses have fibrous root systems.

What Are Stems?

A stem is an organ that connects a plant's roots to its leaves and reproductive structures. A stem does the following jobs: ☑

* Stems support the plant body. Leaves are arranged along stems so that each leaf can get sunlight.
* Stems hold up reproductive structures such as flowers. This helps bees and other pollinators find the flowers.
* Stems carry materials between the root system and the leaves and reproductive structures. Xylem carries water and minerals from the roots to the rest of the plant. Phloem carries the food made in the leaves to roots and other parts of the plant.
* Some stems store materials. For example, the stems of cactuses can store water.

HERBACEOUS STEMS

There are two different types of stems: herbaceous and woody. *Herbaceous* stems are thin, soft, and flexible. Flowers, such as daisies and clover, have herbaceous stems. Many crops, such as tomatoes, corn, and beans, also have herbaceous stems.

Phloem

Xylem

Herbaceous stems are thin and flexible

> ✔ **READING CHECK**
>
> **5. Define** What is a stem?
>
> _____
>
> _____
>
> _____

TAKE A LOOK
6. Compare Examine this figure and the pictures of woody stems on the next page. How are herbaceous and woody stems similar?

_____

_____

Copyright © by Holt, Rinehart and Winston; a Division of Houghton Mifflin Harcourt Publishing Company. All rights reserved.

WOODY STEMS

Other plants have woody stems. *Woody* stems are stiff and are often covered by bark. Trees and shrubs have woody stems. The trunk of a tree is actually its stem!

Trees or shrubs that live in areas with cold winters grow mostly during the spring and summer. During the winter, these plants are *dormant*. This means they are not growing or reproducing. Plants that live in areas with wet and dry seasons are dormant during the dry season.

When a growing season starts, the plant produces large xylem cells. These large cells appear as a light-colored ring when the plant stem is cut. In the fall, right before the dormant period, the plant produces smaller xylem cells. The smaller cells produce a dark ring in the stem. A ring of dark cells surrounding a ring of light cells makes up a *growth ring*. The number of growth rings can show how old the tree is.

Critical Thinking

7. Infer How do you think growth rings can be used to tell how old a tree is?

_____

_____

_____

_____

_____

_____

_____

TAKE A LOOK
8. Compare How are herbaceous and woody stems different?

_____

_____

_____

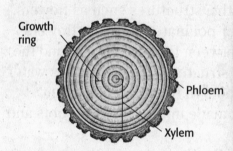

Growth ring

Phloem

Xylem

Woody stems are usually thick and stiff.

What Are Leaves?

FUNCTION OF LEAVES

Leaves are organs, too. The main function of leaves is to make food for the plant. The leaves are where most photosynthesis happens. Chloroplasts in the leaf cells trap energy from sunlight. The leaves also absorb carbon dioxide from the air. They use this energy, carbon dioxide, and water to make food. ☑

All leaf structures are related to the leaf's main job, photosynthesis. A *cuticle* covers the surfaces of the leaf. It prevents the leaf from losing water. The *epidermis* is a single layer of cells beneath the cuticle. Tiny openings in the epidermis, called *stomata* (singular, *stoma*), let carbon dioxide enter the leaf. *Guard cells* open and close the stomata.

 READING CHECK

9. Identify What is the main function of a leaf?

_____

_____

Copyright © by Holt, Rinehart and Winston; a Division of Houghton Mifflin Harcourt Publishing Company. All rights reserved.

SECTION 4 Structures of Seed Plants *continued*

Structure of a Leaf

Cuticle
— Upper epidermis
— Palisade layer
— Spongy layer
— Lower epidermis
Xylem
Phloem — Vascular tissue
Cuticle
Stoma
Guard cells

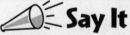

TAKE A LOOK
10. Explain Is this plant vascular or nonvascular? Explain your answer.

_____

_____

_____

LEAF LAYERS

Most photosynthesis takes place in the two layers in the middle of the leaf. The upper layer, called the *palisade layer*, contains many chloroplasts. Sunlight is captured in this layer. The lower layer, called the *spongy layer*, has spaces between the cells, where carbon dioxide can move. The spongy layer also has the vascular tissues that bring water to the leaves and move food away.

LEAF SHAPES

Different kinds of plants can have different shaped leaves. Leaves may be round, narrow, heart-shaped, or fan-shaped. Leaves can also be different sizes. The raffia palm has leaves that may be six times longer than you are tall! Duckweed is a tiny plant that lives in water. Its leaves are so small that several of them could fit on your fingernail. Some leaves, such as those of poison ivy below, can be made of several leaflets.

This is one poison ivy leaf. It is made up of three leaflets

Say It

Describe Some people are allergic to poison ivy. They can get a rash from touching its leaves. Some other plants can be poisonous to eat. Are there any other plants you know of that can be poisonous to touch or eat? Describe some of these plants to a partner.

Copyright © by Holt, Rinehart and Winston; a Division of Houghton Mifflin Harcourt Publishing Company. All rights reserved.

SECTION 4 Structures of Seed Plants *continued*

What Are Flowers?

All plants have reproductive structures. In angiosperms, or flowering plants, flowers are the reproductive structures. Flowers produce eggs and sperm for sexual reproduction. ☑

PARTS OF A FLOWER

There are four basic parts of a flower: sepals, petals, stamens, and one or more pistils. These parts are often arranged in rings, one inside the other. However, not all flowers have every part.

Different species of flowering plants can have different flower types. Flowers with all four parts are called *perfect flowers*. Flowers that have stamens but no pistils are male. Flowers that have pistils but no stamens are female.

Parts of a Flower

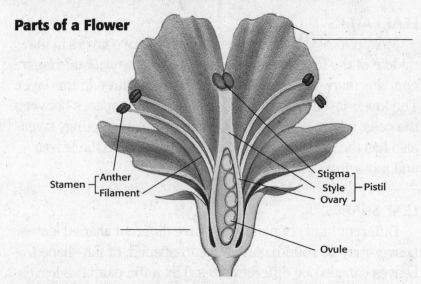

Stamen ⌈Anther
       ⌊Filament

Stigma⌉
Style ⌉─Pistil
Ovary⌋

Ovule

SEPALS

Sepals are leaves that make up the outer ring of flower parts. They are often green like leaves, but they may have other colors. Sepals protect and cover the flower while it is still a bud. When the flower begins to open, the sepals fold back, so the petals can be seen.

PETALS

Petals are leaflike parts of a flower. They make up the next ring inside of the sepals. Petals are sometimes brightly colored, like the petals of poppy flowers or roses. Many plants need animals to help spread their pollen. These colors help attract insects and other animals.

READING CHECK

11. Identify For which group of plants are flowers the reproductive structures?

_____
_____

TAKE A LOOK

12. Label As you read, fill in the missing labels on the diagram.

13. Identify What two parts make up the stamen?

_____
_____
_____

14. Identify What three parts make up the pistil?

_____
_____
_____

Copyright © by Holt, Rinehart and Winston; a Division of Houghton Mifflin Harcourt Publishing Company. All rights reserved.

STAMENS

A **stamen** is the male reproductive structure of a flower. Structures on the stamen called *anthers* produce pollen. Pollen contains the male gametophyte, which produces sperm. The anther rests on a thin stalk called a *filament*. ☑

PISTILS

A **pistil** is the female reproductive structure. The tip of the pistil is called the *stigma*. The long, thin part of the pistil is called the *style*. The rounded base of the pistil is called the **ovary**. The ovary contains one or more ovules. Each ovule contains an egg.

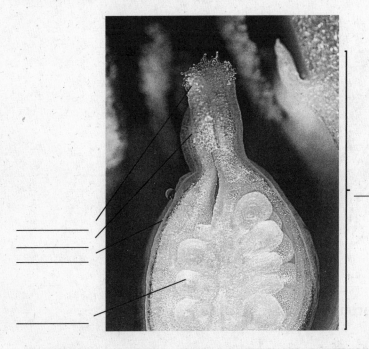

READING CHECK

15. Identify What is the male reproductive structure of a flower?

_____

TAKE A LOOK

16. Label Label the female reproductive structures in this picture.

Pollinators brush pollen onto the style, and sperm from inside the pollen travel down the style to the ovary. One sperm can fertilize the egg of one ovule. After fertilization, an ovule develops into a seed. The ovary surrounding the ovule develops into a fruit.

IMPORTANCE OF FLOWERS

Flowers are important to plants because they help plants reproduce. They are also important to animals, such as insects and bats, that use parts of flowers for food. Humans also use flowers. Some flowers, such as broccoli and cauliflower, can be eaten. Others, such as chamomile, are used to make tea. Flowers are also used in perfumes, lotions, and shampoos.

 **Say It**

Discuss What is your favorite flower? Have you ever seen any unusual flowers in nature? In groups of two or three, discuss your experiences with flowers.

Copyright © by Holt, Rinehart and Winston; a Division of Houghton Mifflin Harcourt Publishing Company. All rights reserved.

Section 4 Review

GLE 0707.Inq.3, GLE 0707.1.2, GLE 0707.4.2 **TN**

SECTION VOCABULARY

ovary in flowering plants, the lower part of a pistil that produces eggs in ovules	**sepal** in a flower, one of the outermost rings of modified leaves that protect the flower bud
petal one of the usually brightly colored, leaf-shaped parts that make up one of the rings of a flower	**stamen** the male reproductive structure of a flower that produces pollen and consists of an anther at the tip of a filament
phloem the tissue that conducts food in vascular plants	**xylem** the type of tissue in vascular plants that provides support and conducts water and nutrients from the roots
pistil the female reproductive part of a flower that produces seeds and consists of an ovary, style, and stigma	

1. Label Label the parts of this flower.

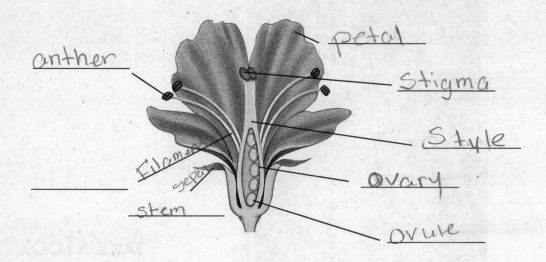

anther   petal   stigma   Style   Ovary   ovule   Filament   sepal   stem

2. Compare How do taproot and fibrous root systems differ?

_____

_____

_____

3. Describe What are the three functions of a stem?

_____

_____

_____

4. List What are the four main organs of a flowering seed plant?

_____

Copyright © by Holt, Rinehart and Winston; a Division of Houghton Mifflin Harcourt Publishing Company. All rights reserved.

CHAPTER 7 | Plant Processes

SECTION 1 Photosynthesis

TN Tennessee Science Standards
GLE 0707.Inq.3
GLE 0707.Inq.5
GLE 0707.1.1
GLE 0707.3.1
GLE 0707.3.2

BEFORE YOU READ

After you read this section, you should be able to answer these questions:

- How do plants make food?
- How do plants get energy from food?
- How do plants exchange gases with the environment?

What Is Photosynthesis?

Many organisms, including humans, have to eat to get energy. Plants, however, are able to make their own food. Plants make their food by a process called **photosynthesis**. During photosynthesis, plants use carbon dioxide, water, and energy from sunlight to make sugars.

How Do Plants Get Energy from Sunlight?

Plant cells have organelles called *chloroplasts*. Chloroplasts capture the energy from sunlight. Inside a chloroplast, membranes called *grana* contain chlorophyll. **Chlorophyll** is a green pigment that absorbs light energy. Many plants look green because chlorophyll reflects the green wavelengths of light. ☑

STUDY TIP

Outline As you read, outline the steps of photosynthesis. Use the questions in the section titles to help you make your outline.

READING CHECK

1. Define What is chlorophyll?

_____

_____

_____

TAKE A LOOK

2. Identify Where is chlorophyll found in a plant cell?

_____

_____

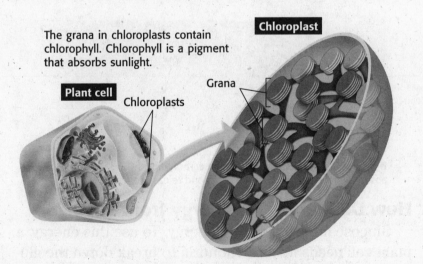

The grana in chloroplasts contain chlorophyll. Chlorophyll is a pigment that absorbs sunlight.

Chloroplast

Plant cell

Chloroplasts

Grana

Copyright © by Holt, Rinehart and Winston; a Division of Houghton Mifflin Harcourt Publishing Company. All rights reserved.

SECTION 1 Photosynthesis *continued*

How Do Plants Make Sugar?

During photosynthesis, plants take in water and carbon dioxide and absorb light energy. Plants use the light energy captured by chlorophyll to help form glucose molecules. *Glucose* is the sugar that plants use for food. In addition to producing sugar, plants give off oxygen during photosynthesis. ☑

The following chemical equation summarizes photosynthesis:

$$6CO_2 \quad + \quad 6H_2O \quad \xrightarrow{\text{light energy}} \quad C_6H_{12}O_6 \quad + \quad 6O_2$$

(carbon dioxide)   (water)   (glucose)   (oxygen)

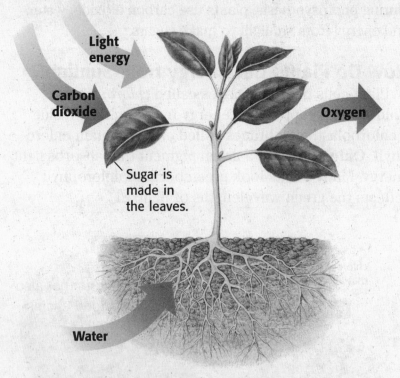

Light energy

Carbon dioxide

Oxygen

Sugar is made in the leaves.

Water

During photosynthesis, plants take in carbon dioxide and water and absorb light energy. They make sugar and release oxygen.

How Do Plants Get Energy from Sugar?

Glucose molecules store energy. To use this energy, a plant cell needs its mitochondria to break down the glucose. This process of breaking down food molecules to get energy is called **cellular respiration**. During cellular respiration, cells use oxygen to break down food molecules. Like all cells, plant cells then use the energy from food to do work.

READING CHECK

3. Identify What are two products of photosynthesis?

_____

_____

TN TENNESSEE STANDARDS CHECK

GLE 0707.1.1 Make observations and describe the <u>structure</u> and <u>function</u> of organelles found in plant and animal cells.

Word Help: <u>structure</u> the whole that is built or put together from parts

Word Help: <u>function</u> use or purpose

4. Identify Which cell structures release the energy stored in sugar?

_____

_____

Copyright © by Holt, Rinehart and Winston; a Division of Houghton Mifflin Harcourt Publishing Company. All rights reserved.

SECTION 1 Photosynthesis *continued*

How Does a Plant Take in the Gases It Needs?

Plants take in carbon dioxide and give off oxygen. These gases move into and out of the leaf through openings called **stomata** (singular, *stoma*). Stomata allow gases to move through the plant's *cuticle*, the waxy layer that prevents water loss. Each stoma is surrounded by two guard cells. The guard cells act like double doors by opening and closing a stoma.

Water vapor also moves out of the leaf through stomata. The loss of water from leaves is called **transpiration**. Stomata open to allow carbon dioxide to enter a leaf. They close to prevent too much water loss.

Critical Thinking

5. Predict What do you think would happen if a plant had no stomata?

_____

_____

_____

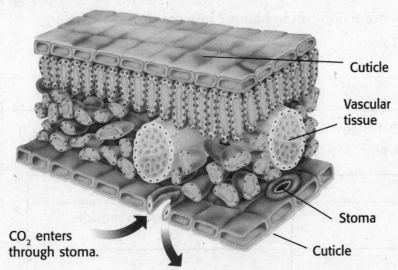

Cuticle

Vascular tissue

Stoma

Cuticle

CO_2 enters through stoma.

H_2O and O_2 exit through stoma.

TAKE A LOOK
6. Identify Circle the guard cells in this picture. What is their function?

_____

_____

Why Is Photosynthesis Important?

Plants, along with many bacteria and protists that also use photosynthesis, form the bases of most food chains on Earth. During photosynthesis, plants store light energy as chemical energy. Animals get this energy when they eat plants. Other animals get energy from plants indirectly. They eat the animals that eat plants. Most organisms could not survive without photosynthetic organisms. ☑

Photosynthesis is also important because it produces oxygen. Recall that cellular respiration requires oxygen to break down food. Most organisms, including plants and animals, depend on cellular respiration to get energy from their food. Without the oxygen produced during photosynthesis, most organisms could not survive.

Say It

Describe Think of all the ways in which photosynthesis is important to you. Describe to the class three ways you depend on photosynthesis.

☑ **READING CHECK**

7. Complete During photosynthesis, plants store light energy as

_____

_____

Copyright © by Holt, Rinehart and Winston; a Division of Houghton Mifflin Harcourt Publishing Company. All rights reserved.

Section 1 Review

GLE 0707.Inq.3, GLE 0707.Inq.5, GLE 0707.1.1,
GLE 0707.3.1, GLE 0707.3.2

SECTION VOCABULARY

cellular respiration the process by which cells use oxygen to produce energy from food **chlorophyll** a green pigment that captures light energy for photosynthesis **photosynthesis** the process by which plants, algae, and some bacteria use sunlight, carbon dioxide, and water to make food	**stoma** one of many openings in a leaf or a stem of a plant that enable gas exchange to occur (plural, *stomata*) **transpiration** the process by which plants release water vapor into the air through stomata

1. Explain Why does chlorophyll look green?

_____

_____

2. Identify What is the role of mitochondria in plants? In what process do they take part?

_____

_____

3. Compare Complete the chart below to show the relationship between photosynthesis and cellular respiration.

Photosynthesis	Cellular respiration
	Cells break down food to provide energy.
Oxygen is produced.	

4. Identify What two structures in plant leaves help prevent the loss of water?

_____

5. Explain Why are photosynthetic organisms, such as plants, so important to life on Earth?

_____

_____

_____

6. Explain If plants need to take in carbon dioxide, why don't they keep their stomata open all the time?

_____

_____

Copyright © by Holt, Rinehart and Winston; a Division of Houghton Mifflin Harcourt Publishing Company. All rights reserved.

SECTION 2
Reproduction of Flowering Plants

BEFORE YOU READ

After you read this section, you should be able to answer these questions:

• What are pollination and fertilization?

• How do seeds and fruits form?

• How can flowering plants reproduce asexually?

Tennessee Science Standards
GLE 0707.Inq.2
GLE 0707.Inq.3
GLE 0707.Inq.5
GLE 0707.4.1
GLE 0707.4.2

What Are Pollination and Fertilization?

Flowering plants are most noticeable to us when they are in bloom. As flowers bloom, they surround us with bright colors and sweet fragrances. However, flowers are not just for us to enjoy. They are the structures for sexual reproduction in flowering plants. Pollination and fertilization take place in flowers.

STUDY TIP

Summarize As you read, write out or draw the steps of pollination and fertilization.

Pollination and Fertilization

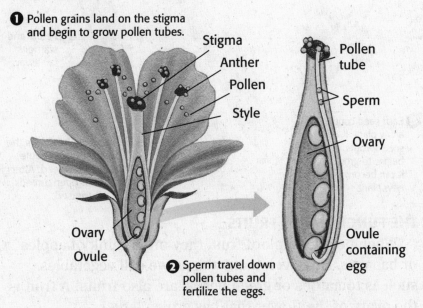

❶ Pollen grains land on the stigma and begin to grow pollen tubes.

Stigma
Anther
Pollen
Style
Pollen tube
Sperm
Ovary
Ovary
Ovule
Ovule containing egg

❷ Sperm travel down pollen tubes and fertilize the eggs.

TAKE A LOOK

1. Identify Circle the part of the flower where pollination occurs.

2. Identify Draw an arrow to show where fertilization will take place.

Sexual reproduction begins in flowers when wind or animals move pollen from one flower to another. *Pollination* occurs when pollen from an anther lands on a stigma. Each pollen grain grows a tube through the style to the ovary. The ovary has ovules, each of which contains an egg. *Fertilization* occurs when a sperm joins with the egg inside an ovule.

Copyright © by Holt, Rinehart and Winston; a Division of Houghton Mifflin Harcourt Publishing Company. All rights reserved.

What Happens After Fertilization?

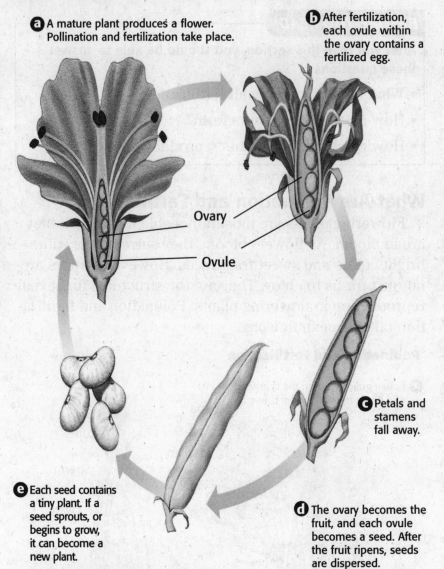

a A mature plant produces a flower. Pollination and fertilization take place.

b After fertilization, each ovule within the ovary contains a fertilized egg.

Ovary

Ovule

c Petals and stamens fall away.

d The ovary becomes the fruit, and each ovule becomes a seed. After the fruit ripens, seeds are dispersed.

e Each seed contains a tiny plant. If a seed sprouts, or begins to grow, it can become a new plant.

TN **TENNESSEE STANDARDS CHECK**

GLE 0707.4.2 Demonstrate an understanding of sexual reproduction in flowering plants.

3. Explain Where do seeds and fruits come from?

_____

_____

_____

_____

_____

TAKE A LOOK
4. Identify In step C, circle the structures that will become seeds.

 **READING CHECK**

5. List What are two functions of a fruit?

_____

_____

_____

THE FUNCTIONS OF FRUITS

When people think of fruit, they often think of apples or bananas. However, many things we call vegetables, such as tomatoes or green beans, are also fruits! A fruit is the ovary of the flower that has grown larger.

Fruits have two major functions. They protect seeds while the seeds develop. Fruits also help a plant spread its seeds to new environments. For example, an animal might eat a fruit and drop the seeds far from the parent plant. Fruits such as burrs spread when they get caught in an animal's fur. Other fruits are carried to new places by the wind or even by water. ☑

Copyright © by Holt, Rinehart and Winston; a Division of Houghton Mifflin Harcourt Publishing Company. All rights reserved.

SECTION 2 Reproduction of Flowering Plants *continued*

How Do Seeds Grow into New Plants?

The new plant inside a seed, called the *embryo*, stops growing once the seed is fully developed. However, the seed might not sprout right away. To sprout, most seeds need water, air, and warm temperatures. A seed might become **dormant**, or inactive, if the conditions are not right for a new plant to grow. For example, if the environment is too cold or too dry, a young plant will not survive. ☑

Dormant seeds often survive for long periods of time during droughts or freezing weather. Some seeds actually need extreme conditions, such as cold winters or forest fires, to *germinate*, or sprout.

☑ **READING CHECK**

6. Explain Why would a seed become dormant?

_____

_____

_____

Seeds grow into new plants. First, the roots begin to grow. Then, the shoots grow up through the soil.

TAKE A LOOK
7. Identify Which part of a new plant grows first?

_____

How Else Can Flowering Plants Reproduce?

Flowering plants can also reproduce asexually, or without flowers. In asexual reproduction, sperm and eggs do not join. A new plant grows from a plant part such as a root or stem. These plant parts include plantlets, tubers, and runners.

Three Structures for Asexual Reproduction

Kalanchoe produces plantlets along the edges of their leaves. The plantlets will fall off and take root in the soil.

A potato is a tuber, or underground stem. The "eyes" of potatoes are buds that can grow into new plants.

The strawberry plant produces runners, or stems that grow along the ground. Buds along the runners take root and grow into new plants.

Critical Thinking

8. Infer When would asexual reproduction be important for the survival of a flowering plant?

_____

_____

_____

_____

_____

_____

Copyright © by Holt, Rinehart and Winston; a Division of Houghton Mifflin Harcourt Publishing Company. All rights reserved.

Section 2 Review

GLE 0707.Inq.2, GLE 0707.Inq.3, GLE 0707.Inq.5,
GLE 0707.4.1, GLE 0707.4.2

SECTION VOCABULARY

dormant describes the inactive state of a seed or other plant part when conditions are unfavorable to growth	

1. Apply Concepts Is fertilization part of asexual reproduction or sexual reproduction? Explain your answer.

_____

_____

2. Compare What is the difference between pollination and fertilization?

_____

_____

3. Summarize Complete the Process Chart below to summarize how sexual reproduction produces new plants.

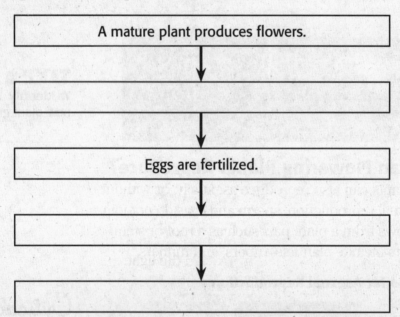

A mature plant produces flowers.

⬇

Eggs are fertilized.

⬇

4. Identify Name two environmental conditions that can cause a seed to become dormant.

_____

5. List What are three structures a flowering plant can use to reproduce asexually?

_____

6. Infer Why do you think roots are the first part of a plant to grow?

_____

Copyright © by Holt, Rinehart and Winston; a Division of Houghton Mifflin Harcourt Publishing Company. All rights reserved.

CHAPTER 7 | Plant Processes
SECTION
3 # Plant Development and Responses

TN **Tennessee Science Standards**
GLE 0707.Inq.3
GLE 0707.Inq.5

BEFORE YOU READ

After you read this section, you should be able to answer these questions:

• How do plants respond to the environment?

• Why do some trees lose their leaves?

How Do Plants Respond to the Environment?

What happens when you get cold? Do you shiver? Do your teeth chatter? These are your responses to an environmental stimulus such as cold air. A *stimulus* (plural, *stimuli*) is anything that causes a reaction in your body. Plants also respond to environmental stimuli, but not to the same ones we do and not in the same way. Plants respond to stimuli such as light and the pull of gravity. ☑

Some plants respond to a stimulus, such as light, by growing in a particular direction. Growth in response to a stimulus is called a **tropism**. A tropism is either positive or negative. Plant growth toward a stimulus is a positive tropism. Plant growth away from a stimulus is a negative tropism.

PLANT GROWTH IN RESPONSE TO LIGHT

Recall that plants need sunlight in order to make food. What would happen to a plant that could get light from only one direction, such as through a window? To get as much light as possible, it would need to grow toward the light. This growth in response to light is called *phototropism*.

STUDY TIP **Summarize** With a partner, take turns summarizing the text under each header.

READING CHECK

1. **Define** What is a stimulus?

_____

_____

_____

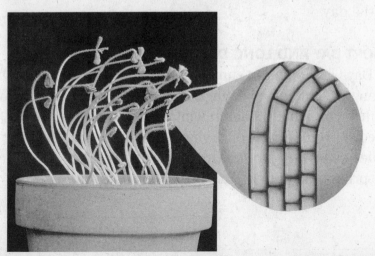

TAKE A LOOK

2. **Explain** Place an **X** on the picture to show where the light must be coming from. Explain your answer.

_____

_____

_____

Copyright © by Holt, Rinehart and Winston; a Division of Houghton Mifflin Harcourt Publishing Company. All rights reserved.

PLANT GROWTH IN RESPONSE TO GRAVITY

Gravity can change the direction in which a plant's roots and shoots grow. Most shoot tips grow upward, away from the center of Earth. Most root tips grow downward, toward the center of Earth. If a plant is placed on its side or turned upside down, the roots and shoots will change direction. Shoots will turn to grow away from the Earth. Roots will turn to grow toward the Earth. This response is called *gravitropism*.

Critical Thinking

3. Apply Concepts Look at the plant on the left. Draw an arrow on the flower pot to show the direction the roots are probably growing.

TAKE A LOOK

4. Explain Look at the plant on the right. What do you think made its stem bend?

_____

_____

_____

_____

_____

Gravitropism

To grow away from the pull of gravity, this plant has grown upward.

What Happens to Plants When Seasons Change?

Have you ever noticed that some plants will drop their leaves in the fall even before the weather turns cool? How do the plants know that fall is coming? We often notice the changing seasons because the temperature changes. Plants, however, respond to change in the length of the day.

Math Focus

5. Calculate It must be dark for 70% of a 24-hour period before a certain plant will bloom. How many hours of daylight does this plant need to bloom?

_____

SHORT DAY AND LONG DAY PLANTS

Days are longer in summer and shorter in winter. The change in amount of daylight is a stimulus for many plants. Some plants that bloom in winter, such as poinsettias, need shorter periods of daylight to reproduce. They are called *short-day plants*. Others, such as clover, reproduce in spring or summer. They are called *long-day plants*.

Copyright © by Holt, Rinehart and Winston; a Division of Houghton Mifflin Harcourt Publishing Company. All rights reserved.

SECTION 3 Plant Development and Responses *continued*

EFFECT OF SEASONS ON LEAF COLOR

The leaves of some trees may change color as seasons change. As the days shorten in fall, the chlorophyll in leaves breaks down. This makes the orange and yellow pigments in the leaves easier to see. During the summer, chlorophyll hides other pigments.

Amount of Leaf Pigment Based on Season

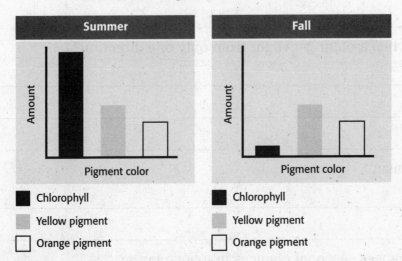

TAKE A LOOK
6. Identify Which pigment's level decreases between summer and fall?

_____

_____

7. Identify Which pigments' levels will stay the same between summer and fall?

_____

_____

LOSS OF LEAVES

Every tree loses leaves throughout its life. Leaves are shed when they become old. For example, pine trees lose some of their leaves, or *needles*, year-round. Because leaves are lost and replaced throughout the year, the tree always has some leaves. These trees are called *evergreen*. A leaf of an evergreen tree is covered with a thick cuticle. The cuticle protects the leaf from cold and dry weather.

Deciduous trees lose all their leaves at about the same time each year. This generally happens as days shorten. The loss of leaves helps these plants survive cold or dry weather. In colder areas, deciduous trees usually lose their leaves before winter begins. In areas that have wet and dry seasons, deciduous trees lose their leaves before the dry season.

 **Say It**

Describe What is your favorite kind of tree? Use the Internet or reference books to find out if the tree is evergreen or deciduous. Describe to the class what the tree looks like and where it lives.

Copyright © by Holt, Rinehart and Winston; a Division of Houghton Mifflin Harcourt Publishing Company. All rights reserved.

Name _____ Class _____ Date _____

Section 3 Review

GLE 0707.Inq.3, GLE 0707.Inq.5 TN

SECTION VOCABULARY

tropism growth of all or part of an organism in response to an external stimulus, such as light	

1. Compare What is the difference between a negative tropism and a positive tropism?

_____

_____

_____

2. Explain What happens when a plant gets light from only one direction?

_____

_____

_____

3. Define What is gravitropism?

_____

_____

4. Identify What stimulus causes seasonal changes in many plants?

_____

5. Compare What is the difference between short-day plants and long-day plants?

_____

_____

_____

6. Explain Why do leaves look green during the summer even though they have orange and yellow pigments?

_____

_____

7. Explain Many evergreen trees live in areas with long, cold winters. How can they keep their leaves all year?

_____

_____

_____

Copyright © by Holt, Rinehart and Winston; a Division of Houghton Mifflin Harcourt Publishing Company. All rights reserved.

CHAPTER 8 Body Organization and Structure

SECTION 1 **Body Organization**

TN **Tennessee Science Standards**
GLE 0707.1.2

BEFORE YOU READ

After you read this section, you should be able to answer these questions:

• What is homeostasis?

• How is the human body organized?

• What are the 11 different human organ systems?

How Is the Body Organized?

The different parts of your body all work together to maintain, or keep, the conditions in your body stable. Your body works to keep itself stable even when things outside your body change. This is called **homeostasis**. For example, your body temperature needs to stay the same even when temperatures outside are very cold or very hot. If your body could not keep its inside conditions the same, many processes in your body would not work.

Conditions inside and outside your body are always changing. Your body can maintain homeostasis because each cell does not have to do everything your body needs. Instead, your body is organized into different levels. The parts at each level work together to help your body maintain homeostasis.

There are four levels of organization in the body: cells, tissues, organs, and organ systems. Cells are the smallest level of organization. A group of similar cells working together forms a **tissue**. Your body has four main kinds of tissue: epithelial, nervous, muscle, and connective.

STUDY TIP 📝

Discuss Read this section silently. When you finish reading, work with a partner to answer any questions you may have about the section.

TN **TENNESSEE STANDARDS CHECK**

GLE 0707.1.2 Summarize how the different levels of organization (cells, tissues, organs, organ systems and organisms) are integrated within living systems.

1. Identify What are the four levels of organization in the body?

_____

_____

Four Kinds of Tissue

Epithelial tissue covers and protects other tissues.

Nervous tissue sends electrical signals through the body.

Muscle tissue is made of cells that contract and relax to produce movement.

Connective tissue joins, supports, protects, insulates, nourishes, and cushions organs. It also keeps organs from falling apart.

Copyright © by Holt, Rinehart and Winston; a Division of Houghton Mifflin Harcourt Publishing Company. All rights reserved.

SECTION 1 Body Organization *continued*

ORGANS

When different kinds of tissues work together, they can do more than any one tissue can do alone. A group of two or more tissues working together to do a job is an **organ**. For example, your stomach is an organ that helps you digest your food. None of the stomach's tissues could digest food alone. Sometimes, if an organ is weak, an artificial organ can be used instead. These man-made organs allow people to live longer and healthier lives. ☑

✓ **READING CHECK**

2. Define What is an organ?

_____

_____

_____

Four Kinds of Tissue in the Stomach

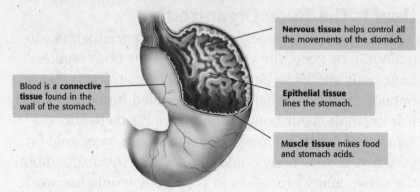

Blood is a **connective tissue** found in the wall of the stomach.

Nervous tissue helps control all the movements of the stomach.

Epithelial tissue lines the stomach.

Muscle tissue mixes food and stomach acids.

ORGAN SYSTEMS

Organs that work together to do a job make up an *organ system*. For example, your stomach works with other organs in the digestive system, such as the intestines, to digest food. Organ systems can do jobs that one organ alone cannot do. Each organ system has a special function.

There are 11 different organ systems that make up the human body. No organ system works alone. For example, the respiratory system and cardiovascular system work together to move oxygen through your body.

Critical Thinking

3. Apply Concepts How does the stomach work as part of an organ system?

_____

_____

_____

_____

_____

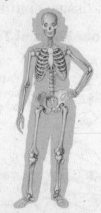

Integumentary System Your skin, hair, and nails protect the tissue that lies beneath them.

Muscular System Your muscular system works with the skeletal system to help you move.

Skeletal System Your bones provide a frame to support and protect your body parts.

Copyright © by Holt, Rinehart and Winston; a Division of Houghton Mifflin Harcourt Publishing Company. All rights reserved.

SECTION 1 Body Organization *continued*

Cardiovascular System
Your heart pumps blood through all of your blood vessels.

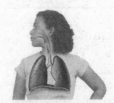

Respiratory System
Your lungs absorb oxygen and release carbon dioxide.

Urinary System Your urinary system removes wastes from the blood and regulates your body's fluids.

Male Reproductive System The male reproductive system produces and delivers sperm.

Female Reproductive System The female reproductive system produces eggs and nourishes and protects the fetus.

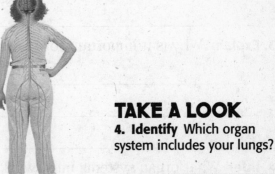

Nervous System Your nervous system receives and sends electrical messages throughout your body.

TAKE A LOOK
4. Identify Which organ system includes your lungs?

_____

5. Identify Which organ system is different in males and females?

_____

Digestive System Your digestive system breaks down the food you eat into nutrients that your body can absorb.

Lymphatic System The lymphatic system returns leaked fluids to blood vessels and helps get rid of bacteria and viruses.

Endocrine System Your glands send out chemical messages. Ovaries and testes are part of this system.

 **Say It**

Discuss With a partner, see how many organs you can name from each organ system.

Copyright © by Holt, Rinehart and Winston; a Division of Houghton Mifflin Harcourt Publishing Company. All rights reserved.

Section 1 Review

GLE 0707.1.2 **TN**

SECTION VOCABULARY

homeostasis the maintenance of a constant internal state in a changing environment **organ** a collection of tissues that carry out a specialized function of the body	**tissue** a group of similar cells that perform a common function

1. Compare How is an organ different from a tissue?

_____

_____

2. List Name five organ systems in the human body.

_____

_____

3. Explain Why is it important for your body to maintain homeostasis?

_____

_____

_____

4. Infer What organ systems must work together to help a person eat and digest a piece of pizza? Give at least three systems.

_____

_____

5. Infer What organ systems must work together to help a person play a soccer game? Give at least four systems.

_____

_____

6. Apply Concepts Can an organ do the same job as an organ system? Explain your answer.

_____

_____

_____

7. Identify Relationships How is the lymphatic system related to the cardiovascular system?

_____

_____

Copyright © by Holt, Rinehart and Winston; a Division of Houghton Mifflin Harcourt Publishing Company. All rights reserved.

CHAPTER 8 | Body Organization and Structure
SECTION 2 **The Skeletal System**

BEFORE YOU READ

After you read this section, you should be able to answer these questions:

- What are the major organs of the skeletal system?
- What are the functions of the skeletal system?
- What are the three kinds of joints in the body?

TN Tennessee Science Standards

GLE 0707.Inq.2
GLE 0707.Inq.3
GLE 0707.Inq.5
GLE 0707.1.2

What Are Bones?

Many people think that bones are dry and brittle, but your bones are actually living organs. Bones are the major organs of the skeletal system. The **skeletal system** is made up of bones, cartilage, and connective tissue.

STUDY TIP

Organize As you read this section, make a chart listing the functions of bones and the tissue or bone structure that does each job.

What Are the Functions of the Skeletal System?

An average adult human skeleton has 206 bones. Bones have many jobs. For example, they help support and protect your body. They work with your muscles so you can move. Bones also help your body maintain homeostasis by storing minerals and making blood cells. The skeletal system does the following jobs for your body:

- It protects other organs. For example, your rib cage protects your heart and lungs.

- It stores minerals that help your nerves and muscles work properly. Long bones store fat that can be used as energy.

- Skeletal muscles pull on bones to cause movement. Without bones, you would not be able to sit, stand, or run.

- Some bones make blood cells. *Marrow* is a special material that makes blood cells.

Critical Thinking

1. Predict Name one organ system, other than the skeletal system, that would be affected if you had no bones. Explain your answer.

_____

_____

_____

_____

```
        ┌─────────────────┐
        │ Skeletal system │
        │    functions    │
        └─────────────────┘
```

To protect other organs	To store minerals	To allow you to move	To make blood cells

Copyright © by Holt, Rinehart and Winston; a Division of Houghton Mifflin Harcourt Publishing Company. All rights reserved.

What Is the Structure of a Bone?

A bone may seem lifeless. Like other organs, however, bone is a living organ made of several different tissues. Bone is made of connective tissue and minerals. Living cells in the bone deposit the minerals.

BONE TISSUE

If you look inside a bone, you will see two kinds of bone tissue: spongy bone and compact bone. Spongy bone has many large open spaces that help the bone absorb shocks. Compact bone has no large open spaces, but it does have tiny spaces filled with blood vessels. Compact bone forms the outer layer of a bone and protects the bone. ☑

MARROW

Some bones contain a tissue called marrow. There are two types of marrow. Red marrow makes red and white blood cells. Yellow marrow stores fat.

CARTILAGE

Did you know that most of your skeleton used to be soft and rubbery? Most bones start out as a flexible tissue called *cartilage*. When you were born, you didn't have much true bone. As you grow, your cartilage is replaced by bone. However, bone will never replace cartilage in a few small areas of your body. For example, the end of your nose and the tops of your ears will always be made of cartilage. ☑

Bone Tissues

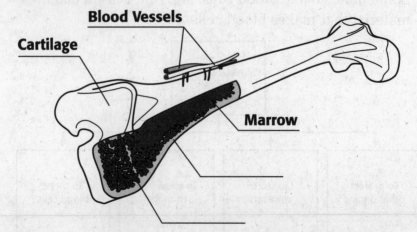

Cartilage

Blood Vessels

Marrow

✔ **READING CHECK**

2. Identify What are the two kinds of bone tissue?

_____

_____

✔ **READING CHECK**

3. Define What is cartilage?

_____

_____

_____

TAKE A LOOK
4. Label Fill in the missing labels for tissues that are found in this bone.

Copyright © by Holt, Rinehart and Winston; a Division of Houghton Mifflin Harcourt Publishing Company. All rights reserved.

What Is a Joint?

A place where two or more bones meet is called a **joint**. Some joints, called *fixed joints*, do not let bones move very much. Many of the joints in the skull are fixed joints. However, most joints let your bones move when your muscles *contract*, or shorten. Joints can be grouped based on how the bones in the joint move. ☑

Gliding Joint Gliding joints let bones in the wrist slide over each other. This type of joint makes a body part flexible.

Ball-and-Socket Joint In the same way that a video-game joystick lets you move your character around, the shoulder lets your arm move freely in all directions.

Hinge Joint A hinge lets a door open and close. Your knee joint lets your leg bend in only one direction.

Joints can handle a lot of wear and tear because of how they are made. Joints are held together by ligaments. *Ligaments* are strong bands of connective tissue. Cartilage covers the ends of many bones and helps cushion the areas where bones meet. ☑

SKELETAL SYSTEM INJURIES AND DISEASES

Sometimes, parts of the skeletal system are injured. For example, bones may be broken. Joints and ligaments can also be injured. Many of these injuries happen when too much stress is placed on the skeletal system.

There are also some diseases that affect the skeletal system. For example, the disease *osteoporosis* makes bones brittle and easy to break. Some diseases make bones soft or affect bone marrow. *Arthritis* is a disease that makes joints stiff, so they are painful and hard to move.

READING CHECK

5. **Define** What is a joint?

_____

_____

TAKE A LOOK

6. **List** What are the three types of joints in the human body?

_____

_____

_____

READING CHECK

7. **Identify** How does cartilage help protect bones and joints?

_____

_____

_____

Copyright © by Holt, Rinehart and Winston; a Division of Houghton Mifflin Harcourt Publishing Company. All rights reserved.

Section 2 Review

GLE 0707.Inq.2, GLE 0707.Inq.3, GLE 0707.Inq.5, GLE 0707.1.2 TN

SECTION VOCABULARY

joint a place where two or more bones meet	**skeletal system** the organ system whose primary function is to support and protect the body and to allow the body to move

1. List What are four functions of the skeletal system?

_____

_____

2. Identify What three things make up the skeletal system?

_____

_____

3. Describe Fill in the chart below to describe the three types of joints. Give an example of each.

Type of joint	Example
	wrist
hinge	

4. Compare What is the difference between red marrow and yellow marrow?

_____

_____

_____

5. Explain What happens to the cartilage in your body as you grow up?

_____

_____

6. Identify What are two diseases that can affect the skeletal system?

_____

_____

7. Describe Describe a joint and its structure.

_____

_____

_____

8. Explain What causes most injuries to the skeletal system?

_____

Copyright © by Holt, Rinehart and Winston; a Division of Houghton Mifflin Harcourt Publishing Company. All rights reserved.

CHAPTER 8 Body Organization and Structure
SECTION 3 The Muscular System

BEFORE YOU READ

After you read this section, you should be able to answer these questions:

- What are the three kinds of muscle tissue?
- How do skeletal muscles work?
- How can exercise help keep you healthy?

TN Tennessee Science Standards
GLE 0707.Inq.2
GLE 0707.Inq.3
GLE 0707.1.2

What Is the Muscular System?

The **muscular system** is made up of the muscles that let you move. There are three kinds of muscle in your body: smooth muscle, cardiac muscle, and skeletal muscle.

Skeletal muscle makes bones move.

Smooth muscle moves food through the digestive system.

Cardiac muscle pumps blood around the body.

STUDY TIP

Circle As you read this section, circle any new science terms. Make sure you know what these words mean before moving to the next chapter.

Muscle action can be voluntary or involuntary. Muscle action that you can control is *voluntary*. Muscle action that you cannot control is *involuntary*. For example, cardiac muscle movements in your heart are involuntary. They happen without you having to think about it. Skeletal muscles, such as those in your eyelids, can be both voluntary and involuntary. You can blink your eyes anytime you want, but your eyes also blink automatically.

Critical Thinking

1. Apply Concepts Your diaphragm is a muscle that helps you breathe. Do you think this muscle is voluntary or involuntary? Explain.

_____

_____

_____

_____

Kind of muscle	Where in your body is it found?	Are its actions voluntary or involuntary?
Cardiac	heart	involuntary
Smooth	digestive tract, blood vessels	involuntary
Skeletal	attached to bones and other organs	both

Copyright © by Holt, Rinehart and Winston; a Division of Houghton Mifflin Harcourt Publishing Company. All rights reserved.

SECTION 3 The Muscular System *continued*

How Do Skeletal Muscles Work?

Skeletal muscles let you move. When you want to move, signals travel from your brain to your skeletal muscle cells. The muscle cells then contract, or get shorter. ☑

HOW MUSCLES AND BONES WORK TOGETHER

Strands of tough connective tissue connect your skeletal muscles to your bones. These strands are called tendons. When a muscle that connects two bones contracts, the bones are pulled closer to each other. For example, tendons attach the biceps muscle to bones in your shoulder and forearm. When the biceps muscle contracts, your forearm bends toward your shoulder.

PAIRS OF MUSCLES

Your skeletal muscles often work in pairs to make smooth, controlled motions. Generally, one muscle in the pair bends part of the body. The other muscle straightens that part of the body. A muscle that bends part of your body is called a *flexor*. A muscle that straightens part of your body is an *extensor*. ☑

In the figure below, the biceps muscle is the flexor. When the biceps muscle contracts, the arm bends. The triceps muscle is the extensor. When it contracts, the arm straightens out.

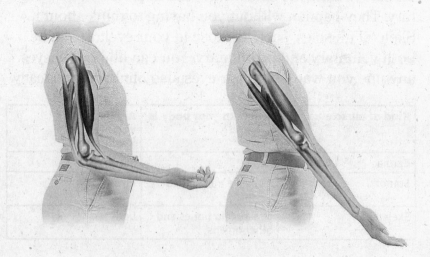

READING CHECK

2. Explain What causes skeletal muscle cells to contract?

_____

_____

READING CHECK

3. Complete A muscle that bends part of your body is a

_____.

TAKE A LOOK

4. Identify On the figure, label the flexor muscle and the extensor muscle.

Copyright © by Holt, Rinehart and Winston; a Division of Houghton Mifflin Harcourt Publishing Company. All rights reserved.

SECTION 3 The Muscular System *continued*

How Can You Keep Your Muscles Healthy?

Muscles get stronger when you exercise them. Strong muscles can help other organs to work better. For example, when your heart is strong, it can pump more blood to the rest of your organs. More blood brings more oxygen and nutrients to your organs.

Certain kinds of exercises can give muscles more strength and endurance. More endurance means that your muscles can work longer before they get tired.

Resistance exercise is a good way to make skeletal muscles stronger. During *resistance exercise*, the muscles work against the resistance, or weight, of an object. Some resistance exercises use weights. Others, such as sit-ups, use your own body weight as resistance.

Aerobic exercise can increase skeletal muscle strength and endurance. Aerobic exercise can also make your heart muscles stronger. During *aerobic exercise*, the muscles work steadily for a fairly long period of time. Jogging, skating, swimming, and walking are all aerobic exercises.

Math Focus
5. Calculate A student is doing resistance exercise. After one week, she can lift a weight of 2 kg. After four weeks, she can lift a weight of 3 kg. By what percentage has the weight that she can lift increased?

_____

Type of exercise	Description	Example
Resistance		weight-lifting, sit-ups
	Muscles work steadily for a long time.	

TAKE A LOOK
6. Describe Complete the table to describe types of exercise.

MUSCLE INJURY

Most muscle injuries happen when people try to do too much exercise too quickly. For example, a *strain* is an injury in which a muscle or tendon is overstretched or torn. To avoid muscle injuries, you should start exercising slowly. Don't try to do too much too fast.

Exercising too much can also harm your muscles and tendons. For example, if you exercise a tendon that has a strain, the tendon cannot heal. It can become swollen and painful. This condition is called *tendonitis*.

Some people try to make their muscles stronger by taking drugs called *anabolic steroids*. These drugs can cause serious health problems. They can cause high blood pressure and can damage the heart, liver, and kidneys. They can also cause bones to stop growing.

 **Say It**

Discuss In a small group, talk about some of the ways that exercise can help keep you healthy.

Copyright © by Holt, Rinehart and Winston; a Division of Houghton Mifflin Harcourt Publishing Company. All rights reserved.

Section 3 Review

GLE 0707.Inq.2, GLE 0707.Inq.3, GLE 0707.1.2 TN

SECTION VOCABULARY

muscular system the organ system whose primary function is movement and flexibility	

1. List What three kinds of muscle make up the muscular system?

_____

_____

2. Identify Which kind of muscle movement happens without you having to think about it? Give two kinds of muscle that show this kind of movement.

_____

_____

3. Describe How are muscles attached to bones?

_____

4. Explain How do muscles cause bones to move?

_____

_____

5. Describe What happens to muscle when you exercise it?

_____

6. Compare How is aerobic exercise different from resistance exercise?

_____

_____

7. Identify What are two kinds of injuries to the muscular system?

_____

8. Compare How is a flexor different from an extensor?

_____

_____

9. Explain What kinds of problems can anabolic steroids cause?

_____

_____

Copyright © by Holt, Rinehart and Winston; a Division of Houghton Mifflin Harcourt Publishing Company. All rights reserved.

SECTION 4 The Integumentary System

TN Tennessee Science Standards
GLE 07071.2

BEFORE YOU READ

After you read this section, you should be able to answer these questions:

- What is the integumentary system?
- What are the functions of the skin, hair, and nails?

What Is the Integumentary System?

Your **integumentary system** is made up of your skin, hair, fingernails, and toenails. The Latin word *integere* means "to cover." Your integumentary system covers your body and helps to protect it. Your integumentary system also helps your body to maintain homeostasis.

THE SKIN

Your skin is the largest organ in your body. It is an important part of the integumentary system. The skin has four main functions.

- Skin protects your body. It keeps water inside your body, and it keeps many harmful particles outside your body.

- Skin keeps you in touch with the world. Nerve endings in your skin let you feel things around you.

- Skin helps to keep your body temperature from getting too high. Small organs in the skin called *sweat glands* make sweat, which flows onto the skin. When sweat evaporates, your body cools down.

- Skin helps your body get rid of some wastes. Sweat can carry these wastes out of your body.

As you know, skin can be many different colors. The color of your skin is determined by a chemical called *melanin*. If your skin contains a lot of melanin, it is dark. If your skin contains very little melanin, it is light.

Melanin helps to protect your skin from being damaged by the ultraviolet radiation in sunlight. People's skin may darken if they are exposed to a lot of sunlight. This happens because the cells in your skin make extra melanin to help protect themselves from ultraviolet radiation.

STUDY TIP

Compare As you read, make a table comparing skin, hair, and nails. In the table, describe each structure and list its functions.

Critical Thinking

1. Apply Concepts Why are you more likely to get sick if you touch a dirty surface with damaged skin than if you touch it with healthy skin?

_____

_____

_____

_____

_____

_____

Copyright © by Holt, Rinehart and Winston; a Division of Houghton Mifflin Harcourt Publishing Company. All rights reserved.

LAYERS OF SKIN

Your skin has two main layers: the epidermis and the dermis. The **epidermis** is the outermost layer of skin. It is the layer that you see when you look at your skin. The prefix *epi-* means "above." Therefore, the epidermis lies above the dermis. The **dermis** is the thick layer of skin that lies underneath the epidermis. ☑

The epidermis is made of *epithelial tissue*. These tissues are made of many layers of cells. However, on most parts of your body, the epidermis is only a few millimeters thick.

Most of the cells in the epidermis are dead. The dead cells are filled with a protein called *keratin*. Keratin helps to make your skin tough.

The dermis is much thicker than the epidermis. It contains many fibers made of a protein called *collagen*. Collagen fibers make the dermis strong and let the skin bend without tearing. The dermis also contains many small structures, as shown in the figure below.

☑ **READING CHECK**

2. List What are the two main layers of skin?

_____

_____

TAKE A LOOK

3. Identify Give three structures that attach to hair follicles.

_____

_____

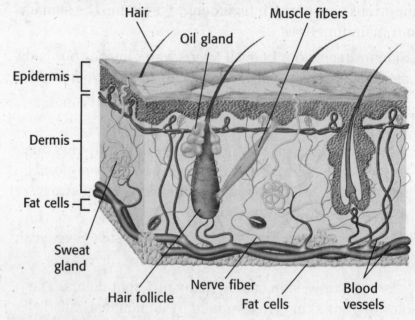

Epidermis — Hair — Oil gland — Muscle fibers
Dermis
Fat cells
Sweat gland — Hair follicle — Nerve fiber — Fat cells — Blood vessels

SKIN INJURIES

Your skin is always coming into contact with the outside world. Therefore, it is often damaged. Fortunately, your skin can repair itself. The figure on the top of the next page shows how a cut in the skin heals.

Some skin problems are caused by conditions inside your body. For example, hormones can cause your skin to make too much oil. The oil can combine with bacteria and dead skin cells to form acne.

Copyright © by Holt, Rinehart and Winston; a Division of Houghton Mifflin Harcourt Publishing Company. All rights reserved.

SECTION 4 The Integumentary System *continued*

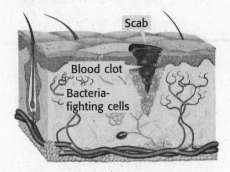

Scab
Blood clot
Bacteria-fighting cells

❶ A blood clot forms over a cut to stop bleeding and to keep bacteria from entering the wound. Bacteria-fighting cells then come to the area to kill bacteria.

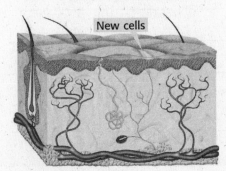

New cells

❷ Damaged cells are replaced through cell division. Eventually, all that is left on the surface is a scar.

TAKE A LOOK
4. Explain How do blood clots help protect your body?

_____

_____

HAIR AND NAILS

Your hair and nails are also important parts of your integumentary system. Like skin, hair and nails are made of both living and dead cells. Hair and nails grow from your skin.

A hair forms at the bottom of a tiny sac called a *hair follicle*. The hair grows as new cells are added at the hair follicle. Older cells get pushed upward. The only living cells in hair are found in the hair follicle. ☑

Like skin, hair gets its color from melanin. Hair helps protect skin from being damaged. The hair in and around your nose, eyes, and ears helps keep dust and other particles out of your body. Your hair also helps to keep you warm. When you feel cold, tiny muscles cause your hair to stand up. The raised hairs act like a sweater. They trap warm air near your body.

A nail forms at the *nail root*. A nail root is found at the base of the nail. As new cells form in the nail root, the nail grows longer. The hard part of the nail is made of dead cells that are filled with keratin. Nails protect the ends of your fingers and toes. They allow your fingers and toes to be soft and sensitive to touch.

☑ READING CHECK
5. Describe How does hair grow?

_____

_____

_____

_____

Copyright © by Holt, Rinehart and Winston; a Division of Houghton Mifflin Harcourt Publishing Company. All rights reserved.

Section 4 Review

GLE 0707.1.2 TN

SECTION VOCABULARY

dermis the layer of skin below the epidermis **epidermis** the surface layer of cells on a plant or animal	**integumentary system** the organ system that forms a protective covering on the outside of the body

1. Identify Name three functions of the integumentary system.

_____

_____

_____

2. Compare Give three differences between the dermis and the epidermis.

_____

_____

_____

3. Infer The epidermis on the palms of your hands and the soles of your feet is thicker than it is on other parts of your body. What do you think is the reason for this?

_____

_____

_____

4. Explain Why can skin get darker if it is exposed to a lot of sunlight?

_____

_____

5. Identify Give two ways that hair helps to protect your body.

_____

_____

6. Infer Blood clots help to prevent bacteria from entering your body through a cut. Why do bacteria-fighting cells travel to a cut, even though there is a blood clot there?

_____

_____

Copyright © by Holt, Rinehart and Winston; a Division of Houghton Mifflin Harcourt Publishing Company. All rights reserved.

CHAPTER 9 Circulation and Respiration

SECTION 1 The Cardiovascular System

BEFORE YOU READ

After you read this section, you should be able to answer these questions:

- What is the cardiovascular system?
- What are some cardiovascular problems?

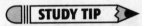
Tennessee Science Standards
GLE 0707.Inq.5
GLE 0707.1.3

What Is the Cardiovascular System?

Your heart, blood, and blood vessels make up your **cardiovascular system**. The word *cardio* means heart. The word *vascular* means blood vessels. *Blood vessels* are hollow tubes that your blood flows through. The cardiovascular system is also sometimes called the *circulatory system*. This is because it *circulates*, or moves, blood through your body.

The cardiovascular system helps your body maintain homeostasis. *Homeostasis* is the state your body is in when its internal conditions are stable. The cardiovascular system helps maintain homeostasis in many ways:

- it carries oxygen and nutrients to your cells
- it carries wastes away from your cells
- it carries heat throughout your body
- it carries chemical signals called *hormones* throughout your body ☑

THE HEART

Your heart is an organ about the same size as your fist. It is near the center of your chest. There is a thick wall in the middle of your heart that divides it into two halves. The right half pumps oxygen-poor blood to your lungs. The left half pumps oxygen-rich blood to your body.

Each side of your heart has two chambers. Each upper chamber is called an *atrium* (plural, *atria*). Each lower chamber is called a *ventricle*. These chambers are separated by flap-like structures called *valves*. Valves keep blood from flowing in the wrong direction. The closing of valves is what makes the "lub-dub" sound when your heart beats. The figure at the top of the next page shows how blood moves through your heart.

STUDY TIP

Summarize As you read, underline the main ideas in each paragraph. When you finish reading, write a short summary of the section using the ideas you underlined.

READING CHECK

1. Identify What are two functions of the cardiovascular system?

_____

_____

_____

Copyright © by Holt, Rinehart and Winston; a Division of Houghton Mifflin Harcourt Publishing Company. All rights reserved.

SECTION 1 The Cardiovascular System *continued*

Math Focus

2. Calculate A person's heart beats about 70 times per minute. How many times does a person's heart beat in one day? How many times does it beat in one year?

_____

_____

_____

TAKE A LOOK

3. Identify Where does the left ventricle receive blood from? Where does the right atrium receive blood from?

_____

_____

_____

_____

❶ Blood enters the atria first. The left atrium receives blood that has a lot of oxygen in it from the lungs. The right atrium receives blood that has little oxygen in it from the body.

❸ While the atria relax, the ventricles contract and push blood out of the heart. Blood from the right ventricle goes to the lungs. Blood from the left ventricle goes to the rest of the body.

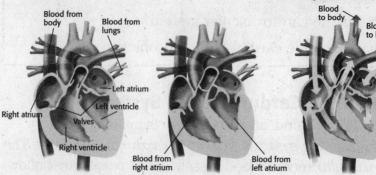

Blood from body

Blood from lungs

Left atrium

Right atrium

Left ventricle

Valves

Right ventricle

Blood from right atrium

Blood from left atrium

Blood to body

Blood to lungs

❷ When the atria contract, blood moves into the ventricles.

BLOOD VESSELS

Blood travels throughout your body in your blood vessels. There are three types of blood vessels: arteries, capillaries, and veins.

An **artery** is a blood vessel that carries blood away from the heart. Arteries have thick walls that contain a layer of muscle. Each heartbeat pumps blood into your arteries. The blood is under high pressure. Artery walls are strong and can stretch to handle this pressure. Your *pulse* is caused by the pumping of blood into your arteries. ☑

A **capillary** is a tiny blood vessel. Capillary walls are very thin. Therefore, substances can move across them easily. Capillaries are also very narrow. They are so narrow that blood cells have to pass through them in single file. Nutrients and oxygen move from the blood in your capillaries into your body's cells. Carbon dioxide and other wastes move from your body's cells into the blood.

A **vein** is a blood vessel that carries blood toward the heart. Veins have valves to keep the blood from flowing backward. When skeletal muscles contract, they squeeze nearby veins and help push blood toward the heart.

✓ **READING CHECK**

4. Describe What causes your pulse?

_____

_____

_____

Copyright © by Holt, Rinehart and Winston; a Division of Houghton Mifflin Harcourt Publishing Company. All rights reserved.

SECTION 1 The Cardiovascular System *continued*

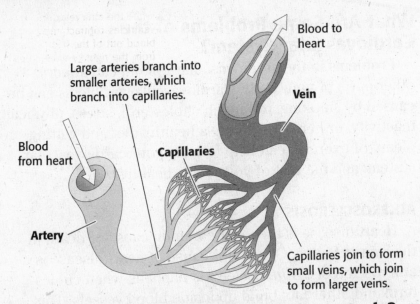

Large arteries branch into smaller arteries, which branch into capillaries.

Blood to heart

Vein

Blood from heart

Capillaries

Artery

Capillaries join to form small veins, which join to form larger veins.

TAKE A LOOK

5. Compare What is one main difference between arteries and veins?

_____

_____

_____

How Does Blood Flow Through Your Body?

Where does blood get the oxygen to deliver to your body? From your lungs! Your heart contracts and pumps blood to the lungs. In the lungs, carbon dioxide leaves the blood and oxygen enters the blood. The oxygen-rich blood then flows back to your heart. This circulation of blood between your heart and lungs is called **pulmonary circulation**. ☑

The oxygen-rich blood returning to your heart from your lungs is then pumped to the rest of your body. The circulation of blood between your heart and the rest of your body is called **systemic circulation**. The figure below shows how blood moves through your body.

✓ READING CHECK

6. Define What is pulmonary circulation?

_____

_____

Pulmonary circulation

The right ventricle pumps oxygen-poor blood into arteries that lead to the lungs. These are the only arteries in the body that carry oxygen-poor blood.

In the lungs, blood gives off carbon dioxide and takes up oxygen. Oxygen-rich blood flows through veins to the left atrium. These are the only veins in the body that carry oxygen-rich blood.

Oxygen-poor blood travels back to the heart through veins. These veins deliver the blood to the right atrium.

The heart pumps oxygen-rich blood from the left ventricle into arteries. The arteries branch into capillaries.

Systemic circulation

Oxygen, nutrients, and water move into the cells of the body as blood moves through capillaries. At the same time, carbon dioxide and other waste materials move out of the cells and into the blood.

TAKE A LOOK

7. Color Use a blue pen or colored pencil to color the vessels carrying oxygen-poor blood. Use a red pen or colored pencil to color the vessels carrying oxygen-rich blood.

Copyright © by Holt, Rinehart and Winston; a Division of Houghton Mifflin Harcourt Publishing Company. All rights reserved.

Critical Thinking

8. Infer How can a problem in your cardiovascular system affect the rest of your body?

_____
_____
_____
_____
_____
_____

What Are Some Problems of the Cardiovascular System?

Problems in the cardiovascular system can affect other parts of your body. Cardiovascular problems can be caused by smoking, too much cholesterol, stress, physical inactivity, or heredity. Eating a healthy diet and getting plenty of exercise can help to keep your cardiovascular system, and the rest of your body, healthy.

ATHEROSCLEROSIS

Heart disease is the most common cause of death in the United States. One major cause of heart disease is atherosclerosis. *Atherosclerosis* happens when cholesterol and other fats build up inside blood vessels. This buildup causes the blood vessels to become narrower and less stretchy. When the pathway through a blood vessel is blocked, blood cannot flow through. ☑

✓ **READING CHECK**

9. Identify What is the most common cause of death in the United States?

_____
_____

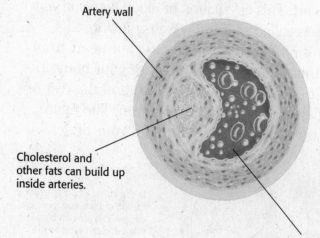

Artery wall

Cholesterol and other fats can build up inside arteries.

If there is a buildup of cholesterol, the artery becomes narrower. Not as much blood can flow through it at a time. If the cholesterol blocks the artery completely, no blood can flow through.

TAKE A LOOK

10. Explain How can too much cholesterol cause problems in your cardiovascular system?

_____
_____
_____

Copyright © by Holt, Rinehart and Winston; a Division of Houghton Mifflin Harcourt Publishing Company. All rights reserved.

SECTION 1 The Cardiovascular System *continued*

HIGH BLOOD PRESSURE

Hypertension is high blood pressure. Hypertension can make it more likely that a person will have cardio-vascular problems. For example, atherosclerosis may be caused by hypertension.

High blood pressure can also cause a stroke. A *stroke* happens when a blood vessel in the brain is blocked or breaks open. Blood cannot flow through the vessel to the brain cells. Without blood, the brain cells cannot get oxygen, so the cells die. ☑

HEART ATTACKS AND HEART FAILURE

Hypertension can also cause heart attacks and heart failure. A *heart attack* happens when heart muscle cells do not get enough blood. Arteries that deliver oxygen to the heart may be damaged. Without oxygen from the arteries, heart muscle cells can be damaged. If enough heart muscle cells are damaged, the heart may stop.

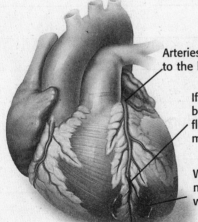

Arteries carry blood and oxygen to the heart muscle.

If an artery is blocked, blood and oxygen cannot flow to part of the heart muscle.

Without oxygen from blood, the heart muscle can be damaged. It can become weak or die.

Heart failure happens when the heart is too weak to pump enough blood to meet the body's needs. Organs may not receive enough oxygen or nutrients to function correctly. Waste products can build up in the organs and damage them.

Say It

Discuss Learn about two ways to maintain healthy blood pressure. In a small group, talk about how you can apply these ideas in your life.

✓ READING CHECK

11. Identify What is a stroke?

_____

_____

TAKE A LOOK

12. Explain How can blocking an artery in the heart cause heart damage?

_____

_____

_____

Copyright © by Holt, Rinehart and Winston; a Division of Houghton Mifflin Harcourt Publishing Company. All rights reserved.

Section 1 Review

GLE 0707.Inq.5, GLE 0707.1.3 TN

SECTION VOCABULARY

artery a blood vessel that carries blood away from the heart to the body's organs	**pulmonary circulation** the flow of blood from the heart to the lungs and back to the heart through the pulmonary arteries, capillaries, and veins
capillary a tiny blood vessel that allows an exchange between blood and cells in tissue	**systemic circulation** the flow of blood from the heart to all parts of the body and back to the heart
cardiovascular system a collection of organs that transport blood throughout the body	**vein** in biology, a vessel that carries blood to the heart

1. Identify What are the three main parts of the cardiovascular system?

_____

_____

2. Describe Beginning and ending in the left atrium, describe the path that blood takes through your body and lungs.

_____

_____

_____

_____

_____

3. Compare How is a heart attack different from heart failure?

_____

_____

4. Explain What is the function of valves in the heart and the veins?

_____

_____

5. Compare How are the arteries that lead from your heart to your lungs different from the other arteries in your body?

_____

_____

_____

Copyright © by Holt, Rinehart and Winston; a Division of Houghton Mifflin Harcourt Publishing Company. All rights reserved.

CHAPTER 9 | Circulation and Respiration

SECTION
2 **Blood**

BEFORE YOU READ

After you read this section, you should be able to answer these questions:

• What is blood?

• What is blood pressure?

• What are blood types?

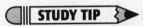

Tennessee Science Standards
GLE 0707.Inq.5
GLE 0707.1.3

What Is Blood?

Your cardiovascular system is made up of your heart, your blood vessels, and blood. **Blood** is a connective tissue made up of plasma, red blood cells, platelets, and white blood cells. Blood travels in blood vessels and carries oxygen and nutrients to all parts of your body. An adult human has only about 5 L of blood. All the blood in your body would not even fill up three 2-L soda bottles! ☑

PLASMA

The fluid part of the blood is called plasma. *Plasma* is made up of water, minerals, nutrients, sugars, proteins, and other substances.

PLATELETS

Platelets are pieces of larger cells found in bone marrow. When you get a cut, you bleed because blood vessels have been opened. Platelets clump together in the damaged area to form a plug. They also give off chemicals that cause fibers to form. The fibers and clumped platelets form a blood clot and stop the bleeding.

STUDY TIP

Ask Questions As you read this section, write down the questions that you have. Then, discuss your questions with a small group.

READING CHECK

1. **Define** What is blood?

_____

_____

_____

Critical Thinking

2. **Infer** If a person does not have enough platelets in her blood, what will happen if she gets a cut?

_____

_____

_____

_____

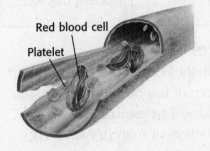

Red blood cell

Platelet

Fibers

Copyright © by Holt, Rinehart and Winston; a Division of Houghton Mifflin Harcourt Publishing Company. All rights reserved.

Math Focus

3. Calculate One cubic millimeter of blood contains 5 million RBCs and 10,000 WBCs. How many times more RBCs are there than WBCs?

_____

_____

✓ **READING CHECK**

4. Identify Where are most white blood cells made?

_____

_____

TN **TENNESSEE STANDARDS CHECK**

GLE 0707.1.3 Describe the function of different organ systems and how collectively they enable complex multicellular organisms to survive.

Word Help: function use or purpose

5. Explain How do wider or narrower blood vessels help your body stay at a constant temperature?

_____

_____

_____

RED BLOOD CELLS

Most blood cells are *red blood cells*, or RBCs. RBCs carry oxygen to all the cells in your body. Cells need oxygen to do their jobs. *Hemoglobin* is the protein in red blood cells that carries the oxygen. It is what makes RBCs look red.

WHITE BLOOD CELLS

A *pathogen* is a virus, bacteria, or other tiny particle that can make you sick. When pathogens get into your body, *white blood cells*, or WBCs, help kill them. WBCs can fight pathogens by:

- leaving blood vessels to destroy pathogens in tissues

- making chemicals called *antibodies* to help destroy pathogens

- destroying body cells that have died or been damaged

Most WBCs are made in bone marrow. Some mature in the lymphatic system. ✓

How Does Blood Control Body Temperature?

Your blood also helps keep your body temperature constant. Your blood vessels can open wider or get narrower to control how much heat is lost through your skin.

Your body is too hot.	→	Blood vessels get wider. More blood flows near the skin.	→	Heat moves from the blood into the air. Your body cools.

Your body is too cold.	→	Blood vessels get narrower. Less blood flows near the skin.	→	Less heat moves from the blood into the air. Your body stays warm.

What Is Blood Pressure?

When your heart beats, it pushes blood out of your heart and into your arteries. The force of the blood on the inside walls of the arteries is called **blood pressure**. Blood pressure is measured in millimeters of mercury (mm Hg).

Copyright © by Holt, Rinehart and Winston; a Division of Houghton Mifflin Harcourt Publishing Company. All rights reserved.

SECTION 2 Blood *continued*

SYSTOLIC AND DIASTOLIC PRESSURE

Blood pressure is given by two numbers, such as 120/80. The first, or top, number is systolic pressure. *Systolic pressure* is the pressure in arteries when the ventricles contract. The rush of blood causes arteries to bulge and produce a pulse. The second, or bottom, number is diastolic pressure. *Diastolic pressure* is the pressure in arteries when the ventricles relax.

For adults, a blood pressure of 120/80 mm Hg or less is healthy. High blood pressure can cause heart or kidney damage.

What Are Blood Types?

Every person has one of four blood types: A, B, AB, or O. Chemicals called *antigens* on the outside of your RBCs determine whch blood type you have. The plasma of different blood types may have different antibodies. *Antibodies* are chemicals that react with antigens of other blood types as if the antigens were pathogens. ☑

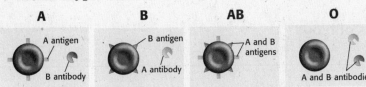

This figure shows which antigens and antibodies may be present in each blood type.

IMPORTANCE OF BLOOD TYPES

A person can lose blood from an injury, illness or surgery. To replace lost blood, a person can receive a blood transfusion. A *transfusion* is when a person is given blood from another person.

However, a person cannot receive blood from just anyone. If someone who is type A gets type B blood, the type B antibodies can make the RBCs clump together. The clumps can block blood vessels. A reaction to the wrong blood type can kill you.

Blood type	Can receive blood from:	Can donate blood to:
A	types A and O	types A and AB
B	types B and O	types B and AB
AB	types A, B, AB, and O	type AB only
O	type O only	types A, B, AB, and O

Copyright © by Holt, Rinehart and Winston; a Division of Houghton Mifflin Harcourt Publishing Company. All rights reserved.

READING CHECK

6. Identify What determines your blood type?

_____

_____

TAKE A LOOK

7. Identify What kinds of antigens are found on the RBCs of a person with type AB blood?

_____

_____

TAKE A LOOK

8. Identify Which blood type can receive blood from the most other blood types? Which type can donate blood to the most other types?

_____

_____

Section 2 Review

GLE 0707.Inq.5, GLE 0707.1.3 TN

SECTION VOCABULARY

blood the fluid that carries gases, nutrients, and wastes through the body and that is made up of platelets, white blood cells, red blood cells, and plasma	**blood pressure** the force that blood exerts on the walls of arteries

1. Identify What are two functions of white blood cells?

_____

_____

2. Describe Complete the table to describe the two parts of blood pressure.

Type of pressure	Description	Where it is found in a blood-pressure measurement
systolic		top number
	pressure in the arteries when ventricles relax	

3. List What are three functions of blood?

_____

_____

4. Infer Why does your face get redder when you are hot?

_____

_____

5. Explain Why is it important that a person with type O blood only receive a blood transfusion from another person with type O blood?

_____

_____

_____

6. Predict If a person has a disease that causes hemoglobin to break down, what can happen to his RBCs?

_____

_____

Copyright © by Holt, Rinehart and Winston; a Division of Houghton Mifflin Harcourt Publishing Company. All rights reserved.

CHAPTER 9 Circulation and Respiration

SECTION
3 **The Lymphatic System**

TN Tennessee Science
Standards
GLE 0707.Inq.5
GLE 0707.1.3

BEFORE YOU READ

After you read this section, you should be able to answer these questions:

• What is the function of the lymphatic system?

• What are the parts of the lymphatic system?

What Does the Lymphatic System Do?

Every time your heart pumps, small amounts of plasma are forced out of the thin walls of the capillaries. What happens to this fluid? Most of it is reabsorbed into your blood through the capillaries. Some of the fluid moves into your lymphatic system.

The **lymphatic system** is the group of vessels, organs, and tissues that collects excess fluid and returns it to the blood. The lymphatic system also helps your body fight pathogens.

What Are the Parts of the Lymphatic System?

Fluid collected by the lymphatic system is carried in vessels. The smallest of these vessels are called lymph capillaries. Larger lymph vessels are called lymphatic vessels. These vessels, along with bone marrow, lymph nodes, the thymus, and the spleen, make up the lymphatic system.

LYMPH CAPILLARIES

Lymph capillaries absorb some of the fluid and particles from between cells in the body. Some of the particles are dead cells or pathogens. These particles are too large to enter blood capillaries. The fluid and particles absorbed into lymph capillaries are called **lymph**.

STUDY TIP

Describe Work with a partner to quiz each other on the names and functions of each structure in the lymphatic system.

TN TENNESSEE STANDARDS CHECK

GLE 0707.1.3 Describe the function of different organ systems and how collectively they enable complex multicellular organisms to survive.

Word Help: function use or purpose

1. Identify Relationships How do the lymphatic and circulatory systems work together?

_____

_____

_____

_____

_____

_____

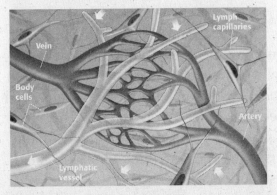

The white arrows show how lymph moves into lymph capillaries and through lymphatic vessels.

Copyright © by Holt, Rinehart and Winston; a Division of Houghton Mifflin Harcourt Publishing Company. All rights reserved.

LYMPHATIC VESSELS

Lymph capillaries carry lymph into larger vessels called *lymphatic vessels*. Skeletal muscles and valves help push the lymph through the lymphatic system. Lymphatic vessels drain the lymph into large veins in the neck. This returns the fluid to the cardiovascular system. ☑

✓ **READING CHECK**

2. Identify Where is fluid returned to the cardiovascular system?

_____

_____

The Lymphatic System

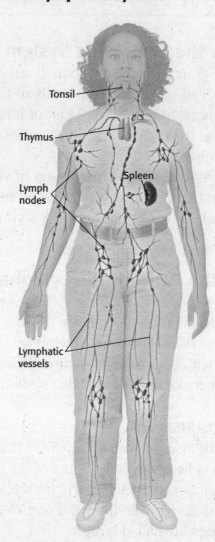

Tonsil

Thymus

Lymph nodes

Spleen

Lymphatic vessels

TAKE A LOOK

3. Describe As you read, write on the diagram the function of each labelled structure.

✓ **READING CHECK**

4. Identify What is the function of bone marrow?

_____

_____

BONE MARROW

Bone marrow is the soft tissue inside bones that makes red and white blood cells. Recall that platelets, which help blood clot, are made in marrow. White blood cells called *lymphocytes* are part of the lymphatic system. They help fight infection. Killer T cell lymphocytes surround and destroy pathogens. B cell lymphocytes make antibodies that cause pathogens to stick together. This marks them for destruction. ☑

Copyright © by Holt, Rinehart and Winston; a Division of Houghton Mifflin Harcourt Publishing Company. All rights reserved.

SECTION 3 The Lymphatic System *continued*

LYMPH NODES

As lymph travels through lymphatic vessels, it passes through lymph nodes. **Lymph nodes** are small masses of tissue that remove pathogens and dead cells from the lymph. When bacteria or other pathogens cause an infection, white blood cells multiply and fill the lymph nodes. This may cause lymph nodes to become swollen and painful.

THYMUS

T cells are made in the bone marrow. Before these cells are ready to fight infections, however, they develop further in the **thymus** gland. The thymus is located just above the heart. Mature T cells leave the thymus and travel through the lymphatic system.

SPLEEN

The **spleen** is the largest lymphatic organ. It stores lymphocytes and fights infection. It is a purplish organ located in the upper left side of the abdomen. As blood flows through the spleen, lymphocytes attack or mark pathogens in the blood. The spleen may release lymphocytes into the bloodstream when there is an infection. The spleen also monitors, stores, and destroys old blood cells. ☑

TONSILS

The **tonsils** are lymphatic tissue at the back of the mouth. Tonsils help defend the body against infection by trapping pathogens. Sometimes, however, tonsils can become infected. Infected tonsils may be red, swollen, and sore. They may be covered with patches of white, infected tissue and make swallowing difficult. Tonsils may be removed if there are frequent, severe tonsil infections that make breathing difficult.

Critical Thinking

5. Infer Sometimes you can easily feel your lymph nodes when they are swollen. If you had swollen lymph nodes, what could you infer?

_____

_____

READING CHECK

6. List Name three functions of the spleen.

_____

_____

_____

_____

Inflamed tonsils

Tonsils help protect your throat and lungs from infection by trapping pathogens.

Copyright © by Holt, Rinehart and Winston; a Division of Houghton Mifflin Harcourt Publishing Company. All rights reserved.

Section 3 Review

GLE 0707.Inq.5, GLE 0707.1.3 **TN**

SECTION VOCABULARY

lymph the fluid that is collected by the lymphatic vessels and nodes	**spleen** the largest lymphatic organ in the body
lymph node an organ that filters lymph and that is found along the lymphatic vessels	**thymus** the main gland of the lymphatic system; it releases mature T lymphocytes
lymphatic system a collection of organs whose primary function is to collect extracellular fluid and return it to the blood	**tonsils** organs that are small, rounded masses of lymphatic tissue located in the pharynx and in the passage from the mouth to the pharynx

1. Describe How does the lymphatic system fight infection?

_____

_____

_____

2. Summarize Complete the Process Chart below to show how fluid travels between the cardiovascular system and the lymphatic system.

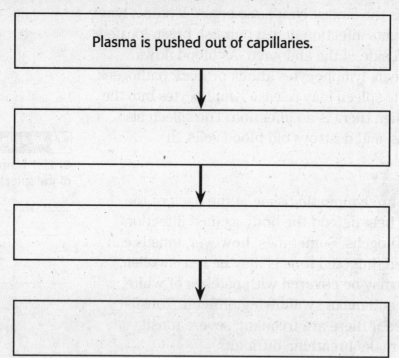

Plasma is pushed out of capillaries.

3. List What are three things that can be found in lymph?

_____

4. Analyze Why is it important that lymphatic tissue is spread throughout the body?

_____

_____

Copyright © by Holt, Rinehart and Winston; a Division of Houghton Mifflin Harcourt Publishing Company. All rights reserved.

CHAPTER 9 | Circulation and Respiration
SECTION 4 | # The Respiratory System

TN **Tennessee Science Standards**
GLE 0707.Inq.5
GLE 0707.1.3
GLE 0707.3.2

BEFORE YOU READ

After you read this section, you should be able to answer these questions:

• What is the respiratory system?

• What are some respiratory disorders?

What Is the Respiratory System?

Breathing: you do it all the time. You're doing it right now. You probably don't think about it unless you can't breathe. Then, it becomes very clear that you have to breathe in order to live. Why is breathing important? Breathing helps your body get oxygen. Your body needs oxygen in order to get energy from the foods you eat.

The words *breathing* and *respiration* are often used to mean the same thing. However, breathing is only one part of respiration. **Respiration** is the way the body gains and uses oxygen and gets rid of carbon dioxide. ☑

Respiration is divided into two parts. The first part involves inhaling and exhaling, or breathing. The second part is cellular respiration. *Cellular respiration* involves the chemical reactions that let you get energy from food.

The **respiratory system** is the group of organs and structures that take in oxygen and get rid of carbon dioxide. The nose, throat, lungs, and passageways that lead to the lungs make up the respiratory system.

STUDY TIP

Compare Make a chart showing the features of the different parts of the respiratory system.

READING CHECK

1. Define What is respiration?

_____

_____

_____

Parts of the Respiratory System

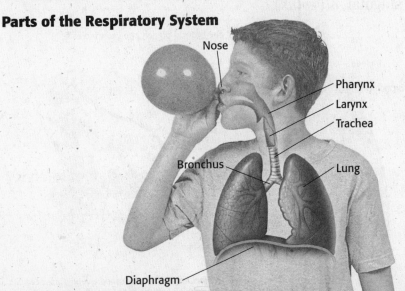

Nose

Pharynx

Larynx

Trachea

Bronchus

Lung

Diaphragm

TAKE A LOOK
2. List What are the parts of the respiratory system?

_____

_____

_____

_____

Copyright © by Holt, Rinehart and Winston; a Division of Houghton Mifflin Harcourt Publishing Company. All rights reserved.

SECTION 4 The Respiratory System *continued*

THE NOSE, PHARYNX, LARYNX, AND TRACHEA

Your *nose* is the main passageway into and out of the respiratory system. You breathe air in through your nose. You also breathe air out of your nose. Air can also enter and leave through your mouth. ☑

From the nose or mouth, air flows through the **pharynx**, or throat. Food and drink also move through the pharynx on the way to the stomach. The pharynx branches into two tubes. One tube, the *esophagus*, leads to the stomach. The other tube leads to the lungs. The larynx sits at the start of this tube.

The **larynx** is the part of the throat that contains the vocal cords. The *vocal cords* are bands of tissue that stretch across the larynx. Muscles connected to the larynx control how much the vocal cords are stretched. When air flows between the vocal cords, the cords vibrate. These vibrations make sound.

The larynx guards the entrance to a large tube called the **trachea**, or windpipe. The trachea is the passageway for air traveling from the larynx to the lungs. ☑

THE BRONCHI AND ALVEOLI

Inside your chest, the trachea splits into two branches called **bronchi** (singular, *bronchus*). One bronchus connects to each lung. Each bronchus branches into smaller and smaller tubes. These branches form a smaller series of airways called *bronchioles*. In the lungs, each bronchiole branches to form tiny sacs that are called **alveoli** (singular, *alveolus*).

☑ **READING CHECK**

3. Describe What is the main function of the nose?

_____

_____

_____

☑ **READING CHECK**

4. Identify What is the trachea?

_____

_____

TAKE A LOOK

5. Infer Which do you have the most of in your lungs: bronchi, bronchioles, or alveoli?

_____

_____

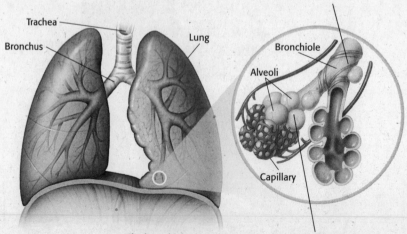

Inside your lungs, the bronchi branch into bronchioles.

Trachea

Bronchus

Lung

Bronchiole

Alveoli

Capillary

The bronchioles branch into tiny sacs called alveoli. Gases can move between the alveoli and the blood that is in the capillaries.

Copyright © by Holt, Rinehart and Winston; a Division of Houghton Mifflin Harcourt Publishing Company. All rights reserved.

How Does Breathing Work?

Your lungs have no muscles of their own. Instead, your diaphragm and rib muscles do the work that helps you breathe. The *diaphragm* is a dome-shaped muscle underneath the lungs. When the diaphragm contracts and moves down, you inhale. At the same time, some of your rib muscles contract and lift your rib cage. The volume of your chest gets larger. As a result, air is sucked in.

Exhaling is this process in reverse. Your diaphragm relaxes, your rib muscles relax, and air moves out.

BREATHING AND CELLULAR RESPIRATION

In cellular respiration, cells use oxygen to release the energy that is stored in molecules of a sugar. This sugar is called *glucose*. When cells break down glucose, they give off carbon dioxide.

Critical Thinking

6. Predict Consequences What would happen to a person whose diaphragm could not contract?

_____

_____

Oxygen moves into your blood. Your red blood cells carry the oxygen to other parts of your body.

Alveoli

Blood

CO_2 enters the blood. The blood carries the carbon dioxide back to the lungs.

Carbon dioxide gas can move from your blood into your lungs to be exhaled.

Capillary

Tissues and cells pick up O_2 from the blood.

When you breathe in, air enters your lungs. The air contains oxygen gas. When you breathe out, air moves out of your lungs. The air carries carbon dioxide out of your body.

TAKE A LOOK
7. Explain How does oxygen gas get from the air into the cells in your body?

_____

_____

_____

_____

_____

What Are Some Respiratory Disorders?

People who have *respiratory disorders* have trouble getting the oxygen they need. Their cells cannot release all the energy they need from the food they eat. Therefore, these people may feel tired all the time. They may also have problems getting rid of carbon dioxide. The carbon dioxide can build up in their cells and make them sick.

Respiratory Disorder	What it is
Asthma	A disorder that causes bronchioles to narrow, making it hard to breathe.
Emphysema	A disorder caused when alveoli are damaged.
Severe Acute Respiratory Syndrome (SARS)	A disorder caused by a virus that makes it hard to breathe.

Copyright © by Holt, Rinehart and Winston; a Division of Houghton Mifflin Harcourt Publishing Company. All rights reserved.

Section 4 Review

GLE 0707.Inq.5, GLE 0707.1.3, GLE 0707.3.2   **TN**

SECTION VOCABULARY

alveoli any of the tiny air sacs of the lungs where oxygen and carbon dioxide are exchanged	**respiration** the exchange of oxygen and carbon dioxide between living cells and their environment; includes breathing and cellular respiration
bronchus one of the two tubes that connect the lungs with the trachea	**respiratory system** a collection of organs whose primary function is to take in oxygen and expel carbon dioxide
larynx the area of the throat that contains the vocal cords and produces vocal sounds	**trachea** the tube that connects the larynx to the lungs
pharynx the passage from the mouth to the larynx and esophagus	

1. List What are three respiratory disorders?

_____

2. Define What is cellular respiration?

_____

_____

3. Compare How is respiration different from breathing?

_____

_____

_____

4. Explain The nose is the main way for air to get into and out of your body. How can a person still breathe if his or her nose is blocked?

_____

5. Describe How do vocal cords produce sound?

_____

_____

_____

6. Explain What are two ways that a respiratory disorder can make a person sick?

_____

_____

_____

Copyright © by Holt, Rinehart and Winston; a Division of Houghton Mifflin Harcourt Publishing Company. All rights reserved.

CHAPTER 10 The Digestive and Urinary Systems

SECTION 1 The Digestive System

 **Tennessee Science Standards**
GLE 0707.Inq.5
GLE 0707.1.3

BEFORE YOU READ

After you read this section, you should be able to answer these questions:

• What are the parts of the digestive system?

• How does each part of the digestive system work?

What Are the Parts of the Digestive System?

The **digestive system** is a group of organs that break down, or digest, food so your body can get nutrients. The main organs of the digestive system make one long tube through the body. This tube is called the *digestive tract*. The digestive tract includes the mouth, pharynx, esophagus, stomach, small intestine, and large intestine.

The digestive system has several organs that are not part of the digestive tract. The liver, gallbladder, pancreas, and salivary glands add materials to the digestive tract to help break down food. However, food does not go into these organs. ☑

STUDY TIP

Organize As you read, make combination notes about each digestive organ. Write the function of the organ in the left column of the notes. Draw or describe the structure in the right column.

READING CHECK

1. Identify What is the name of the tube that food passes through?

_____

_____

The Digestive System

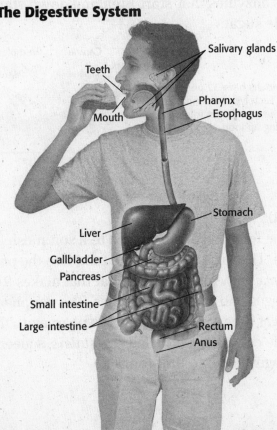

Salivary glands
Teeth
Pharynx
Mouth
Esophagus
Stomach
Liver
Gallbladder
Pancreas
Small intestine
Large intestine
Rectum
Anus

TAKE A LOOK

2. Color Use a colored pencil to shade all of the digestive organs that food passes through.

Copyright © by Holt, Rinehart and Winston; a Division of Houghton Mifflin Harcourt Publishing Company. All rights reserved.

How Is Food Broken Down?

The sandwich you eat for lunch has to be broken into tiny pieces to be absorbed into your blood. First, food is crushed and mashed into smaller pieces. This is called *mechanical digestion*. However, the food is still too large to enter your blood. Next, the small pieces of food are broken into their chemical parts, or molecules. This is called *chemical digestion*. The molecules can now be taken in and used by the body's cells.

Most food is made up of three types of nutrients: carbohydrates, proteins, and fats. The digestive system uses proteins called *enzymes* to break your food into molecules. Enzymes act as chemical scissors to cut food into smaller particles that the body can use.

MOUTH

Digestion begins in the mouth where food is chewed. You use your teeth to mash and grind food. Chewing creates small, slippery pieces of food that are easy to swallow. As you chew, the food mixes with a liquid called *saliva*. Saliva is made in salivary glands in the mouth. Saliva has enzymes that start breaking down starches into simple sugars.

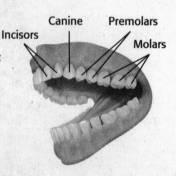

Most adults have 32 permanent teeth. Each type of permanent tooth has a different role in breaking up food.

Canine   Premolars

Incisors   Molars

ESOPHAGUS

Once the food has been chewed to a soft mush, it can be swallowed. The tongue pushes the food into the *pharynx*. The *pharynx* is the part of the throat that makes food go to the esophagus and air go to the lungs. The **esophagus** is a long, straight tube that leads from the pharynx to the stomach. Muscle contractions, called *peristalsis*, squeeze food in the esophagus down to the stomach.

TN TENNESSEE STANDARDS CHECK

GLE 0707.1.3 Describe the function of different organ systems and how collectively they enable complex multicellular organisms to survive.

Word Help: function use or purpose.

3. Infer How does the circulatory system work with the digestive system?

_____

_____

_____

_____

_____

Critical Thinking

4. Infer Why do you think you should chew your food well before you swallow it?

_____

_____

_____

_____

_____

Math Focus

5. Compute Ratios Young children get a first set of 20 teeth called *baby teeth*. These teeth usually fall out and are replaced by 32 permanent teeth. What is the ratio of baby teeth to permanent teeth?

Copyright © by Holt, Rinehart and Winston; a Division of Houghton Mifflin Harcourt Publishing Company. All rights reserved.

SECTION 1 The Digestive System *continued*

STOMACH

The stomach is a muscular, saclike organ. The stomach uses its muscles to continue mechanical digestion. It squeezes and mashes food into smaller and smaller pieces.

Squeezing also mixes the food with enzymes and acids produced by tiny glands in the stomach. These chemicals help break down food into nutrients. Stomach acid also kills most bacteria in the food. After a few hours of chemical and mechanical digestion, the food is a soupy mixture called *chyme*. ☑

The stomach squeezes and mixes food for hours before it releases the mixture into the small intestine.

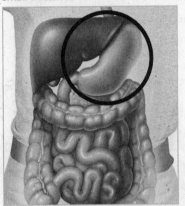
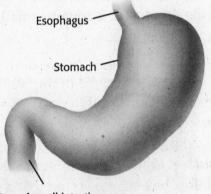

Esophagus

Stomach

Part of small intestine

READING CHECK
6. Identify What kinds of digestion occur in the stomach?

_____

_____

What Happens to Food in the Small and Large Intestines?

Most chemical digestion takes place after food leaves the stomach. Chyme slowly leaves the stomach and enters the small intestine. The pancreas, liver, and gallbladder add enzymes and other fluids to the small intestine to help finish digestion. The large intestine absorbs water and gets rid of waste.

PANCREAS

The **pancreas** is an organ located between the stomach and small intestine. Food does not enter the pancreas. Instead, the pancreas makes a fluid that flows into the small intestine. The chart below shows the chemicals that make up this pancreatic fluid, and the function of each.

TAKE A LOOK
7. Identify What is the role of enzymes in pancreatic fluid?

_____

_____

_____

Chemical in pancreatic fluid	Function
sodium bicarbonate	to protect the small intestine from acid in the chyme
enzymes	to chemically digest chyme

Copyright © by Holt, Rinehart and Winston; a Division of Houghton Mifflin Harcourt Publishing Company. All rights reserved.

SECTION 1 The Digestive System *continued*

LIVER

The **liver** is a large, reddish brown organ found on the right side of the body under the ribs. The liver helps with digestion in the following ways.

• It makes bile.

• It stores extra nutrients.

• It breaks down toxins, such as alcohol.

GALLBLADDER

Bile is made in the liver, and stored in a small, saclike organ called the **gallbladder**. The gallbladder squeezes bile into the small intestine when there is food to digest. Bile breaks fat into very small droplets so that enzymes can digest it. ☑

READING CHECK

8. Identify What is the function of bile?

_____

_____

_____

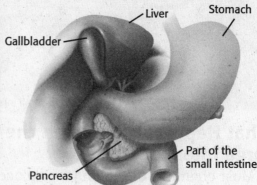

The liver, gall bladder, and pancreas are linked to the small intestine. However, food does not pass through these organs.

Liver

Stomach

Gallbladder

Pancreas

Part of the small intestine

TAKE A LOOK

9. Identify Which organs in this diagram does food pass through?

_____

_____

SMALL INTESTINE

The **small intestine** is a long, thin, muscular tube where nutrients are absorbed. If you stretched out your small intestine, it would be much longer than you are tall—about 6 m! If you flattened out the surfaces of the small intestine, it would be larger than a tennis court.

The inside wall of the small intestine is covered with many small folds. The folds are covered with cells called *villi*. Villi absorb nutrients. The large number of folds and villi increase the surface area of the small intestine. A large surface area helps the body get as many nutrients from food as possible. ☑

 **READING CHECK**

10. Explain Why is a large surface area in the small intestine important?

_____

_____

_____

_____

_____

_____

Copyright © by Holt, Rinehart and Winston; a Division of Houghton Mifflin Harcourt Publishing Company. All rights reserved.

SECTION 1 The Digestive System *continued*

The inside of the small intestine is folded. The folds are covered with finger-like structures called **villi**.

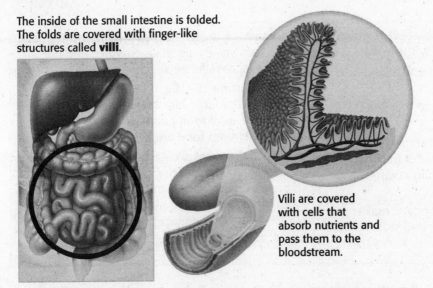

Villi are covered with cells that absorb nutrients and pass them to the bloodstream.

TAKE A LOOK
11. Explain How do nutrients get into the bloodstream?

_____
_____
_____

Most nutrient molecules are taken into the blood from the small intestine. However, the body can't use everything you eat. The soupy mixture of water and food that cannot be absorbed moves into the large intestine.

LARGE INTESTINE

The **large intestine** is the last part of the digestive tract. It stores, compacts, and rids the body of waste. The large intestine is wider than the small intestine, but shorter. It takes most of the water out of the mixture from the small intestine. By removing water, the large intestine changes the liquid into mostly solid waste called *feces*, or *stool*.

The *rectum* is the last part of the large intestine. The rectum stores feces until your body can get rid of them. Feces leave the body through an opening called the *anus*. It takes about 24 hours for food to make the trip from your mouth to the end of the large intestine.

You can help keep your digestive system healthy by eating whole grains, fruits, and vegetables. These foods contain a carbohydrate called *cellulose*, or *fiber*. Humans cannot digest fiber. However, fiber keeps the stool soft and keeps materials moving well through the large intestine.

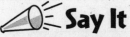 **Say It**

Name With a partner, name as many foods as you can that are sources of fiber.

Copyright © by Holt, Rinehart and Winston; a Division of Houghton Mifflin Harcourt Publishing Company. All rights reserved.

SECTION VOCABULARY

digestive system the organs that break down food so that it can be used by the body

esophagus a long, straight tube that connects the pharynx to the stomach

gallbladder a sac-shaped organ that stores bile produced by the liver

large intestine the wider and shorter portion of the intestine that removes water from mostly digested food and that turns the waste into semisolid feces, or stool

liver the largest organ in the body; it makes bile, stores and filters blood, and stores excess sugars as glycogen

pancreas the organ that lies behind the stomach and that makes digestive enzymes and hormones that regulate sugar levels

small intestine the organ between the stomach and the large intestine where most of the breakdown of food happens and most of the nutrients from food are absorbed

stomach the saclike, digestive organ between the esophagus and the small intestine and that breaks down food by the action of muscles, enzymes, and acids

1. List What organs in the digestive system are not part of the digestive tract?

_____

_____

2. Describe What is the function of saliva?

_____

_____

3. Compare How is chemical digestion different from mechanical digestion?

_____

_____

_____

4. List Name three places in the digestive tract where chemical digestion takes place.

_____

5. Explain How does the structure of the small intestine help it absorb nutrients?

_____

_____

_____

6. Apply Concepts How would digestion change if the liver didn't make bile?

_____

_____

_____

Copyright © by Holt, Rinehart and Winston; a Division of Houghton Mifflin Harcourt Publishing Company. All rights reserved.

CHAPTER 10 The Digestive and Urinary Systems

SECTION 2
The Urinary System

BEFORE YOU READ

After you read this section, you should be able to answer these questions:

- What is the function of the urinary system?
- How do the kidneys filter the blood?
- What are common problems with the urinary system?

TN **Tennessee Science Standards**
GLE 0707.Inq.5
GLE 0707.1.3

What Does the Urinary System Do?

As your cells break down food for energy, they produce waste. Your body must get rid of this waste or it could poison you! As blood moves through the body and drops off oxygen, it picks up the waste from cells. The **urinary system** is made of organs that take wastes from the blood and send them out of the body.

The chart below describes the main organs of the urinary system.

STUDY TIP

Underline As you read, underline any unfamiliar words. Use the glossary or a dictionary to find out what these words mean. Write the definitions in the margins of the text.

Organ	What it looks like	Function
Kidneys	a pair of organs	to clean the blood and produce urine
Ureters	pair of thin tubes leading from the kidneys to the bladder	to carry urine from the kidneys to the bladder
Urinary bladder	a sac	to store urine
Urethra	tube leading from the bladder to outside your body	to carry urine out of the body

TAKE A LOOK
1. **Identify** Which organs clean the blood?

_____

The Urinary System

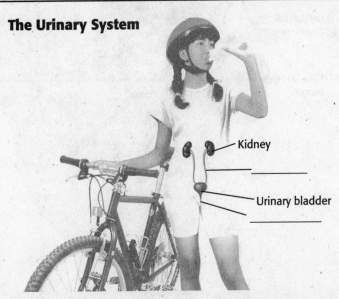

Kidney

Urinary bladder

TAKE A LOOK
2. **Identify** Use the chart above to help you label the organs of the urinary system.

Copyright © by Holt, Rinehart and Winston; a Division of Houghton Mifflin Harcourt Publishing Company. All rights reserved.

Math Focus

3. Calculate Your kidneys filter about 2,000 L of blood each day. Your body has about 5.6 L of blood. About how many times does your blood cycle through your kidneys each day?

_____

How Do the Kidneys Clean the Blood?

The **kidneys** are a pair of organs that clean the blood. Each kidney is made of many small filters called **nephrons**. Nephrons remove waste from the blood. One of the most important substances that nephrons remove is urea. *Urea* is formed when your cells use protein for energy. The figure below shows how kidneys clean the blood.

How the Kidneys Filter Blood

❶ Blood enters the kidney through an artery and moves into nephrons.

❷ Water, sugar, salts, urea, and other waste products move out of the blood and into the nephron.

❸ The nephron returns most of the water and nutrients to the blood. The wastes are left in the nephron.

❹ The cleaned blood leaves the kidney and goes back to the body.

❺ The waste left in the nephron is a yellow fluid called *urine*. Urine leaves the kidneys through tubes called *ureters*.

❻ The urine is stored in the bladder. Urine leaves the body through the urethra.

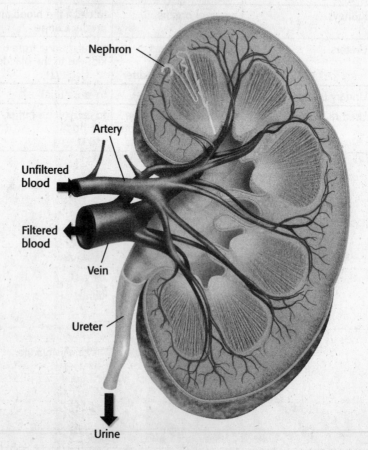

Nephron
Artery
Unfiltered blood
Filtered blood
Vein
Ureter
Urine

TAKE A LOOK

4. Explain What happens to blood after it is cleaned in the kidneys?

_____
_____
_____

5. Identify What structures carry urine from the kidneys?

_____
_____

Copyright © by Holt, Rinehart and Winston; a Division of Houghton Mifflin Harcourt Publishing Company. All rights reserved.

How Does the Urinary System Control Water?

You lose water every day in sweat and urine, and you replace water when you drink. You need to get rid of as much water as you drink. If you don't, your body will swell up. Chemical messengers called *hormones* help control this balance.

One of these hormones is called *antidiuretic hormone* (ADH). ADH keeps your body from losing too much water. When there is not much water in your blood, ADH tells the nephrons to put water back in the blood. The body then makes less urine. If there is too much water in your blood, your body releases less ADH.

Critical Thinking

6. Infer What do you think happens when your body releases less ADH?

_____

_____

_____

_____

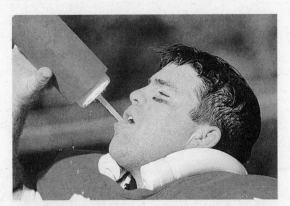

Drinking water when you exercise helps replace the water you lose when you sweat.

Some drinks contain caffeine, which is a diuretic. *Diuretics* cause kidneys to make more urine. This means that a drink with caffeine can actually cause you to lose water.

What Problems Can Happen in the Urinary System?

The chart below shows some common problems of the urinary system.

Problem	Description	Treatment
Bacterial infections	Bacteria can infect the urinary system and cause pain or permanent damage.	Antibiotics
Kidney stones	Wastes can be trapped in the kidney and form small stones. They can stop urine flow and cause pain.	Most can pass out of the body on their own. Some may need to be removed by a doctor.
Kidney disease	Damage to the nephrons can stop the kidneys from working.	A kidney machine can be used to filter the blood.

TAKE A LOOK

7. Explain What problems do kidney stones cause?

_____

_____

Copyright © by Holt, Rinehart and Winston; a Division of Houghton Mifflin Harcourt Publishing Company. All rights reserved.

Section 2 Review

GLE 0707.Inq.5, GLE 0707.1.3 TN

SECTION VOCABULARY

kidney one of the pair of organs that filter water and wastes from the blood and that excrete products as urine	**urinary system** the organs that make, store, and eliminate urine.
nephron the unit in the kidney that filters blood	

1. List What are the main organs that make up the urinary system?

_____

2. Summarize Complete the process chart to show how blood is filtered.

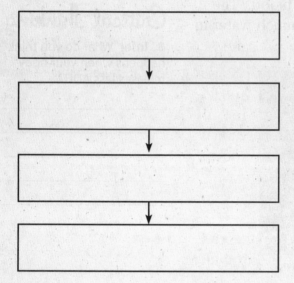

    a. Water is put back in the blood.

    b. Blood goes into the kidney.

    c. Water and waste go into the nephron.

    d. Nephrons separate water and waste.

3. Apply Concepts Which of the following has more water: the blood going into the kidney, or the blood leaving it? Explain your answer.

_____

_____

_____

4. Explain How does the urinary system control the amount of water in the body?

_____

_____

_____

5. Infer Is it a good idea to drink beverages with caffeine when you are exercising? Explain your answer.

_____

_____

_____

Copyright © by Holt, Rinehart and Winston; a Division of Houghton Mifflin Harcourt Publishing Company. All rights reserved.

CHAPTER 11 Communication and Control

SECTION 1

The Nervous System

TN Tennessee Science Standards
GLE 0707.Inq.5
GLE 0707.1.3

BEFORE YOU READ

After you read this section, you should be able to answer these questions:

• What does the nervous system do?

• What is the structure of the nervous system?

What Are the Two Main Parts of the Nervous System?

What is one thing that you have done today that did not involve your nervous system? This is a trick question! Your nervous system controls almost everything you do.

The nervous system has two basic functions. First, it collects information and decides what the information means. This information comes from inside your body and from the world outside your body. Second, the nervous system responds to the information it has collected.

The nervous system has two parts: the central nervous system and the peripheral nervous system. The **central nervous system** (CNS) includes the brain and the spinal cord. The CNS takes in and responds to information from the peripheral nervous system. ☑

The **peripheral nervous system** (PNS) includes all the parts of the nervous system except the brain and the spinal cord. The PNS connects all parts of the body to the CNS. Special structures called nerves in the PNS carry information between the body and the CNS.

Part of the nervous system	What it includes	What it does
Central nervous system (CNS)		takes in and responds to messages from the PNS
Peripheral nervous system (PNS)		

STUDY TIP

Organize As you read, make a chart that describes different structures in the nervous system.

READING CHECK

1. **Identify** What are the two main parts of the nervous system?

_____

_____

_____

TAKE A LOOK

2. **Summarize** Complete the chart to describe the main parts of the nervous system.

Copyright © by Holt, Rinehart and Winston; a Division of Houghton Mifflin Harcourt Publishing Company. All rights reserved.

SECTION 1 The Nervous System *continued*

Math Focus

3. Calculate To calculate how long an impulse takes to travel a certain distance, you can use the following equation:

$$time = \frac{distance}{speed}$$

If an impulse travels 100 m/s, about how long will it take the impulse to travel 10 meters?

How Does Information Move Through the Nervous System?

Special cells called **neurons** carry the information that travels through your nervous system. Neurons carry information in the form of electrical energy. These electrical messages are called *impulses*. Impulses may travel up to 150 m/s!

Like any other cell in your body, a neuron has a nucleus and organelles. The nucleus and organelles are found in the *cell body* of the neuron. However, neurons also have structures called dendrites and axons that are not found in other kinds of cells.

Dendrites are parts of the neuron that branch from the cell body. Most dendrites are very short compared to the rest of the neuron. A single neuron may have many dendrites. Dendrites bring messages from other cells to the cell body.

Axons are longer than dendrites. Some axons can be as long as 1 m! Axons carry information away from the cell body to other cells. The end of an axon is called an *axon terminal*.

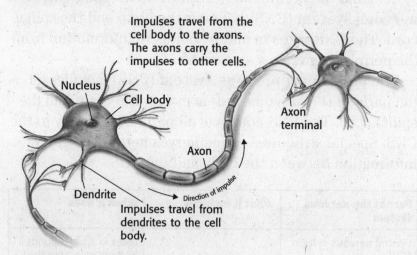

Impulses travel from the cell body to the axons. The axons carry the impulses to other cells.

Nucleus

Cell body

Axon terminal

Axon

Dendrite

Direction of impulse

Impulses travel from dendrites to the cell body.

TAKE A LOOK

4. Identify Add an arrow to the diagram that shows the direction of impulses at the dendrites.

 **Say It**

Identify In a small group, discuss what you think sensory receptors in your eyes, nose, ears, and finger tips respond to.

SENSORY NEURONS

Some neurons are sensory neurons. *Sensory neurons* carry information about what is happening in and around your body. Some sensory neurons are called receptors. *Receptors* can detect changes inside and outside your body. For example, receptors in your eyes can sense light. Sensory neurons carry information from the receptors to the CNS.

Copyright © by Holt, Rinehart and Winston; a Division of Houghton Mifflin Harcourt Publishing Company. All rights reserved.

MOTOR NEURONS

Motor neurons carry impulses from the CNS to other parts of your body. Most motor neurons carry impulses to muscle cells. When muscles cells receive impulses from motor neurons, the muscle cells contract. Some motor neurons carry impulses to glands, such as sweat glands. These messages tell sweat glands when to make sweat.

NERVES

In many parts of your body, groups of axons are wrapped together with blood vessels and connective tissue to form bundles. These bundles are called **nerves**. Your central nervous system is connected to the rest of your body by nerves.

Nerves are found everywhere in your PNS. Most nerves contain axons from both sensory neurons and motor neurons. Many nerves carry impulses from your CNS to your PNS. Other nerves carry impulses from your PNS to your CNS.

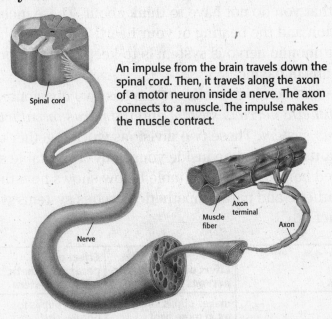

An impulse from the brain travels down the spinal cord. Then, it travels along the axon of a motor neuron inside a nerve. The axon connects to a muscle. The impulse makes the muscle contract.

Spinal cord

Nerve

Muscle fiber

Axon terminal

Axon

TN TENNESSEE STANDARDS CHECK

GLE 0707.1.3 Describe the <u>function</u> of different organ systems and how collectively they enable complex multicellular organisms to survive.

Word Help: <u>function</u> use or purpose

5. Predict What would happen if your nerves stopped working?

_____

_____

_____

_____

_____

What Are the Parts of the Peripheral Nervous System?

Recall that the PNS connects the CNS to the rest of the body. Sensory neurons and motor neurons are both found in the PNS. Sensory neurons carry information to the CNS. Motor neurons carry information from the CNS to the PNS. The motor neurons in the PNS make up two groups: the somatic nervous system and the autonomic nervous system. ☑

✓ READING CHECK

6. Identify What are the two main groups of motor neurons in the PNS?

_____

_____

_____

Copyright © by Holt, Rinehart and Winston; a Division of Houghton Mifflin Harcourt Publishing Company. All rights reserved.

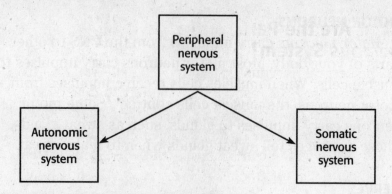

THE SOMATIC NERVOUS SYSTEM

The *somatic nervous* system is made up of motor neurons that you can control. These neurons are connected to skeletal muscles. They control voluntary movements, or movements that you have to think about. These movements include walking, writing, and talking.

THE AUTONOMIC NERVOUS SYSTEM

The *autonomic nervous system* controls body functions that you do not have to think about. These include digestion and the beating of your heart. The main job of the autonomic nervous system is to keep all of the body's functions in balance.

The autonomic nervous system has two divisions: the *sympathetic nervous system* and the *parasympathetic nervous system*. These two divisions work together to maintain a stable state inside your body. This stable state is called *homeostasis*. The table below shows how the sympathetic and parasympathetic nervous systems work together. ☑

Critical Thinking

7. Explain How does eating a piece of pizza involve both the somatic and autonomic nervous systems?

_____
_____
_____
_____
_____
_____

☑ **READING CHECK**

8. Identify What are the two divisions of the autonomic nervous system?

_____
_____

Organ	Effect of sympathetic nervous system	Effect of parasympathetic nervous system
Eye	makes pupils larger to let in more light	returns pupils to normal size
Heart	raises heart rate to increase blood flow	lowers heart rate to decrease blood flow
Lungs	makes bronchioles larger to get more oxygen into the blood	returns bronchioles to normal size
Blood vessels	makes blood vessels smaller to increase blood pressure	has little or no effect
Intestines	reduces blood flow to stomach and intestines to slow digestion	returns digestion to normal pace

Copyright © by Holt, Rinehart and Winston; a Division of Houghton Mifflin Harcourt Publishing Company. All rights reserved.

What Are the Parts of the Central Nervous System?

The central nervous system receives information from sensory neurons. It responds by sending messages to the body through motor neurons. The CNS is made of two important organs: the brain and the spinal cord.

The **brain** is the control center of the nervous system. It is also the largest organ in the nervous system. Many processes that the brain controls are involuntary. However, the brain also controls many voluntary processes. The brain has three main parts: the cerebrum, the cerebellum, and the medulla. Each part of the brain has its own job.

THE CEREBRUM

The *cerebrum* is the largest part of your brain. This dome-shaped area is where you think and where most memories are kept. The cerebrum controls voluntary movements. It also lets you sense touch, light, sound, odors, tastes, pain, heat, and cold. ☑

The cerebrum is made up of two halves called *hemispheres*. The left hemisphere controls most movements on the right side of the body. The right hemisphere controls most movements on the left side of the body. The two hemispheres also control different types of activities, as shown in the figure below. However, most brain activities use both hemispheres.

✓ READING CHECK

9. Identify What kind of movements does the cerebrum control?

_____

_____

TAKE A LOOK

10. Explain Which hemisphere of your brain are you mainly using as you read this book? Explain your answer.

_____

_____

_____

_____

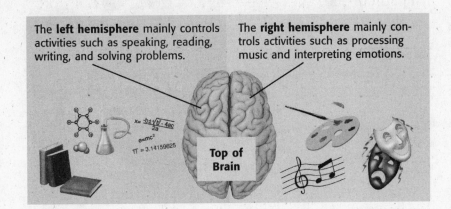

The **left hemisphere** mainly controls activities such as speaking, reading, writing, and solving problems.

The **right hemisphere** mainly controls activities such as processing music and interpreting emotions.

Top of Brain

Copyright © by Holt, Rinehart and Winston; a Division of Houghton Mifflin Harcourt Publishing Company. All rights reserved.

SECTION 1 The Nervous System *continued*

THE CEREBELLUM

The *cerebellum* is found beneath the cerebrum. The cerebellum receives and processes information from your body, such as from your skeletal muscles and joints. This information lets the brain keep track of your body's position. For example, your cerebellum lets you know when you are upside-down. Your cerebellum also sends messages to your muscles to help you keep your balance. ☑

THE MEDULLA

The *medulla* is the part of your brain that connects to your spinal cord. It is only about 3 cm long, but you cannot live without it. It controls involuntary processes, such as breathing and regulating heart rate.

Your medulla is always receiving sensory impulses from receptors in your blood vessels. It uses this information to control your blood pressure. If your blood pressure gets too low, your medulla sends impulses that cause blood vessels to tighten. This makes your blood pressure rise. The medulla also sends impulses to the heart to make it beat faster or slower.

✓ READING CHECK

11. Identify What are two functions of the cerebellum?

_____

_____

_____

_____

TAKE A LOOK

12. Identify Which part of the brain controls breathing?

_____

13. Identify Which part of the brain senses smells?

_____

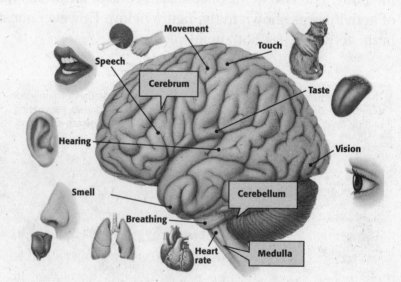

Movement, Speech, Touch, Cerebrum, Taste, Hearing, Vision, Smell, Cerebellum, Breathing, Heart rate, Medulla

Copyright © by Holt, Rinehart and Winston; a Division of Houghton Mifflin Harcourt Publishing Company. All rights reserved.

SECTION 1 The Nervous System *continued*

THE SPINAL CORD

Your spinal cord is part of your central nervous system. It is made up of neurons and bundles of axons that send impulses to and from the brain. The spinal cord is surrounded by bones called *vertebrae* (singular, *vertebra*) to protect it.

The axons in your spinal cord let your brain communicate with your PNS. The axons of sensory neurons carry impulses from your skin and muscles to your spinal cord. The impulses travel through the spinal cord to your brain. The brain then processes the impulses and sends signals back through the spinal cord. The impulses travel from the spinal cord to motor neurons. The axons of motor neurons carry the signals to your body.

Critical Thinking

14. Explain What is one way your nervous system and skeletal system work together?

_____
_____
_____
_____
_____

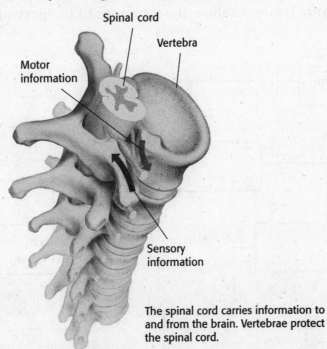

Spinal cord

Vertebra

Motor information

Sensory information

The spinal cord carries information to and from the brain. Vertebrae protect the spinal cord.

SPINAL CORD INJURIES

A spinal cord injury may block the messages to and from the brain. For example, a spinal cord injury might block signals to and from the feet and legs. People with such an injury cannot sense pain, touch, or temperature in their feet or legs. They may not be able to move their legs.

Each year, thousands of people are paralyzed by spinal cord injuries. Most spinal cord injuries in young people happen during sports. You can help prevent such injuries by using proper safety equipment. ☑

✓ **READING CHECK**

15. Identify How can you help prevent spinal cord injuries while playing sports?

_____
_____
_____

Copyright © by Holt, Rinehart and Winston; a Division of Houghton Mifflin Harcourt Publishing Company. All rights reserved.

Section 1 Review

GLE 0707.Inq.5, GLE 0707.1.3 TN

SECTION VOCABULARY

brain the mass of nerve tissue that is the main control center of the nervous system	**neuron** a nerve cell that is specialized to receive and conduct electrical impulses
central nervous system the brain and the spinal cord	**peripheral nervous system** all of the parts of the nervous system except for the brain and the spinal cord
nerve a collection of nerve fibers through which impulses travel between the central nervous system and other parts of the body	

1. Compare How do the functions of the CNS and the PNS differ?

_____

_____

_____

2. Summarize Complete the diagram below to show the structure of the nervous system.

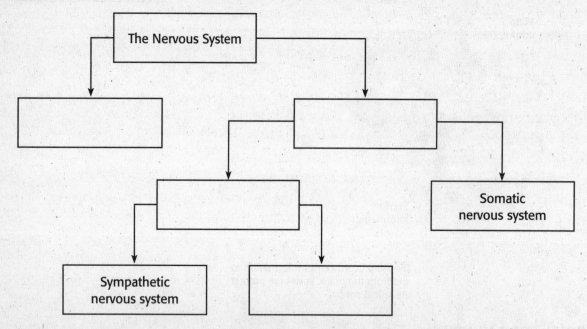

3. Compare How do the functions of dendrites and axons differ?

_____

_____

4. Explain What can happen to someone with a spinal cord injury?

_____

_____

Copyright © by Holt, Rinehart and Winston; a Division of Houghton Mifflin Harcourt Publishing Company. All rights reserved.

SECTION 2 Responding to the Environment

BEFORE YOU READ

After you read this section, you should be able to answer these questions:

- How do the integumentary system and nervous system work together?
- What is a feedback mechanism?
- How do your five senses work?

TN Tennessee Science Standards
GLE 0707.Inq.5
GLE 0707.1.3

How Does Your Sense of Touch Work?

When a friend taps you on the shoulder or when you feel a breeze, how does your brain know what has happened? Receptors throughout your body gather information about the environment and send this information to your brain.

You skin is part of the integumentary system. The **integumentary system** is an organ system that protects the body. This system also includes hair and nails. Your skin is the main organ that helps you sense touch. It has many different *sensory receptors* that are part of the nervous system. Each kind of receptor responds mainly to one kind of stimulation. For example, *thermoreceptors* respond to temperature changes.

Sensory receptors detect a stimulus and create impulses. These impulses travel to your brain. In your brain, the impulses produce a sensation. A *sensation* is the awareness of a stimulus.

STUDY TIP
List As you read, make a list of the five senses. In your list, include the type of receptors used by those senses.

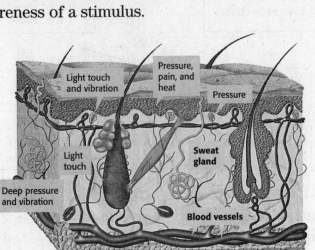

Your skin is made of several layers of tissues and contains many structures. Different kinds of receptors in your skin can sense different things.

TAKE A LOOK
1. List What are three types of sensations that your skin can detect?

_____

_____

Copyright © by Holt, Rinehart and Winston; a Division of Houghton Mifflin Harcourt Publishing Company. All rights reserved.

 **Say It**

Discuss With a partner, name some other examples of reflexes. What part of the body is involved? When does the reflex happen? How does the reflex protect your body?

✓ READING CHECK

2. Complete Feedback mechanisms in your nervous system are controlled by the

_____.

REFLEXES

When you step on something sharp, pain receptors in your foot send messages to your spinal cord. The spinal cord sends a message back to move your foot. This immediate reaction that you can't control is a **reflex**. Messages that cause reflexes do not travel all the way to your brain. If you had to wait for your brain to act, you could be badly hurt.

FEEDBACK MECHANISMS

Most of the time, the brain decides what to do with the messages from the skin receptors. Your brain helps to control many of your body's functions by using feedback mechanisms. A **feedback mechanism** is a cycle of events in which one step controls or affects another step. In the example below, your brain senses a change in temperature. It tells your sweat glands and blood vessels to react. ☑

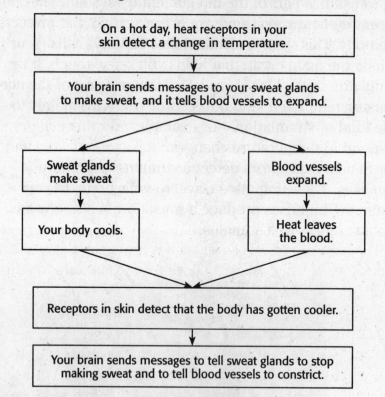

How Does Your Sense of Sight Work?

Sight is the sense that lets you know the size, shape, motion, and color of objects. Light bounces off an object and enters your eyes. Your eyes send impulses to the brain that produce the sensation of sight.

Copyright © by Holt, Rinehart and Winston; a Division of Houghton Mifflin Harcourt Publishing Company. All rights reserved.

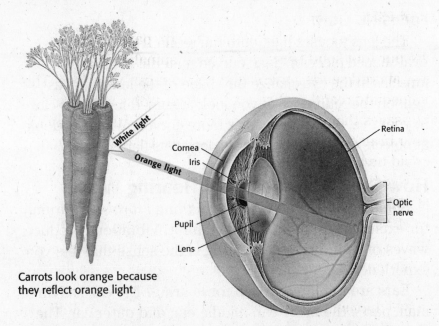

Carrots look orange because
they reflect orange light.

Labels in figure: White light, Orange light, Cornea, Iris, Pupil, Lens, Retina, Optic nerve

TAKE A LOOK
3. Apply Concepts Why do
strawberries look red?

_____

_____

Eyes are complex sensory organs. A clear membrane
called the *cornea* protects the eye but lets light through. Light
enters an opening called the *pupil*. The light then travels
through the lens and hits the retina at the back of the eye.

The **retina** contains neurons called *photoreceptors*
that sense light. When light hits a photoreceptor, it sends
electrical impulses to the brain. The brain interprets
these impulses as light.

The retina has two kinds of photoreceptors: rods and
cones. *Rods* are very sensitive to dim light. They are
important for night vision and for seeing in black and
white. *Cones* are very sensitive to bright light. They let
you see colors and fine details. ☑

The impulses from rods and cones travel along axons. The
axons form the optic nerve that carries the impulses to your
brain.

REACTING TO LIGHT

Your pupil looks like a black dot in the center of your
eye. Actually, it is an opening that lets light enter the eye.
Around the pupil is a ring of muscle called the *iris*. The
iris controls how much light enters your eye. It also gives
your eye its color. ☑

✓ READING CHECK
4. Identify What are the two
kinds of photoreceptors in
the retina?

_____

_____

✓ READING CHECK
5. Identify What is the
function of the iris?

_____

_____

In bright light the iris
contracts. This makes
the pupil smaller.

In dim light the iris relaxes.
This makes the pupil dilate,
or get larger

Copyright © by Holt, Rinehart and Winston; a Division of Houghton Mifflin Harcourt Publishing Company. All rights reserved.

FOCUSING LIGHT

The lens focuses light onto the retina. The *lens* is an oval-shaped piece of clear, curved material behind the iris. Muscles in the eye change the shape of the lens to focus light on the retina. When you look at something that is close to your eye, the lens becomes more curved. When you look at objects that are far away, the lens gets flatter. ☑

How Does Your Sense of Hearing Work?

Sound is produced when something *vibrates*. A drum, for example, vibrates when you hit it. Vibrations produce waves of sound energy. Hearing is the sense that lets you experience sound energy.

Ears are the organs used for hearing. The ear has three main parts: the inner ear, middle ear, and outer ear. The chart below shows the structures and functions of each.

Part of ear	Main structures	Function
Outer ear	ear canal	funnels sound waves into the middle ear
Middle ear	eardrum; three ear bones	sends vibrations to the inner ear
Inner ear	cochlea and auditory nerve	sends impulses to the brain

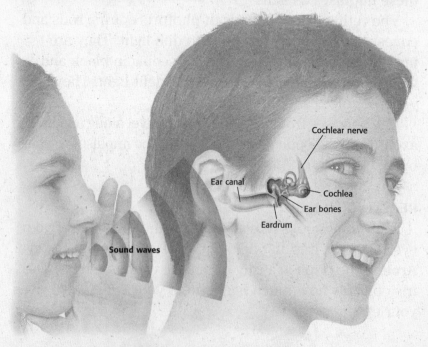

A sound wave travels through the air into the outer ear. The wave produces vibrations in the middle ear and inner ear. These vibrations produce impulses in the cochlear nerve that travel to the brain

---

READING CHECK

6. Identify What is the function of the lens?

_____

_____

TAKE A LOOK

7. Identify Use colored pencils to color the outer ear blue, the middle ear green, and the inner ear red.

Copyright © by Holt, Rinehart and Winston; a Division of Houghton Mifflin Harcourt Publishing Company. All rights reserved.

SECTION 2 Responding to the Environment *continued*

SENDING SOUND ENERGY TO YOUR BRAIN

Sound waves create vibrations throughout your ear that your brain can interpret as sound. The outer ear funnels sound waves to the middle ear. Sound waves hit the eardrum and make it vibrate. These vibrations make tiny ear bones vibrate. One of these bones vibrates against the **cochlea**, an organ filled with fluid. The vibrations make waves in the fluid. This causes neurons in the cochlea to send impulses along the auditory nerve to the brain. ☑

How Does Your Sense of Taste Work?

Taste is the sense that lets you detect chemicals and tell one flavor from another flavor. Your tongue is covered with tiny bumps called *papillae* (singular, *papilla*) that contain taste buds. Taste buds have receptors called taste cells that respond to dissolved food molecules in your mouth. Taste cells react to four basic tastes: sweet, sour, salty, and bitter.

How Does Your Sense of Smell Work?

Neurons called *olfactory cells* are located in the nose and have receptors for smell. Your brain combines the information from taste buds and olfactory cells to let you sense flavor.

☑ **READING CHECK**

8. Explain What does your brain interpret as sound?

_____

_____

Critical Thinking

9. Apply Concepts Why do you have a hard time tasting things when you have a cold?

_____

_____

_____

_____

_____

_____

Brain

Olfactory cell

Nasal passage

Molecules that you inhale dissolve in the moist lining of the nasal cavity and trigger an impulse. The brain interprets the impulse as the sensation of smell.

Copyright © by Holt, Rinehart and Winston; a Division of Houghton Mifflin Harcourt Publishing Company. All rights reserved.

Section 2 Review

GLE 0707.Inq.5, GLE 0707.1.3 TN

SECTION VOCABULARY

cochlea a coiled tube that is found in the inner ear and that is essential to hearing	**reflex** an involuntary and almost immediate movement in response to a stimulus
feedback mechanism a cycle of events in which information from one step controls or affects a previous step	**retina** the light-sensitive inner layer of the eye that receives images formed by the lens and transmits them through the optic nerve to the brain
integumentary system the organ system that forms a protective covering on the outside of the body	

1. List What are the five senses?

_____

2. Explain How do the integumentary system and nervous system work together?

_____

_____

_____

3. Explain What are reflexes and why are they important for the body?

_____

_____

_____

4. Explain Why is important for your eyes to have both rods and cones?

_____

_____

_____

5. Summarize Complete the chart to summarize the major senses and sense receptors.

Sense	Receptors	What the receptors respond to
Touch	many different kinds	
Sight		
	neurons in the cochlea	
Taste		dissolved molecules
	olfactory cells	

Copyright © by Holt, Rinehart and Winston; a Division of Houghton Mifflin Harcourt Publishing Company. All rights reserved.

CHAPTER 11 Communication and Control
SECTION 3 # The Endocrine System

BEFORE YOU READ

After you read this section, you should be able to answer these questions:

• Why is the endocrine system important?

• How do feedback systems work?

• What are common hormone imbalances?

TN **Tennessee Science Standards**
GLE 0707.Inq.5
GLE 0707.1.3

What Is the Endocrine System?

The **endocrine system** controls body functions using chemicals made by endocrine glands. A **gland** is a group of cells that makes special chemicals for your body. The chemicals made by endocrine glands are called hormones. A **hormone** is a chemical messenger. It is made in one cell or tissue and causes a change in another cell or tissue. Hormones flow through the bloodstream to all parts of the body. ☑

Glands and Organs of the Endocrine System

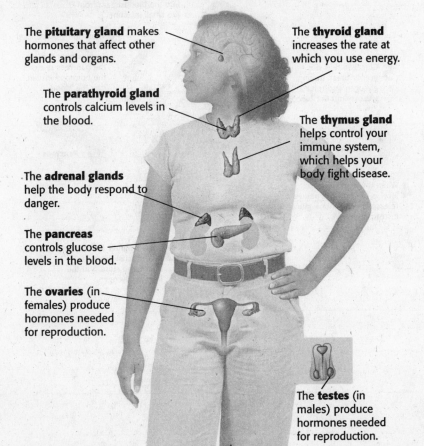

The **pituitary gland** makes hormones that affect other glands and organs.

The **parathyroid gland** controls calcium levels in the blood.

The **adrenal glands** help the body respond to danger.

The **pancreas** controls glucose levels in the blood.

The **ovaries** (in females) produce hormones needed for reproduction.

The **thyroid gland** increases the rate at which you use energy.

The **thymus gland** helps control your immune system, which helps your body fight disease.

The **testes** (in males) produce hormones needed for reproduction.

STUDY TIP

Describe As you read, fill in the chart at the end of the section to name the major endocrine glands and what each one does.

READING CHECK

1. Explain How do hormones move from one part of the body to another?

_____

_____

_____

TAKE A LOOK

2. Identify What two structures produce hormones needed for reproduction?

_____

_____

3. Identify Which gland makes hormones that affect organs and other glands?

_____

Copyright © by Holt, Rinehart and Winston; a Division of Houghton Mifflin Harcourt Publishing Company. All rights reserved.

ADRENAL GLANDS

Endocrine glands may affect many organs at one time. For example, the adrenal glands release the hormone *epinephrine*, sometimes called *adrenaline*. Epinephrine increases your heartbeat and breathing rate. This response is called the fight-or-flight response. When you are scared, angry, or excited, the fight-or-flight response prepares you either to fight the danger or to run from it.

How Do Feedback Systems Work?

Recall how feedback mechanisms work in the nervous system. Feedback mechanisms are cycles in which information from one step controls another step in the cycle. In the endocrine system, endocrine glands control similar feedback mechanisms.

The pancreas has specialized cells that make two different hormones, *insulin* and *glucagon*. These two hormones control the level of glucose in the blood.

Critical Thinking

4. Infer Why do you think increasing your breathing rate helps prepare you to fight or run away?

_____

_____

_____

_____

_____

_____

TAKE A LOOK

5. Explain How could you raise your blood-glucose level without involving hormones?

_____

_____

6. Identify Which hormone tells the liver to release glucose into the blood?

_____

_____

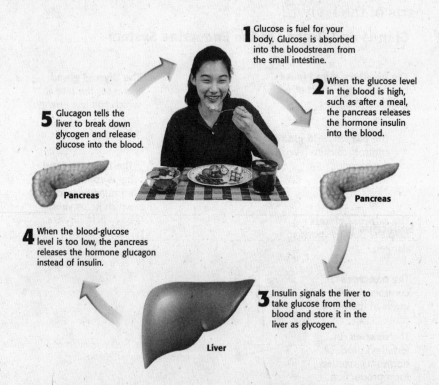

1 Glucose is fuel for your body. Glucose is absorbed into the bloodstream from the small intestine.

2 When the glucose level in the blood is high, such as after a meal, the pancreas releases the hormone insulin into the blood.

5 Glucagon tells the liver to break down glycogen and release glucose into the blood.

Pancreas

Pancreas

4 When the blood-glucose level is too low, the pancreas releases the hormone glucagon instead of insulin.

3 Insulin signals the liver to take glucose from the blood and store it in the liver as glycogen.

Liver

Copyright © by Holt, Rinehart and Winston; a Division of Houghton Mifflin Harcourt Publishing Company. All rights reserved.

What Is a Hormone Imbalance?

Sometimes, an endocrine gland makes too much or not enough of a hormone. For example, a person's body may not make enough insulin or be able use it properly. This is a condition called *diabetes mellitus*. A person who has diabetes may need daily injections of insulin. These injections help keep his or her blood-glucose levels within safe limits. Some patients get their insulin automatically from a small machine worn on the body. ☑

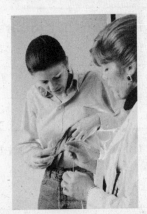

This woman has diabetes. She is wearing a device that delivers insulin to her body.

READING CHECK

7. Define What is a hormone inbalance?

_____

_____

_____

Another hormone imbalance is when a child's pituitary gland doesn't make enough growth hormone. As a result, the child does not grow as quickly or as much as he or she should. If the problem is found early in childhood, a doctor can prescribe growth hormone.

In some cases, the pituitary gland may make too much growth hormone. This may cause a child to grow taller than expected.

Endocrine gland	Function
Pituitary	
	increases the rate at which you use energy

TAKE A LOOK

8. Summarize Use this chart to help you summarize the major endocrine glands and their functions.

Copyright © by Holt, Rinehart and Winston; a Division of Houghton Mifflin Harcourt Publishing Company. All rights reserved.

Section 3 Review

GLE 0707.Inq.5, GLE 0707.1.3 TN

SECTION VOCABULARY

endocrine system a collection of glands and groups of cells that secrete hormones that regulate growth, development, and homeostasis; includes the pituitary, thyroid, parathyroid, and adrenal glands, the hypothalamus, the pineal body, and the gonads	**gland** a group of cells that make special chemicals for the body **hormone** a substance that is made in one cell or tissue and that causes a change in another cell or tissue in a different part of the body

1. Explain What is the function of the endocrine system?

_____

2. Compare How are the thymus and thyroid gland similar? How are they different?

_____

_____

_____

3. Identify Relationships How do the circulatory system and the endocrine system work together?

_____

_____

4. Explain What does insulin do?

_____

_____

5. Apply Concepts Many organs in the body are part of more than one organ system. List three examples from this section of organs that are part of both the endocrine system and another organ system.

_____

_____

_____

6. Infer Glucose is a source of energy. Epinephrine quickly raises your blood-glucose level when you are excited or scared. Why is epinephrine important during these times?

_____

_____

_____

Copyright © by Holt, Rinehart and Winston; a Division of Houghton Mifflin Harcourt Publishing Company. All rights reserved.

CHAPTER 12 Reproduction and Development
SECTION 1

Animal Reproduction

TN Tennessee Science
Standards
GLE 0707.Inq.5
GLE 0707.4.1
GLE 0707.4.3

BEFORE YOU READ

After you read this section, you should be able to answer these questions:

- What is asexual reproduction?
- What is sexual reproduction?
- What is the difference between external and internal fertilization?

Why Do Organisms Reproduce?

Organisms live for different amounts of time. For example, a fruit fly lives for about 40 days. Some pine trees can live for almost 5,000 years. However, all living things will die. For a species to surivive, its members must reproduce.

How Are Offspring Made with One Parent?

Some animals reproduce asexually. In **asexual reproduction**, one parent has offspring. These offspring are genetically identical to the parent.

Asexual reproduction can occur through budding, fragmentation, or regeneration. *Budding* happens when the offspring buds or grows out from the parent's body. The new organism then pinches off. *Fragmentation* occurs when part of an organism breaks off and then develops into a new organism. *Regeneration* occurs when part of the organism's body is lost and the organism regrows it. The lost body part can also develop into an offspring. ☑

The hydra bud will separate from its parent. Buds from other organisms, such as some coral, stay attached to the parent. ▶

◀ The largest arm on this sea star was lost from another sea star. A new sea star has grown from this arm.

STUDY TIP

Describe As you read, make a chart describing characteristics and types of asexual and sexual reproduction.

READING CHECK

1. List What are three types of asexual reproduction?

_____

_____

_____

TAKE A LOOK

2. Identify What kind of asexual reproduction does this sea star in this picture show?

_____

Copyright © by Holt, Rinehart and Winston; a Division of Houghton Mifflin Harcourt Publishing Company. All rights reserved.

What Is Sexual Reproduction?

Most animals reproduce sexually. In **sexual reproduction**, offspring form when sex cells from more than one parent combine. The offspring will share traits from both parents. Sexual reproduction usually requires two parents—a male and a female. The female makes sex cells called **eggs**. The male makes sex cells called **sperm**. When a sperm and an egg join together, a fertilized egg, or *zygote* is made. This process is called *fertilization*. ☑

Almost all human cells have 46 chromosomes. However, egg and sperm cells only have 23 chromosomes. These sex cells are formed by a process called *meiosis*. In humans, meiosis occurs when one cell with 46 chromosomes divides to make sex cells with 23 chromosomes. This way, when a sperm and egg join, the zygote will have the typical number of chromosomes.

How Do Offspring Get Their Parent's Traits?

Genetic information is carried by *genes*. Genes are located on *chromosomes*. At fertilization in humans, the egg and sperm both give 23 chromosomes to the zygote. This zygote will then develop into an organism that has traits from both parents. The figure below shows how genes are passed from parent to offspring.

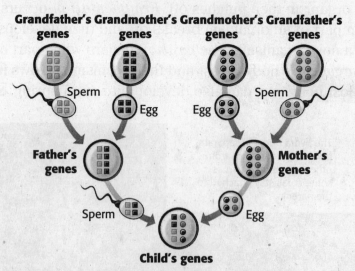

Eggs and sperm contain chromosomes. You inherited half of your chromosomes from each parent.

✔ READING CHECK

3. Define What is fertilization?

_____

_____

_____

TAKE A LOOK

4. Identify What fraction of your genes did you inherit from each grandparent?

_____

_____

Copyright © by Holt, Rinehart and Winston; a Division of Houghton Mifflin Harcourt Publishing Company. All rights reserved.

How Does Fertilization Occur?

Fertilization can occur outside or inside the female's body. **External fertilization** occurs when the sperm fertilizes the eggs outside the female's body. Some fish and amphibians, such as frogs, use external fertilization. The female releases her eggs and the male releases his sperm over the eggs. External fertilization generally takes place in wet environments so that the new zygotes do not dry out.

Internal fertilization occurs when an egg and sperm join inside a female's body. This allows the female to protect the developing zygote inside her body. Reptiles, birds, mammals, and some fishes use internal fertilization. Many mammals give birth to well-developed young. However, other animals, such as chickens, use internal fertilization to lay fertilized eggs.

The female clownfish has laid her eggs on the rock. The male will then fertilize them.

Critical Thinking
5. Infer Name some environments where animals could use external fertilization.

_____

_____

_____

TAKE A LOOK
6. Identify Is this an example of internal or external fertlization?

_____

How Do Mammals Reproduce?

All mammals reproduce sexually by internal fertilization. They also feed their young with milk. Mammals reproduce in one of the following ways. ☑

- *Monotremes* lay eggs. A platypus is a monotreme.

- *Marsupials* give birth to young that are not well developed. After birth, the young crawl into their mother's pouch to develop. A kangaroo is a marsupial.

- *Placental mammals* are fed and protected inside their mother's body before birth. These mammals are born well developed. Humans are placental mammals.

☑ **READING CHECK**
7. Identify Name two ways that all mammals are alike.

_____

_____

Copyright © by Holt, Rinehart and Winston; a Division of Houghton Mifflin Harcourt Publishing Company. All rights reserved.

Section 1 Review

GLE 0707.Inq.5, GLE 0707.4.1, GLE 0707.4.3 TN

SECTION VOCABULARY

asexual reproduction reproduction that does not involve the union of sex cells and in which a single parent produces offspring that are genetically identical to the parent	**internal fertilization** fertilization of an egg by sperm that occurs inside the body of a female
egg a sex cell produced by a female	**sexual reproduction** reproduction in which the sex cells from two parents unite to produce offspring that share traits from both parents
external fertilization the union of sex cells outside the bodies of the parents	**sperm** the male sex cell

1. Compare How is fragmentation different from budding?

_____

_____

_____

2. Infer Why is reproduction as important to a bristlecone pine as it is to a fruit fly?

_____

_____

3. Apply Concepts Why is meiosis important in sexual reproduction?

_____

_____

_____

_____

4. Identify What type of reproduction produces offspring that are genetically identical to the parent?

_____

5. Compare What is the difference between external and internal fertilization?

_____

_____

6. Identify What is one advantage of internal fertilization?

_____

_____

7. List What are three groups of animals that use internal fertilization?

_____

Copyright © by Holt, Rinehart and Winston; a Division of Houghton Mifflin Harcourt Publishing Company. All rights reserved.

CHAPTER 12 Reproduction and Development

SECTION 2 Human Reproduction

■.☞ Tennessee Science Standards
GLE 0707.Inq.5
GLE 0707.1.3

BEFORE YOU READ

After you read this section, you should be able to answer these questions:

• How are sperm and eggs made?

• How does fertilization occur?

• What problems can happen in the reproductive system?

What Happens in the Male Reproductive System?

The male reproductive system has two functions:

• to make sperm
• to deliver sperm to the female reproductive system

To perform these functions, organs in the male reproductive system make sperm, hormones, and fluids. The **testes** (singular, *testis*) are a pair of organs that hang outside the body covered by a skin sac called the *scrotum*. They make sperm and *testosterone*, the main male sex hormone.

A male can make millions of sperm each day. Immature sperm cells divide and change shape as they travel through the testes and epididymis. The *epididymis* is a tube attached to the testes that stores sperm as they mature. ☑

Mature sperm pass into the *vas deferens*, which connects the epididymis and urethra. The *urethra* is a tube that runs from the bladder through the penis. The **penis** is the male organ that delivers sperm to the female. Before leaving the body, sperm mixes with a fluid mixture to form *semen*.

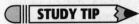

STUDY TIP

Summarize As you read, create two Process Charts. In the first, describe the path an egg takes from ovulation to fertilization. In the second, describe the path of an egg that does not get fertilized.

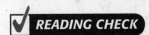

READING CHECK

1. Identify How many sperm can a male make in one day?

Copyright © by Holt, Rinehart and Winston; a Division of Houghton Mifflin Harcourt Publishing Company. All rights reserved.

What Happens in the Female Reproductive System?

The female reproductive system has three functions: to produce eggs, to protect and nourish developing offspring, and to give birth. Unlike males, who produce new sperm throughout their lives, females have all their eggs when they are born.

Eggs are produced in an **ovary**. Ovaries also release the main female sex hormones: estrogen and progesterone. These hormones control the release of eggs from the ovaries and the development of female characteristics. Females generally have two ovaries.

THE EGG'S JOURNEY

During *ovulation* an ovary releases an egg. The egg passes into a *fallopian tube*. The fallopian tube leads from the ovary to the uterus. If sperm are present, fertilization usually takes place in the fallopian tube.

After fertilization, the embryo moves to the uterus and may embed in the thick lining. An embyo develops into a fetus in the **uterus**. When the baby is born, it passes from the uterus and through the vagina. The **vagina** is the canal between the outside of the body and the uterus.

THE MENSTRUAL CYCLE

From puberty through her late 40s or early 50s, a woman's reproductive system goes through the *menstrual cycle*. This cycle of about 28 days prepares the body for pregnancy. An ovary releases an egg at *ovulation*. This happens at around the 14th day of the cycle. If the egg is not fertilized, menstruation begins. *Menstruation* is the monthly discharge of blood and tissue from the uterus.

Math Focus

3. Calculate The average woman ovulates each month from about the age of 12 to about the age of 50. How many mature eggs does she release from age 18 to age 50? Assume that she has never been pregnant.

Copyright © by Holt, Rinehart and Winston; a Division of Houghton Mifflin Harcourt Publishing Company. All rights reserved.

What Problems Can Happen in the Reproductive System?

Problems such as disease can cause the reproductive system to fail. When couples cannot have children, they are considered *infertile*. Men are infertile if they do not make enough healthy sperm. Women are infertile if they do not ovulate normally. Reproductive problems are often caused by sexually transmitted diseases and cancers.

SEXUALLY TRANSMITTED DISEASES

A *sexually transmitted disease* (STD) is a disease that can pass from one person to another during sexual contact. STDs are also called *sexually transmitted infections* (STIs). Sexually-active young people have the highest risk for STDs.

One example of an STD is human immunodeficiency virus (HIV), the virus that leads to AIDS. HIV destroys the immune system of the infected person. People with AIDS generally die from infections that are not fatal to people with healthy immune systems. Below is a table showing the most common STDs and how fast they are spreading in the United States.

STD	Approximate number of new cases each year
Chlamydia	3 to 10 million
Genital HPV (human papilloma virus)	5.5 million
Genital herpes	1 million
Gonorrhea	650,000
Syphilis	70,000
HIV/AIDS	40,000 to 50,000

CANCER

Sometimes cancer happens in reproductive organs. *Cancer* is a disease in which cells grow at an uncontrolled rate. In men, the two most common cancers of the reproductive system happen in the testes and prostate gland. In women, two common reproductive cancers are cancer of the cervix and cancer of the ovaries.

TN TENNESSEE STANDARDS CHECK

GLE 0707.1.3 Describe the function of different organ systems and how collectively they enable complex multicellular organisms to survive.

Word Help: function use or purpose.

5. Identify What are two common causes of reproductive problems?

_____

_____

Critical Thinking

6. Infer In women, some untreated STDs can block the fallopian tubes. How would this affect fertilization?

_____

_____

_____

_____

Say It

Research Use your school library or the internet to research one of the STDs in the chart. What organism or virus causes it? How does it affect the body? What treatments are available? Present your findings to the class.

Copyright © by Holt, Rinehart and Winston; a Division of Houghton Mifflin Harcourt Publishing Company. All rights reserved.

Section 2 Review

GLE 0707.Inq.5, GLE 0707.1.3 **TN**

SECTION VOCABULARY

ovary an organ in the female reproductive system of animals that produces eggs	**uterus** in female placental mammals, the hollow, muscular organ in which an embryo embeds itself and develops into a fetus
penis the male organ that transfers sperm to a female and that carries urine out of the body	**vagina** the female reproductive organ that connects the outside of the body to the uterus
testes the primary male reproductive organs which produce sperm and testosterone (singular, *testis*)	

1. Explain What is the purpose of the menstrual cycle?

_____

_____

2. Organize Complete the chart below to describe the functions or characteristics of structures in the female reproductive system.

Structure	Characteristic or function
	produces eggs; releases female sex hormones
uterus	
fallopian tube	
	canal that connects uterus to the outside

3. Explain What is fertilization and where does it occur?

_____

_____

_____

4. Apply Concepts Fraternal twins are created when two sperm fertilize two different eggs. Paternal, or identical, twins are created when a single egg divides after fertilization. Why are fraternal twins not identical?

_____

_____

_____

_____

Copyright © by Holt, Rinehart and Winston; a Division of Houghton Mifflin Harcourt Publishing Company. All rights reserved.

CHAPTER 12 Reproduction and Development

SECTION 3 Growth and Development

BEFORE YOU READ

After you read this section, you should be able to answer these questions:

- What happens after an egg is fertilized?
- How does a fetus develop?
- How does a person develop after birth?

How Does Fertilization Occur?

A man can release millions of sperm at once. However, only one sperm can fertilize an egg. Why are so many sperm needed?

Only a few hundred sperm survive the journey from the vagina to the uterus and into a fallopian tube. In the fallopian tube only a few sperm find and cover the egg. Once one sperm enters, or *penetrates*, the egg, it causes the egg's covering to change. This change keeps other sperm from entering the egg. When the nuclei of the sperm and egg join, the egg is fertilized.

Fertilization and Implantation

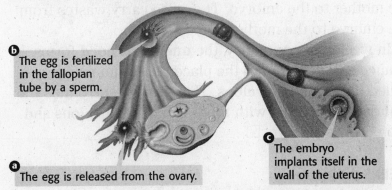

ⓑ The egg is fertilized in the fallopian tube by a sperm.

ⓐ The egg is released from the ovary.

ⓒ The embryo implants itself in the wall of the uterus.

What Stages Does a Fertilized Egg Go Through?

At fertilization, the egg is only a single cell. At this stage, the fertilized egg is called a *zygote*. As the zygote becomes an embryo it moves down the fallopian tube, and divides many times. After about a week, the **embryo** is a ball of cells. This ball of cells implants in the uterus. ☑

As the cells of the embryo continue to divide, some cells start to *differentiate*. They develop special structures for certain jobs in the body. After week 10 of the pregnanacy, the embryo is called a **fetus**.

TN Tennessee Science Standards

GLE 0707.Inq.2
GLE 0707.Inq.3
GLE 0707.Inq.5

STUDY TIP

Summarize As you read, make a a timeline that shows the stages of human development from fertilized egg to adulthood.

Critical Thinking

1. Form Hypothesis Why do you think millions of sperm are released to fertilize one egg?

_____

_____

_____

_____

TAKE A LOOK

2. Identify Where does fertilization usually take place?

_____

_____

READING CHECK

3. Identify What is the egg called right after fertilization?

_____

Copyright © by Holt, Rinehart and Winston; a Division of Houghton Mifflin Harcourt Publishing Company. All rights reserved.

TAKE A LOOK

4. List List the three stages a fertilized egg goes through as it develops.

_____

_____

_____

| The fertilized egg, called a *zygote*, divides and becomes a ball of cells | → | The ball of cells, called an *embryo*, implants in the uterus | → | The cells of the embryo divide and differentiate; the embryo becomes a *fetus* |

How Does an Embryo Develop?

WEEKS 1 AND 2

A woman's pregnancy starts when an egg is fertilized, and ends at birth. Pregnancy is measured from the starting date of a woman's last menstruation. This is easier than trying to determine the exact date fertilization took place. Fertilization takes place at about the end of week 2. ☑

✔ **READING CHECK**

5. Explain Why is pregnancy measured from the starting date of the woman's last period?

_____

_____

_____

WEEKS 3 AND 4

In week 3, the embryo moves to the uterus. As the embryo travels, it divides many times. It becomes a ball of cells that implants itself in the wall of the uterus.

WEEKS 5 TO 8

After an embryo implants in the uterus, the placenta forms. The **placenta** is an organ used by the embryo to exchange materials with the mother. The placenta has many blood vessels that carry nutrients and oxygen from the mother to the embryo. They also carry wastes from the embryo to the mother.

In week 5 of pregnancy, the **umbilical cord** forms. It connects the embryo to the placenta. A thin membrane called the amnion develops. The *amnion* surrounds the embryo and is filled with fluid. This fluid cushions and protects the embryo. ☑

✔ **READING CHECK**

6. Explain What is the function of the umbilical cord?

_____

_____

_____

WEEKS 9 TO 16

At week 9, the embryo may start to make tiny movements. The fetus grows very quickly during this stage. It doubles, then triples in size within a month. In about week 13, the fetus's face begins to look more human. During this stage the fetus's muscles also grow stronger. It can even make a fist.

Copyright © by Holt, Rinehart and Winston; a Division of Houghton Mifflin Harcourt Publishing Company. All rights reserved.

WEEKS 17 TO 24

By week 17 the fetus can make faces. By week 18 the fetus starts to make movements that its mother can feel. It can also hear sounds through the mother's uterus. By week 23 the fetus makes a lot of movements. A baby born during week 24 might survive, but it would need a lot of help.

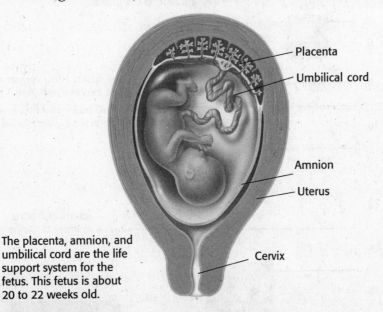

Placenta

Umbilical cord

Amnion

Uterus

Cervix

The placenta, amnion, and umbilical cord are the life support system for the fetus. This fetus is about 20 to 22 weeks old.

TAKE A LOOK
7. Identify What structure surrounds the fetus?

_____

_____

WEEKS 25 TO 36

At about 25 or 26 weeks, the fetus's lungs are well-developed. However, they are not fully mature. The fetus still gets oxygen from its mother through the placenta. The fetus will not take its first breath of air until it is born.

By the 32nd week, the fetus can open and close its eyes. Studies show that the fetus responds to light. Some scientists think fetuses at this stage show brain activity and eye movements like sleeping children or adults. These scientists think a sleeping fetus may dream. After 36 weeks, the fetus is almost ready to be born.

BIRTH

At weeks 37 to 38 the fetus is fully developed. A full pregnancy is usually 40 weeks. As birth begins, the muscles of the uterus begin to squeeze, or contract. This is called *labor*. These contractions push the fetus out of the mother's body through the vagina. ☑

Once the baby is born, the umbilical cord is tied and cut. The navel is all that will remain of the point where the umbilical cord was attached. Once the mother's body has pushed out the placenta, labor is complete.

READING CHECK
8. Define What is labor?

_____

_____

_____

Copyright © by Holt, Rinehart and Winston; a Division of Houghton Mifflin Harcourt Publishing Company. All rights reserved.

SECTION 3 Growth and Development *continued*

Pregnancy Timeline

Week

Fertilization takes place.

2

The fertilized egg becomes a ball of hundreds of cells.

4 — Implantation is complete.

The spinal cord and brain begin to form.

6

8 — Well-defined tiny fingers and toes become apparent.

The embryo may make tiny movements that may be detected by ultrasound.

10

The embryo is now called a *fetus*.

12

14

Bones and bone marrow continue to form.

16

18

20

A layer of fat begins to form under the skin.

22

The fetus is developing taste buds, and its brain is growing rapidly.

24

The fetus's lungs are almost ready to breathe air.

26

The fetus practices breathing and has brain wave activity.

28

The eyes of the fetus are open, and the fetus may turn toward a bright light.

30

32

The fetus's skin turns from red to pink.

34

36 — The fetus's skull has hardened.

38

The baby is born.

40

TAKE A LOOK

9. Identify By what week has the brain of the embryo started to form?

_____

10. Identify At around what week does the fetus start to develop taste buds?

_____

Copyright © by Holt, Rinehart and Winston; a Division of Houghton Mifflin Harcourt Publishing Company. All rights reserved.

How Does a Person Grow and Change?

The human body goes through several stages of development. One noticeable difference is the change in body proportion.

Body Proportions During Stages of Human Development

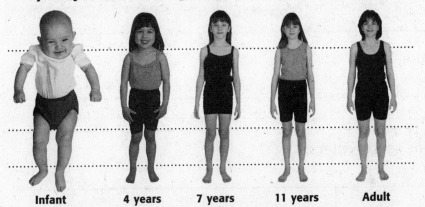

Infant   4 years   7 years   11 years   Adult

The chart below lists the different stages of life and the characteristics of that stage.

Stage	Ages	Characteristics
Infancy	birth to age 2	• Baby teeth appear. • Nervous system and muscles develop. • Coordination improves.
Childhood	age 2 to puberty	• Permanent teeth grow. • Nerve pathways mature, and child can learn new skills. • Muscle coordination increases.
Adolescence	in females, puberty takes place between the ages of 9 and 14	**Females:** • Breasts enlarge. • Body hair appears. • Menstruation begins.
	in males, puberty takes place between the ages of 11 and 16	**Males:** • Body grows more muscular. • Voice deepens. • Facial and body hair appear.
	puberty to adulthood	• Reproductive system matures (puberty).
Adulthood	age 20+	• From ages 20 to 40, physical development is at its peak. • After age 40, hair may turn gray, skin may wrinkle, athletic ability may decrease.

TAKE A LOOK

11. Identify At which stage or age is the head largest in proportion to the rest of the body?

_____

Math Focus

12. Calculate Alice is 80 years old. She started puberty at age 12. Calculate the percentage of her life that she has spent in each stage of development.

_____

_____

_____

TAKE A LOOK

13. Identify Which stage of development are you in?

_____

Copyright © by Holt, Rinehart and Winston; a Division of Houghton Mifflin Harcourt Publishing Company. All rights reserved.

Section 3 Review

GLE 0707.Inq.2, GLE 0707.Inq.3, GLE 0707.Inq.5 **TN**

SECTION VOCABULARY

embryo in humans, a developing individual from first division after fertilization through the 10th week of pregnancy	**placenta** the partly fetal and partly maternal organ by which materials are exchanged between a fetus and the mother
fetus a developing human from the end of the 10th week of pregnancy until birth	**umbilical cord** the ropelike structure through which blood vessels pass and by which a developing mammal is connected to the placenta

1. Define What is fertilization?

_____

_____

2. Explain Why can only one sperm enter an egg?

_____

_____

_____

3. Explain What does it mean that cells differentiate?

_____

_____

4. Identify What is the function of the amnion?

_____

5. Explain Why is the placenta important to a developing embryo?

_____

_____

_____

6. Explain Why is it necessary for cells in an embryo to differentiate?

_____

_____

7. List What are the four stages of human development after birth?

_____

_____

8. Identify What is the main characteristic of adolescence?

_____

Copyright © by Holt, Rinehart and Winston; a Division of Houghton Mifflin Harcourt Publishing Company. All rights reserved.

| CHAPTER 13 | Minerals of the Earth's Crust |
| SECTION 1 | **What Is a Mineral?** |

TN **Tennessee Science Standards**
GLE 0707.7.1

BEFORE YOU READ

After you read this section, you should be able to answer these questions:

• What are minerals?

• What determines the shape of a mineral?

• What are two main groups of minerals?

What Are Minerals?

A **mineral** is a naturally formed, inorganic solid that forms crystals and is always made of the same elements. The figure below shows four questions that you can ask in order to learn whether something is a mineral.

Is it nonliving?
Minerals are inorganic. This means that they are not made of living things or their remains.

Is it a solid?
Minerals are not gases or liquids.

Does it have a crystalline structure?
Minerals are crystals. Each mineral has a certain crystal structure that is always the same.

Does it form naturally?
Minerals are not made by people.

All minerals have four features, as described in the figure.

You might not be familiar with the term "crystalline structure." To understand what crystalline structure is, you need to know a little about how elements form minerals. **Elements** are pure substances that cannot be broken down into simpler substances. Oxygen, chlorine, carbon, and iron are examples of elements. Elements can come together in certain ways to form new substances, such as minerals. All minerals are made of one or more elements.

STUDY TIP

Learn New Words As you read, underline words you don't understand. When you figure out what they mean, write the words and their definitions in your notebook.

TAKE A LOOK

1. Explain Why are diamonds that are made by people not considered minerals?

you might not be familid

Critical Thinking

2. Apply Concepts Coal is made from the remains of dead plants. Is coal a mineral? Explain your answer.

Oxygen, chlorine

Copyright © by Holt, Rinehart and Winston; a Division of Houghton Mifflin Harcourt Publishing Company. All rights reserved.

SECTION 1 What Is a Mineral? *continued*

COMPOUNDS AND ATOMS

Most minerals are made of compounds of several different elements. A **compound** is a substance made of two or more elements that are chemically bonded. For example, the mineral halite is a compound of sodium, Na, and chlorine, Cl. A few minerals, such as gold and silver, are made of only one element. A mineral that is made of only one element is called a *native element*. ☑

Each element is made of only one kind of atom. An *atom* is the smallest part of an element that has the properties of that element. Like other compounds, minerals are made up of atoms of one or more elements.

CRYSTALS

Remember that minerals have a definite crystalline structure. This means that the atoms in the mineral line up in a regular pattern. The regular pattern of the atoms in a mineral causes the mineral to form crystals. **Crystals** are solid, geometric forms of minerals that are formed by repeating a pattern of atoms. ☑

The shape of a crystal depends on how the atoms in it are arranged. The atoms that make up each mineral are different. However, there are only a few ways that atoms can be arranged. Therefore, the crystals of different minerals can have similar shapes.

Although different minerals may form similar shapes, each mineral forms only one shape of crystal. Therefore, geologists say that a mineral has a definite crystalline structure. This means that crystals of a certain mineral always form the same shape.

READING CHECK

3. Define What is a compound?

is a substance made of two or more elements that are chemically bonded.

READING CHECK

4. Explain What causes minerals to form crystals?

remember that minerals have a definite

TAKE A LOOK

5. Identify What shape are gold crystals?

gold form cubes

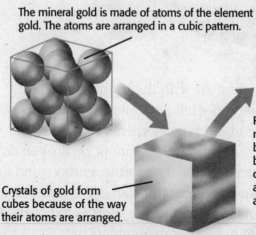

The mineral gold is made of atoms of the element gold. The atoms are arranged in a cubic pattern.

Crystals of gold form cubes because of the way their atoms are arranged.

Real crystals of gold may not be perfect cubes because the crystals may be damaged or not form completely. However, the atoms are still arranged in a cubic pattern.

Copyright © by Holt, Rinehart and Winston; a Division of Houghton Mifflin Harcourt Publishing Company. All rights reserved.

SECTION 1 What Is a Mineral? *continued*

How Do Geologists Classify Minerals?

Geologists classify minerals based on the elements or compounds in the minerals. Two main groups of minerals are silicate minerals and nonsilicate minerals.

SILICATE MINERALS

Silicon and oxygen are two of the most common elements in the Earth's crust. Minerals that contain compounds of silicon and oxygen are called **silicate minerals**. Silicate minerals make up more than 90% of the Earth's crust. Most silicate minerals also contain elements other than silicon and oxygen, such as aluminum, iron, or magnesium.

Common Silicate Minerals

Quartz is a mineral that is found in many rocks of the Earth's crust.

Mica breaks into sheets easily.

Feldspar is also common in the rocks of the Earth's crust. Feldspar can contain many elements other than silicon and oxygen, such as potassium or sodium.

TAKE A LOOK
6. Identify What two elements are found in all of the minerals in the figure? Explain your answer.

<u>More than 90%</u>
<u>of the earth's</u>
<u>crust.</u>

NONSILICATE MINERALS

Minerals that do not contain compounds of silicon and oxygen are called **nonsilicate minerals**. Some of these minerals are made of elements such as carbon, oxygen, fluorine, and sulfur.

Types of Nonsilicate Minerals

Native elements are minerals that are made of only one element. Copper, gold, silver, and diamonds are native elements. Copper

Oxides are minerals that contain compounds of oxygen and another element, such as iron or aluminum. Rubies and sapphires are forms of the mineral corundum, which is an oxide mineral. Corundum

Carbonates are minerals that contain compounds of carbon and oxygen. Calcite is a carbonate mineral. Calcite

Sulfates are minerals that contain compounds of oxygen and sulfur. Gypsum is a sulfate mineral. Gypsum

Halides are minerals that contain the elements fluorine, chlorine, iodine, or bromine. Fluorite and halite are halide minerals. Fluorite

Sulfides are minerals that contain compounds of sulfur and an element other than oxygen, such as lead, iron, or nickel. Galena and pyrite ("fool's gold") are sulfide minerals. Galena

TAKE A LOOK
7. Compare How are sulfate minerals different from sulfide minerals?

<u>Copper and</u>
<u>Calcite</u>
_____
_____

Copyright © by Holt, Rinehart and Winston; a Division of Houghton Mifflin Harcourt Publishing Company. All rights reserved.

Section 1 Review

GLE 0707.7.1 **TN**

SECTION VOCABULARY

compound a substance made up of atoms of two or more different elements joined by chemical bonds	**mineral** a naturally formed, inorganic solid that has a definite chemical structure
crystal a solid whose atoms, ions, or molecules are arranged in a regular, repeating pattern	**nonsilicate mineral** a mineral that does not contain compounds of silicon and oxygen
element a substance that cannot be separated or broken down into simpler substances by chemical means	**silicate mineral** a mineral that contains a combination of silicon, oxygen, and one or more metals

1. Identify What are four features of a mineral?

_____Silicate_____

_____

_____

2. Compare What is the difference between an atom and an element?

_____

_____

3. Infer What determines the shape of a crystal?

_____

_____

4. Apply Concepts Why is the ice in a glacier considered a mineral, but the water in a river is not considered a mineral?

_____

_____

5. Describe What are the features of the two major groups of minerals?

_____

_____

_____

6. List Give four types of nonsilicate minerals.

_____

_____

Copyright © by Holt, Rinehart and Winston; a Division of Houghton Mifflin Harcourt Publishing Company. All rights reserved.

CHAPTER 13 (Minerals of the Earth's Crust)

SECTION 2 **Identifying Minerals**

BEFORE YOU READ

After you read this section, you should be able to answer these questions:

• What seven properties can be used to identify a mineral?

• What are some special properties of minerals?

TN Tennessee Science Standards

GLE 0707.Inq.2
GLE 0707.Inq.5
GLE 0707.7.1

How Can You Identify Minerals?

If you close your eyes and taste different foods, you can usually figure out what the foods are. You can identify foods by noting their properties, such as texture and flavor. Minerals also have properties that you can use to identify them.

COLOR

The same mineral can have many different colors. For example, the mineral quartz can be clear, white, pink, or purple. Minerals can also change colors when they react with air or water. For example, pyrite ("fool's gold") has a golden color. If pyrite is exposed to air and water, it can turn brown or black. Because the color of a mineral can vary a lot, color is not the best way to identify a mineral. ☑

LUSTER

The way a surface reflects light is called **luster**. When you say that something looks shiny, you are describing its luster. A mineral can have a metallic, submetallic, or nonmetallic luster. The table below gives some examples of different kinds of luster.

Luster	Description	Examples
Metallic	bright and shiny, like metal	gold, copper wire
Submetallic	dull, but reflective	graphite (pencil "lead")
Nonmetallic		
Vitreous	glassy, brilliant	glass, quartz
Waxy	greasy, oily	wax, halite
Silky	looks like light is reflecting off long fibers	satin fabric, asbestos
Pearly	creamy	pearls, talc
Resinous	looks like plastic	plastic, sulfur
Earthy	rough, dull	concrete, clay

STUDY TIP

Reading Organizer As you read this section, create an outline of the section. Use the properties of minerals to form the headings of your outline.

READING CHECK

1. Explain How can the color of a mineral change?

_____

_____

_____

 **Say It**

Apply Ideas In a small group, think of a list of 10 to 15 everyday materials. Together, try to describe the luster of each material using the terms in the table.

Copyright © by Holt, Rinehart and Winston; a Division of Houghton Mifflin Harcourt Publishing Company. All rights reserved.

STREAK

The color of a mineral in powdered form is called its **streak**. You can find a mineral's streak by rubbing the mineral against a piece of unglazed porcelain. The piece of unglazed porcelain is called a *streak plate*. The mark left on the streak plate is the streak.

Streak is a more useful property than color for identifying minerals. This is because the color of a mineral's streak is always the same. For example, the color of the mineral hematite may vary, but its streak will always be red-brown. ☑

✓ **READING CHECK**

2. Explain Why is streak more useful than color in identifying a mineral?

_____

_____

CLEAVAGE AND FRACTURE

Different minerals break in different ways. The way that a mineral breaks depends on how its atoms are arranged. When some minerals break, the surfaces that form are smooth and flat. These minerals show the property of **cleavage**. Other minerals break unevenly, along curved or rough surfaces. These minerals show the property of **fracture**.

The mineral biotite, a type of mica, shows the property of cleavage. It breaks easily into thin, flat sheets.

The mineral halite also shows the property of cleavage. Its crystals break into cubes.

The mineral quartz shows the property of fracture. It breaks along a curved surface. This kind of fracture is called *conchoidal* fracture.

TN TENNESSEE STANDARDS CHECK

GLE 0707.7.1 Describe the physical properties of minerals.

Word Help: physical able to be seen or touched

3. Identify What kind of fracture does quartz show?

_____

DENSITY

Density is a measure of how much matter is in a given amount of space. Density is usually measured in grams per cubic centimeter (g/cm³). For example, the density of water is 1 g/cm³.

Geologists often use specific gravity to describe the density of a mineral. A mineral's *specific gravity* is the density of the mineral divided by the density of water. For example, gold has a density of 19 g/cm³. Its specific gravity is 19 g/cm³ ÷ 1 g/cm³ = 19.

Math Focus

4. Calculate How many times denser is gold than water?

_____

Copyright © by Holt, Rinehart and Winston; a Division of Houghton Mifflin Harcourt Publishing Company. All rights reserved.

HARDNESS

A mineral's resistance to being scratched is its **hardness**. Scientists use the *Mohs hardness scale* to describe the hardness of minerals. The harder a mineral is to scratch, the higher its rating on the Mohs scale. Talc, one of the softest minerals, has a rating of 1. Diamond, the hardest mineral, has a rating of 10. ☑

Scientists use reference minerals to find the hardness of unknown minerals. They try to scratch the surface of the unknown mineral with the edge of a reference mineral. If the reference mineral scratches the unknown mineral, the reference mineral is harder than the unknown mineral.

You probably don't have pieces of these reference minerals. However, you can find the hardness of a mineral using common objects. For example, your fingernail has a hardness of about 2 on the Mohs scale. A piece of window glass has a hardness of about 5.5.

Hardness	Mineral		Hardness	Mineral
1	Talc		6	Orthoclase
2	Gypsum		7	Quartz
3	Calcite		8	Topaz
4	Fluorite		9	Corundum
5	Apatite		10	Diamond

SPECIAL PROPERTIES

Some minerals have special properties. These properties can be useful in identifying the minerals.

Special Properties of Some Minerals

 Calcite and fluorite show the property of *fluorescence*. This means that they glow under ultraviolet light.

 Calcite produces a *chemical reaction* when a drop of weak acid is placed on it. It fizzes and produces gas bubbles.

 Some minerals, such as this calcite, show *optical properties*. Images look doubled when they are viewed through calcite.

 Magnetite shows the property of *magnetism*. It is a natural magnet.

 Halite has a salty *taste*. You should not taste a mineral unless your teacher tells you to.

 Minerals that contain radioactive elements may show the property of *radioactivity*. The radiation they give off can be detected by a Geiger counter.

☑ **READING CHECK**

5. Define What is hardness?

_____

_____

Critical Thinking

6. Apply Concepts A scientist tries to scratch a sample of orthoclase with a sample of apatite. Will he be able to scratch the orthoclase? Explain your answer.

_____

_____

_____

_____

TAKE A LOOK

7. Describe Under ultraviolet light, what happens to minerals that show the property of fluorescence?

_____

_____

Copyright © by Holt, Rinehart and Winston; a Division of Houghton Mifflin Harcourt Publishing Company. All rights reserved.

Section 2 Review

GLE 0707.Inq.2, GLE 0707.Inq.5, GLE 0707.7.1

SECTION VOCABULARY

cleavage the splitting of a mineral along smooth, flat surfaces	**hardness** a measure of the ability of a mineral to resist scratching
density the ratio of the mass of a substance to the volume of the substance	**luster** the way in which a mineral reflects light
fracture the manner in which a mineral breaks along either curved or irregular surfaces	**streak** the color of the powder of a mineral

1. Compare How are cleavage and fracture different?

_____

_____

_____

2. Explain Why is color not the best property to use to identify a mineral?

_____

3. Identify Give five properties that you can use to identify a mineral.

_____

_____

_____

_____

4. Apply Concepts A geologist has found an unknown mineral. She finds that a sample of calcite will not scratch the unknown mineral. She also finds that a sample of apatite will scratch the unknown mineral. About what is the unknown mineral's hardness? Explain your answer.

_____

_____

_____

_____

5. Calculate The density of a mineral is 2.6 g/cm³. What is its specific gravity?

_____

Copyright © by Holt, Rinehart and Winston; a Division of Houghton Mifflin Harcourt Publishing Company. All rights reserved.

CHAPTER 13 Minerals of the Earth's Crust

SECTION
3 # The Formation, Mining, and Use of Minerals

TN Tennessee Science Standards
GLE 0707.Inq.3
GLE 0707.Inq.5
GLE 0707.T/E.1
GLE 0707.7.6

BEFORE YOU READ

After you read this section, you should be able to answer these questions:

- How do minerals form?
- How are mineral resources used?

How Do Minerals Form?

Different minerals form in different environments. The table below shows five ways that minerals can form.

Process	Description	Minerals that form this way
Evaporation	When a body of salt water dries up, minerals are left behind. As the water evaporates, the minerals crystallize.	gypsum, halite
Metamorphism	High temperatures and pressures deep below the ground can cause the minerals in rock to change into different minerals.	garnet, graphite, magnetite, talc
Deposition	Surface water and ground water carry dissolved minerals into lakes or seas. The minerals can crystallize on the bottom of the lake or sea.	calcite, dolomite
Reaction	Water underground can be heated by hot rock. The hot water can dissolve some minerals and deposit other minerals in their place.	gold, copper, sulfur, pyrite, galena
Cooling	Melted rock can cool slowly under Earth's surface. As the melted rock cools, minerals form.	mica, feldspar, quartz

✏️ **STUDY TIP**

Describe As you read this section, make a chart showing the uses of different rock and mineral resources.

TAKE A LOOK

1. Identify Give three minerals that form by metamorphism and three minerals that form by reaction.

Metamorphism:

_____

_____

_____

Reaction:

_____

_____

_____

How Are Minerals Removed from the Earth?

People mine many kinds of minerals from the ground and make them into objects we need. Some minerals have more useful materials in them than others. An **ore** is a rock or mineral that contains enough useful materials for it to be mined at a profit.

There are two ways of removing ores from Earth: surface mining and subsurface mining. The type of mining used depends on how close the ore is to the surface.

Copyright © by Holt, Rinehart and Winston; a Division of Houghton Mifflin Harcourt Publishing Company. All rights reserved.

SURFACE MINING

People use surface mining to remove ores that are near Earth's surface. Three types of surface mines include open pits, surface coal mines, and quarries.

Open-pit mining is used to remove large, near-surface deposits of gold and copper. Explosives break up the rock layers above the ore. Then, trucks haul the ore from the mine to a processing plant. ☑

Quarries are open mines that are used to remove sand, gravel, and crushed rock. The layers of rock near the surface are removed and used to make buildings and roads.

Strip mines are often used to mine coal. The coal is removed in large pieces. These pieces are called *strips*. The strips of coal may be up to 50 m wide and 1 km long.

SUBSURFACE MINING

People use subsurface mining to remove ores that are deep underground. Iron, coal, and salt can be mined in subsurface mines. ☑

READING CHECK

2. Identify Give two minerals that are mined using open-pit mining.

_____

_____

READING CHECK

3. List Give three resources that can be mined using subsurface mining.

_____

_____

_____

TAKE A LOOK

4. Identify What are three kinds of tunnels used in subsurface mining?

_____

_____

_____

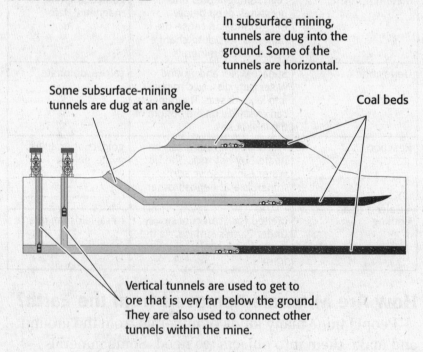

In subsurface mining, tunnels are dug into the ground. Some of the tunnels are horizontal.

Coal beds

Some subsurface-mining tunnels are dug at an angle.

Vertical tunnels are used to get to ore that is very far below the ground. They are also used to connect other tunnels within the mine.

Copyright © by Holt, Rinehart and Winston; a Division of Houghton Mifflin Harcourt Publishing Company. All rights reserved.

RESPONSIBLE MINING

Mining can help us get the resources we need, but it can also create problems. Mining may destroy or harm the places where plants and animals live. The wastes from mining can be poisonous. They can pollute water and air. ☑

One way to reduce these problems is to return the land to nearly its original state after mining is finished. This is called **reclamation**. Since the mid-1970s, laws have required the reclamation of land used for mining.

Another way to reduce the problems with mining is to reduce our need for minerals that are mined. For example, when you recycle materials made from minerals, you reduce the need for further mining. If you recycle the aluminum in your soda can, less aluminum has to be removed from the Earth. ☑

How Are Minerals Used?

We can use some minerals just as they are. However, most minerals must be processed before they can be used. The table below shows how some common minerals are used. The figure on the next page shows some of the processed minerals that are used in a bicycle.

Mineral	Uses
Bauxite (aluminum ore)	source of aluminum for cans, foil, appliances, and utensils
Copper	electrical wire, plumbing, coins
Diamond	jewelry, cutting tools, drill bits
Galena (lead ore)	source of lead for batteries and ammunition
Gold	jewelry, computers, spacecraft
Gypsum	plaster, cement, wallboard
Halite	table salt, road salt, water softener
Quartz	glass, source of silicon for computer chips
Silver	photography, electronic products, jewelry
Sphalerite (zinc ore)	jet aircraft, spacecraft, paint

☑ **READING CHECK**

5. Describe What are two problems with mining?

_____

_____

☑ **READING CHECK**

6. Explain How can recycling help reduce the problems with mining?

_____

_____

_____

_____

TAKE A LOOK

7. Identify Give two uses for the mineral silver and two uses for the mineral bauxite.

Silver:

_____

_____

Bauxite:

_____

_____

Copyright © by Holt, Rinehart and Winston; a Division of Houghton Mifflin Harcourt Publishing Company. All rights reserved.

SECTION 3 The Formation, Mining, and Use of Minerals *continued*

Minerals Used in the Parts of a Bicycle

Handlebars titanium from ilmenite

Frame aluminum from bauxite

Spokes iron from magnetite

Pedals beryllium from beryl

TAKE A LOOK
8. Identify Name four minerals that are used in the parts of a bicycle.

_____

_____

_____

_____

Critical Thinking

9. Infer Electricity can pass through metals easily. How does this make metals useful in computers and other electronic appliances?

_____

_____

_____

_____

_____

METALLIC MINERALS

Many minerals contain metals. Many of the features of metals make them useful in aircraft, automobiles, computer parts, and spacecraft. All metals have the features given below:

• Metals have shiny surfaces.

• Light cannot pass through metals.

• Heat and electricity can pass through metals easily.

• Metals can be rolled into sheets or stretched into wires.

Some metals react easily with air and water. For example, iron can react with oxygen in the air to produce rust. However, many of these metallic minerals can be processed into materials that do not react with air and water. For example, iron can be used to make stainless steel, which does not rust. Other metals do not react very easily. For example, gold is used in parts of aircraft because it does not react with many chemicals.

Many metals are strong. Their strength makes them useful in making ships, automobiles, airplanes, and buildings. For example, tall buildings are too heavy to be supported by a wooden frame. However, steel frames can support skyscrapers that are hundreds of meters tall.

Copyright © by Holt, Rinehart and Winston; a Division of Houghton Mifflin Harcourt Publishing Company. All rights reserved.

SECTION 3 The Formation, Mining, and Use of Minerals *continued*

NONMETALLIC MINERALS

Many minerals also contain nonmetals. Some important features of nonmetals are given below:

• Nonmetals can have shiny or dull surfaces.

• Light can pass through some kinds of nonmetals.

• Heat and electricity cannot pass through nonmetals easily.

Nonmetallic minerals are some of the most widely used minerals in industry. For example, the mineral calcite is used to make concrete. The mineral quartz is used to make glass. Quartz can also be processed to produce the element silicon, which is used in computer chips. ☑

GEMSTONES

Some nonmetallic minerals are considered valuable because of their beauty or rarity. These minerals are called *gemstones*. Important gemstones include diamond, ruby, sapphire, emerald, topaz, and tourmaline.

Color is the feature that determines the value of a gemstone. The more attractive the color, the more valuable the gemstone is. The colors of many gemstones are caused by impurities. An *impurity* is a small amount of an element not usually found in the mineral. For example, rubies and sapphires are both forms of the mineral corundum. Rubies look red because they have chromium impurities. Sapphires look blue because they have iron impurities. ☑

Most gemstones are very hard. This allows them to be cut and polished easily. For example, corundum (rubies and sapphires) and diamond are the two hardest minerals. They are also some of the most valuable gemstones.

✓ **READING CHECK**

10. Identify Give two nonmetallic minerals that are used in industry.

_____

_____

✓ **READING CHECK**

11. Explain What gives many gemstones their color?

_____

Diamonds are some of the most valuable gemstones. They are used in jewelry and in other items, such as this scepter.

Copyright © by Holt, Rinehart and Winston; a Division of Houghton Mifflin Harcourt Publishing Company. All rights reserved.

Section 3 Review

GLE 0707.Inq.3, GLE 0707.Inq.5, GLE 0707.T/E.1, GLE 0707.7.6 **TN**

SECTION VOCABULARY

ore a natural material whose concentration of economically valuable minerals is high enough for the material to be mined profitably	**reclamation** the process of returning land to its original condition after mining is completed

1. Define Write your own definition for ore.

_____

_____

_____

2. Describe Fill in the spaces in table to describe metals and nonmetals.

Type of material	Main features	Common objects made from it

metal	has shiny surfaces
	does not transmit light
	transmits heat and
	    electricity easily
	can be rolled into sheets
	    or stretched into wires	

nonmetal		

3. List What are three ways minerals can form?

_____

_____

_____

4. Identify Give three types of surface mines and an example of the kind of material that each is used to mine.

_____

_____

_____

Copyright © by Holt, Rinehart and Winston; a Division of Houghton Mifflin Harcourt Publishing Company. All rights reserved.

CHAPTER 14 | Rocks: Mineral Mixtures
SECTION
1 **The Rock Cycle**

TN **Tennessee Science Standards**
GLE 0707.Inq.3
GLE 0707.Inq.5
GLE 0707.7.2

BEFORE YOU READ

After you read this section, you should be able to answer these questions:

• What is a rock?

• How are rocks classified?

• What does the texture of a rock reveal about how it was formed?

Why Are Rocks Important?

You know that you can recycle paper, aluminum, and plastic. Did you know that the Earth also recycles? One thing the Earth recycles is rock. A **rock** is a naturally occurring solid mixture of one or more minerals. Some rocks also contain the remains of living things.

Rock is an important resource for human beings. Early humans used rocks as hammers and other tools. They shaped rocks like chert and obsidian into spear points, knives, and scrapers. Rock is also used in buildings, monuments, and roads. The figure below shows how rock has been used as a building material in ancient and modern civilizations.

The ancient Egyptians used a rock called **limestone** to build the pyramids at Giza (left-hand figure). The Texas state capitol building in Austin is constructed of a rock called **granite** (right-hand figure).

It may seem like rocks never change, but this is not true. In fact, rocks are changing all the time. Most of these changes are slow, which is why it seems like rocks do not change. The processes by which new rocks form from older rock material is called the **rock cycle**.

> **STUDY TIP**
> **Describe** As you read this section, make a chart describing the processes of weathering, erosion, and deposition.

> **TAKE A LOOK**
> **1. Identify** What are two kinds of rocks that people have used for constructing buildings?
>
> _____
>
> _____

Copyright © by Holt, Rinehart and Winston; a Division of Houghton Mifflin Harcourt Publishing Company. All rights reserved.

What Processes Shape the Earth's Surface?

Many different processes are part of the rock cycle. These processes shape the features of our planet. They form the mountains and valleys that we see around us. They also affect the types of rock found on the Earth's surface.

WEATHERING, EROSION, AND DEPOSITION

Weathering happens when water, wind, ice, and heat break down rock into smaller fragments. These fragments are called *sediment*. Sediment can move over the Earth's surface through erosion and deposition.

Erosion happens when water, wind, ice, or gravity move sediment over the Earth's surface. Over time, sediment that has been eroded stops moving and is deposited. When sediment stops moving, it is called **deposition**. Sediment can be deposited in bodies of water and other low-lying areas.

TN **TENNESSEE STANDARDS CHECK**

GLE 0707.7.2 Summarize the basic events that <u>occur</u> during the rock <u>cycle</u>.

Word Help: occur to happen

Word Help: cycle a repeating series of changes

2. Explain How does weathering shape the Earth's surface?

_____

_____

_____

_____

TAKE A LOOK
3. Identify Give two things that may have caused the weathering and erosion in Bryce Canyon.

_____

_____

The rocks in Bryce Canyon, Utah, have been shaped by weathering and erosion. Although these processes can be slow, they can cause large changes in the Earth's surface.

HEAT AND PRESSURE

Rock can also form when buried sediment is squeezed by the weight of the layers above it. In addition, temperature and pressure can change the minerals in the rocks. In some cases, the rock gets hot enough to melt. This melting produces liquid rock, or *magma*. When the magma cools, it hardens to form new rock. The new rock contains different minerals than the rock that melted.

Copyright © by Holt, Rinehart and Winston; a Division of Houghton Mifflin Harcourt Publishing Company. All rights reserved.

SECTION 1 The Rock Cycle *continued*

THE ROCK CYCLE

Geologists put rocks into three main groups based on how they form. These groups are igneous rock, sedimentary rock, and metamorphic rock. *Igneous rock* forms when melted rock cools and hardens. *Sedimentary rock* is made of pieces of other rock (sediment). *Metamorphic rock* forms when heat and pressure change the chemical composition of a rock.

Remember that the rock cycle is made of all of the processes that make new rock out of older rock material. Weathering, erosion, deposition, heat, and pressure are some of the processes that are part of the rock cycle. The figure below shows how the processes in the rock cycle can change rocks from one kind to another.

Critical Thinking
4. Compare How are igneous rocks different from metamorphic rocks?

_____

_____

_____

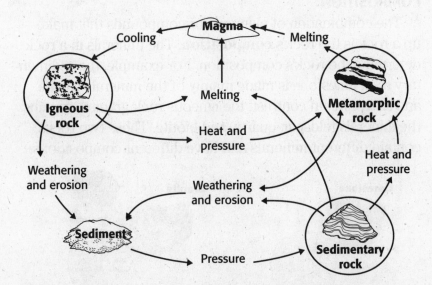

As you can see, rocks do not have to follow a single path through the rock cycle. An igneous rock may be weathered to form sediment, which then forms sedimentary rock. The igneous rock could also melt and cool to form a new igneous rock.

The path that a rock takes through the rock cycle depends on the forces that act on the rock. These forces change depending on where the rock is located. For example, high pressures and temperatures below the Earth's surface can cause metamorphic rock to form.

TAKE A LOOK

5. Use a Model Find two paths through the rock cycle that lead from sedimentary rock to igneous rock. Use a colored pen or marker to trace both paths on the figure.

Copyright © by Holt, Rinehart and Winston; a Division of Houghton Mifflin Harcourt Publishing Company. All rights reserved.

How Do Geologists Classify Rocks?

Remember that rocks can be divided into three groups based on how they form. Each main group of rock can be divided into smaller groups. These divisions are also based on the ways rocks form. For example, all igneous rock forms when magma cools and hardens. However, different kinds of igneous rock form when magma cools above the ground and when it cools underground.

Each kind of rock has specific features that make it different from other kinds of rock. Geologists can learn how a rock formed by studying its features. Two features that are especially helpful for classifying rocks are composition and texture ☑

COMPOSITION

The combination of elements or compounds that make up a rock is the rock's **composition**. The minerals in a rock determine the rock's composition. For example, the sedimentary rock limestone is made mainly of the minerals calcite and aragonite. In contrast, the igneous rock granite contains the minerals feldspar, quartz, and biotite. These two rocks contain different minerals and have different compositions.

Limestone, a sedimentary rock, contains the minerals calcite and aragonite.

Granite, an igneous rock, contains the minerals biotite, feldspar, and quartz.

Composition can help geologists classify rocks. This is because different minerals form under different conditions. For example, remember that the mineral garnet forms under high temperatures and pressures. Therefore, a rock with garnet in it probably formed under high temperature and pressure. Such a rock is probably a metamorphic rock.

READING CHECK

6. Explain How do geologists learn how a rock formed?

_____

_____

Math Focus

7. Calculate Rock A contains 10% quartz and 45% calcite. The rest of the rock is mica. What percentage of the rock is mica?

_____

Copyright © by Holt, Rinehart and Winston; a Division of Houghton Mifflin Harcourt Publishing Company. All rights reserved.

TEXTURE

The sizes, shapes, and positions of the grains that make up a rock are the rock's **texture**. The texture of a rock can be affected by different things. The texture of a sedimentary rock is mainly affected by the sediment that formed it. For example, a sedimentary rock that forms from small sediment pieces will have a fine-grained texture. The figures below show some examples of sedimentary rock textures.

Fine-grained

Siltstone

Siltstone is made of tiny pieces of sediment, such as silt and clay. Therefore, it has a fine-grained texture. It feels smooth when you touch it.

Medium-grained

Sandstone

Sandstone is made of pieces of sand. It has a medium-grained texture. It feels a bit rough, like sandpaper.

Coarse-grained

Conglomerate

Conglomerate is made of sediment pieces that are large, such as pebbles. Therefore, it has a coarse-grained texture. It feels bumpy.

TAKE A LOOK
8. Explain What determines the texture of a sedimentary rock?

_____

_____

The texture of an igneous rock depends on how fast the melted rock cools. As melted rock cools, mineral crystals form. When melted rock cools quickly, only very small mineral crystals can form. Therefore, igneous rocks that cool quickly tend to have a fine-grained texture. When melted rock cools slowly, large crystals can form, which make a coarse-grained igneous rock.

Fine-grained

Basalt

Basalt forms when melted rock cools quickly on the Earth's surface. It has a fine-grained texture because the mineral crystals in it are very small.

Coarse-grained

Granite

Granite forms when melted rock cools slowly underground. It has a coarse-grained texture because the mineral crystals in it are large.

TAKE A LOOK
9. Describe How does granite form?

_____

_____

Copyright © by Holt, Rinehart and Winston; a Division of Houghton Mifflin Harcourt Publishing Company. All rights reserved.

Section 1 Review

GLE 0707.Inq.3, GLE 0707.Inq.5, GLE 0707.7.2 **TN**

SECTION VOCABULARY

composition the chemical makeup of a rock; describes either the minerals or other materials in the rock	**rock** a naturally occurring solid mixture of one or more minerals or organic matter
deposition the process in which material is laid down	**rock cycle** the series of processes in which rock forms, changes from one type to another, is destroyed, and forms again by geologic processes
erosion the process by which wind, water, ice, or gravity transports soil and sediment from one location to another	**texture** the quality of a rock that is based on the sizes, shapes, and positions of the rock's grains

1. **Compare** What is the difference between weathering and erosion?

_____

_____

_____

2. **Identify** Complete the diagram to show how igneous rock can turn into sedimentary rock.

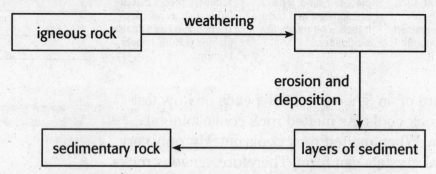

3. **List** What are two features that geologists use to classify rocks?

_____

_____

4. **Describe** What determines the texture of an igneous rock?

_____

_____

5. **Explain** How can a rock's composition help geologists to classify the rock?

_____

_____

Copyright © by Holt, Rinehart and Winston; a Division of Houghton Mifflin Harcourt Publishing Company. All rights reserved.

■ BEFORE YOU READ

After you read this section, you should be able to answer these questions:

• How do igneous rocks form?

• What factors affect the texture of igneous rock?

TN Tennessee Science
Standards
GLE 0707.Inq.3
GLE 0707.Inq.5
GLE 0707.7.2

How Does Igneous Rock Form?

Igneous rocks form when hot, liquid rock, or *magma*, cools and hardens. There are three main ways that magma can form.

• An *increase in temperature:* when temperature increases, the minerals in a rock can melt.

• A *decrease in pressure:* hot rock can remain solid if it is under high pressure deep within the Earth. When the hot rock rises to the surface, the pressure goes down, and the rock can melt.

• An *addition of fluids:* when fluids, such as water, mix with rock, the melting temperature of the rock decreases and the rock can melt. ☑

When magma cools enough, mineral crystals form. This is similar to how water freezes. When you put water into the freezer, the water cools. When its temperature gets low enough, crystals of ice form. In the same way, crystals of different minerals can form as magma cools.

Water is made of a single compound. Therefore, all water freezes at the same temperature (0°C). However, magma is made of many different compounds. These compounds can combine to form different minerals. Each mineral becomes solid at a different temperature. Therefore, as magma cools, different parts of it become solid at different temperatures. Magma can become solid, or freeze, between 700°C and 1,250°C. ☑

STUDY TIP

Compare After you read this section, make a table comparing the properties of intrusive igneous rock and extrusive igneous rock.

✓ READING CHECK

1. Identify Give three ways that magma can form.

Similar to how Freezzer, cool, crystals ice.

✓ READING CHECK

2. Explain Why do different parts of magma become solid at different times?

magma cools

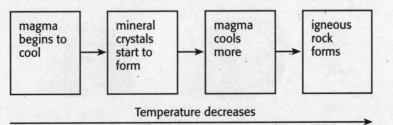

| magma begins to cool | → | mineral crystals start to form | → | magma cools more | → | igneous rock forms |

Temperature decreases ⟶

Copyright © by Holt, Rinehart and Winston; a Division of Houghton Mifflin Harcourt Publishing Company. All rights reserved.

How Do Geologists Classify Igneous Rocks?

Geologists group igneous rocks by how they form. Geologists use clues from the rocks' compositions and textures to guess how they formed.

COMPOSITION

Based on composition, there are two main groups of igneous rocks—felsic rocks and mafic rocks. *Felsic* igneous rocks are rich in elements such as sodium, potassium, and aluminum. These elements combine to form light-colored minerals. Therefore, most felsic igneous rocks are light-colored. Granite and rhyolite are examples of felsic rocks.

Mafic igneous rocks are rich in elements such as iron, magnesium, and calcium. These elements combine to form dark-colored minerals. Therefore, most mafic igneous rocks are dark-colored. Gabbro and basalt are examples of mafic rocks.

TEXTURE

Remember that the texture of a rock is determined by the sizes of the grains in the rock. The texture of an igneous rock depends on how fast the magma cooled.

When magma cools quickly, mineral crystals do not have time to grow very large. Therefore, the rock that forms has a fine-grained texture. When magma cools slowly, large mineral crystals can form. Therefore, the rock that forms has a coarse-grained texture.

Critical Thinking

3. Compare Give two differences between felsic and mafic igneous rocks.

igneous rocks are
rich in elements
_____
_____

TAKE A LOOK
4. Identify Give an example of a felsic, fine-grained igneous rock.

Big rock

5. Identify Give an example of a mafic, coarse-grained igneous rock.

Gabber rock

	Coarse-grained	Fine-grained
Felsic	Granite	Rhyolite
Mafic	Gabbro	Basalt

Copyright © by Holt, Rinehart and Winston; a Division of Houghton Mifflin Harcourt Publishing Company. All rights reserved.

Rock's Texture?

Many people know that volcanoes form from melted rock. Therefore, they may think that igneous rocks only form at volcanoes on the Earth's surface. However, some igneous rocks form deep within the Earth's crust.

INTRUSIVE IGNEOUS ROCKS

Intrusive igneous rock forms when magma cools below the Earth's surface. Because the magma cools slowly, intrusive igneous rock usually has a coarse-grained texture. The minerals can grow into large, visible crystals. Bodies of intrusive igneous rock are grouped by their sizes and shapes. ☑

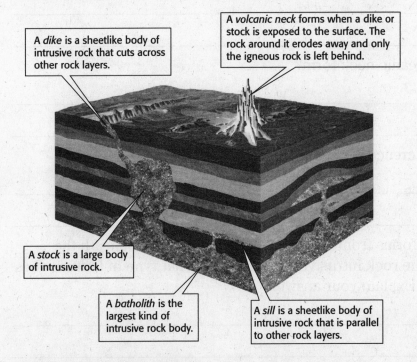

A *dike* is a sheetlike body of intrusive rock that cuts across other rock layers.

A *volcanic neck* forms when a dike or stock is exposed to the surface. The rock around it erodes away and only the igneous rock is left behind.

A *stock* is a large body of intrusive rock.

A *batholith* is the largest kind of intrusive rock body.

A *sill* is a sheetlike body of intrusive rock that is parallel to other rock layers.

EXTRUSIVE IGNEOUS ROCKS

Magma that reaches the Earth's surface is called *lava*. **Extrusive igneous rock** forms when lava cools. Extrusive igneous rock is common around volcanoes. Because extrusive rock cools quickly, it contains very small crystals or no crystals. ☑

When lava erupts from a volcano, it forms a *lava flow*. Lava flows can cover the land and bury objects on the Earth's surface.

Sometimes, lava erupts and flows along long cracks in Earth's crust called *fissures*. Many fissures are found on the ocean floor. Lava can also flow out of fissures onto land and form a *lava plateau*.

✓ READING CHECK

6. Define Write your own definition for intrusive igneous rock.

form when magma cools below the Earth's surface.

TAKE A LOOK

7. Identify Give four kinds of intrusive rock bodies.

rock layers
rock body
large body
rock that cuts

✓ READING CHECK

8. Explain Why do extrusive rocks have very small crystals or no crystals?

very small crystals

Copyright © by Holt, Rinehart and Winston; a Division of Houghton Mifflin Harcourt Publishing Company. All rights reserved.

Section 2 Review

GLE 0707.Inq.3, GLE 0707.Inq.5, GLE 0707.7.2 ▰ TN▰

SECTION VOCABULARY

extrusive igneous rock rock that forms as a result of volcanic activity at or near the Earth's surface	**intrusive igneous rock** rock formed from the cooling and solidification of magma beneath the Earth's surface

1. Compare How are intrusive and extrusive igneous rocks different?

Becuse the rock are very diffent from the rocks in intrusive

_____

_____

2. Identify Give two examples of fine-grained igneous rocks.

like to weich it heary

_____

_____

3. Describe How does a volcanic neck form?

_____

_____

4. Compare What is the difference between a dike and a sill?

_____

_____

5. Predict An igneous rock forms from slowly cooled magma deep beneath the surface of the Earth. Is the rock intrusive or extrusive? What type of texture does the rock probably have? Explain your answer.

_____

_____

_____

6. Apply Concepts Complete the table below. (Hint: What is the texture of each rock?)

Rock Name	Composition	Intrusive or Extrusive?
basalt	mafic	
gabbro	mafic	
granite	felsic	In
rhyolite	felsic	Ex

Copyright © by Holt, Rinehart and Winston; a Division of Houghton Mifflin Harcourt Publishing Company. All rights reserved.

CHAPTER 14 | Rocks: Mineral Mixtures

SECTION 3 **Sedimentary Rock**

TN **Tennessee Science Standards**
GLE 0707.Inq.1
GLE 0707.Inq.2
GLE 0707.Inq.3
GLE 0707.Inq.5
GLE 0707.7.2

BEFORE YOU READ

After you read this section, you should be able to answer these questions:

• How do sedimentary rocks form?

• How do geologists classify sedimentary rocks?

• What are some sedimentary structures?

How Does Sedimentary Rock Form?

Remember that wind, water, ice, and gravity can cause rock to break down into fragments. These fragments are called *sediment*. During erosion, sediment is moved across the Earth's surface. Then the sediment is deposited in layers on the Earth's surface. As new layers are deposited, they cover older layers. The weight of the new layers *compacts*, or squeezes, the sediment in the older layers.

Water within the sediment layers can contain dissolved minerals, such as calcite and quartz. As the sediment is compacted, these minerals can crystallize between the sediment pieces. The minerals act as a natural glue and hold the sediment pieces together. As the loose sediment grains become bound together, a kind of sedimentary rock forms.

Unlike igneous and metamorphic rocks, sedimentary rock does not form at high temperatures and pressures. Sedimentary rock forms at or near the Earth's surface. ☑

Sediment is deposited in layers. Therefore, most sedimentary rocks contain layers called **strata** (singular, *stratum*).

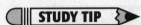
STUDY TIP
Reading Organizer As you read this section, create an outline of this section using the headings from this section.

Say It
Infer and Discuss In what kinds of areas are you likely to find sediment? Write down four places that sediment can be found. Think about the size of the sediment pieces that may be found at each place. Then, discuss your ideas with a small group.

READING CHECK
1. Describe Where does sedimentary rock form?

_____

_____

These "monuments" in Monument Valley, Arizona, formed as sedimentary rock eroded over millions of years.

Copyright © by Holt, Rinehart and Winston; a Division of Houghton Mifflin Harcourt Publishing Company. All rights reserved.

How Do Geologists Classify Sedimentary Rock?

Like other kinds of rock, sedimentary rock is classified by how it forms. Some sedimentary rock forms when rock or mineral fragments are stuck together. Some forms when minerals crystallize out of water. Other sedimentary rock forms from the remains of plants and animals.

CLASTIC SEDIMENTARY ROCK

Most sedimentary rock is clastic sedimentary rock. *Clastic sedimentary rock* forms when fragments of other rocks are cemented together. In most cases, the cement is a mineral such as calcite or quartz. The sediment pieces in different rocks can be of different sizes. Geologists group clastic sedimentary rocks by the sizes of the sediment pieces in them. ☑

READING CHECK

2. Identify Give two minerals that can act as cement in sedimentary rocks.

_____

_____

| Conglomerate | Sandstone | Siltstone | Shale |

Coarse-grained ◄———————————————————► Fine-grained

Coarse-grained sedimentary rocks, such as conglomerate, contain large sediment pieces. Fine-grained rocks, such as shale, are made of tiny sediment pieces.

TAKE A LOOK

3. Describe What is the texture of conglomerate?

_____

_____

CHEMICAL SEDIMENTARY ROCK

Chemical sedimentary rock forms when minerals crystallize out of water. Water moves over rocks on the Earth's surface. As the water moves, it dissolves some of the minerals in the rocks. When the water evaporates, the dissolved minerals can crystallize to form chemical sedimentary rocks. ☑

Many chemical sedimentary rocks contain only one or two kinds of mineral. For example, evaporite is a chemical sedimentary rock. Evaporite is made mainly of the minerals halite and gypsum. These minerals crystallize when water evaporates.

READING CHECK

4. Explain How do chemical sedimentary rocks form?

_____

_____

Copyright © by Holt, Rinehart and Winston; a Division of Houghton Mifflin Harcourt Publishing Company. All rights reserved.

ORGANIC SEDIMENTARY ROCK

Organic sedimentary rock forms from the remains of plants and animals. Coal is one type of organic sedimentary rock. Coal forms from plant material that has been buried deep underground. Over millions of years, the buried plant material turns into coal.

Some organic sedimentary rock forms from the remains of sea creatures. For example, some limestone is made from the skeletons of creatures called *coral*. Coral are tiny creatures that make hard skeletons out of calcium carbonate. These skeletons and the shells of other sea creatures can be glued together to form *fossiliferous limestone*.

The shells of sea creatures can be cemented together to form fossiliferous limestone.

TAKE A LOOK
5. Define What is fossiliferous limestone?

_____

_____

What Are Some Features of Sedimentary Rock?

The features of sedimentary rocks can give you clues about how the rocks formed. For example, many clastic sedimentary rocks show **stratification**. This means that they contain strata. Clastic sedimentary rocks show stratification because sediment is deposited in layers.

Some sedimentary rock features show the motions of wind and water. For example, some sedimentary rocks show ripple marks or mud cracks. *Ripple marks* are parallel lines that show how wind or water has moved sediment. *Mud cracks* form when fine-grained sediment dries out and cracks.

☑ **READING CHECK**
6. Explain Why do many clastic sedimentary rocks show stratification?

_____

_____

These ripple marks show how sediment was moved by flowing water.

Copyright © by Holt, Rinehart and Winston; a Division of Houghton Mifflin Harcourt Publishing Company. All rights reserved.

Section 3 Review

GLE 0707.Inq.1, GLE 0707.Inq.2, GLE 0707.Inq.3,
GLE 0707.Inq.5, GLE 0707.7.2

SECTION VOCABULARY

strata layers of rock (singular, *stratum*)	**stratification** the process in which sedimentary rocks are arranged in layers

1. **Define** Write your own definition for stratification.

_____

_____

2. **List** Give three examples of clastic sedimentary rocks.

_____

_____

_____

3. **Compare** How are clastic and organic sedimentary rocks different?

_____

_____

_____

4. **Describe** How does evaporite form?

_____

_____

5. **Describe** How does fossiliferous limestone form?

_____

_____

_____

6. **Infer** Imagine that a geologist finds a sedimentary rock with ripple marks in it. What can the geologist guess about the environment in which the sediment was deposited? Explain your answer.

_____

_____

_____

_____

Copyright © by Holt, Rinehart and Winston; a Division of Houghton Mifflin Harcourt Publishing Company. All rights reserved.

CHAPTER 14 Rocks: Mineral Mixtures

SECTION 4 # Metamorphic Rock

BEFORE YOU READ

After you read this section, you should be able to answer these questions:

• How do metamorphic rocks form?

• How do geologists classify metamorphic rocks?

TN Tennessee Science Standards

GLE 0707.Inq.3
GLE 0707.Inq.5
GLE 0707.7.2

How Does Metamorphic Rock Form?

Metamorphic rock forms when the chemical composition of a rock changes because of heat and pressure. This change is called *metamorphism*. Metamorphism can happen to any kind of rock.

Most metamorphism happens at temperatures between 150°C and 1,000°C. Some metamorphism happens at even higher temperatures. Many people think that all rocks must melt at such high temperatures. However, these rocks are also under very high pressure, so they do not melt.

High pressure can keep a hot rock from melting. Even very hot rocks may not melt if the pressure is high. Instead of melting, the minerals in the rock react with each other to form new minerals. In this way, the composition of the rock can change, even though the rock remains solid. ☑

High pressure can also affect the minerals in a rock. It can cause minerals to react quickly. It can also cause minerals to move slowly through the rock. In this way, different minerals can separate into stripes in the rock. The figure below shows an example of these stripes.

STUDY TIP

Ask Questions Read this section quietly to yourself. As you read, write down any questions you have. When you finish reading, try to figure out the answer to your questions in a small group.

✓ READING CHECK

1. Describe How does the composition of a rock change during metamorphism?

_____

_____

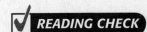

The bands in this metamorphic rock formed as molecules of different minerals moved together.

TAKE A LOOK

2. Identify How did the bands in the rock in the figure form?

_____

_____

_____

Copyright © by Holt, Rinehart and Winston; a Division of Houghton Mifflin Harcourt Publishing Company. All rights reserved.

CONTACT METAMORPHISM

There are two main ways that rock can go through metamorphism—contact metamorphism and regional metamorphism. *Contact metamorphism* happens when rock is heated by nearby magma. As the magma moves through the crust, the rocks in the crust heat up. The minerals in those rocks can react to produce new minerals.

TAKE A LOOK
3. Define What is contact metamorphism?

_____

_____

_____

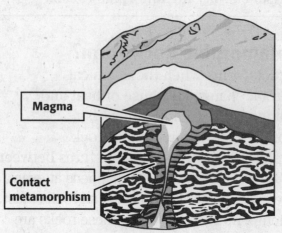

Magma

Contact metamorphism

Contact metamorphism happens when magma heats nearby rock.

Rock that is very near the magma changes the most during contact metamorphism. The farther the rock is from the magma, the smaller the changes. This is because the temperature decreases with distance from the magma. Contact metamorphism usually only affects rock in a small area.

REGIONAL METAMORPHISM

During *regional metamorphism*, high pressures and temperatures cause the rock in a large area to change. Regional metamorphism can happen where rock is buried deep below the surface or where pieces of the Earth's crust collide.

TAKE A LOOK
4. Describe Give two places where regional metamorphism can happen.

_____

_____

_____

_____

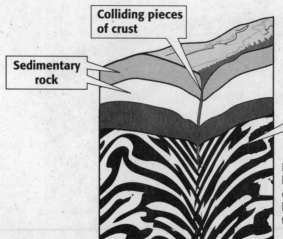

Colliding pieces of crust

Sedimentary rock

Regional metamorphism

Regional metamorphism happens when rock changes because of high pressures and temperatures.

Copyright © by Holt, Rinehart and Winston; a Division of Houghton Mifflin Harcourt Publishing Company. All rights reserved.

SECTION 4 Metamorphic Rock *continued*

METAMORPHIC STRUCTURES

Both contact and regional metamorphism can cause deformation. *Deformation* is a change in the shape of a rock. When forces act on a rock, they may cause the rock to be squeezed or stretched.

Folds are features of a rock that show that the rock has been deformed. Some folds are so small that they can only be seen with a microscope. Other folds, like the ones below, are visible to the naked eye.

These folds formed during metamorphism. The rocks in this picture are found in Labrador, Canada.

TAKE A LOOK
5. Infer Were these folds probably caused by squeezing the rock or by stretching it?

_____

_____

What Are Metamorphic Rocks Made Of?

Remember that different minerals form under different conditions. Minerals that form near the Earth's surface, such as calcite, may not be stable under higher temperatures and pressures. During metamorphism, these minerals are likely to react and produce new minerals. The new minerals are stable under high temperatures and pressures. The figure below shows how new minerals can form from unstable minerals.

Critical Thinking

6. Predict The mineral gypsum forms at low temperatures and pressures. The mineral sillimanite forms at high temperatures and pressures. Which mineral would most likely be found in a metamorphic rock? Explain your answer.

_____

_____

_____

_____

_____

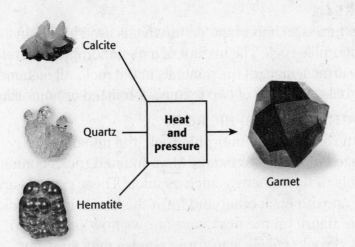

Calcite, quartz, and hematite are not stable under high temperatures and pressures. They react to form garnet in metamorphic rocks.

Copyright © by Holt, Rinehart and Winston; a Division of Houghton Mifflin Harcourt Publishing Company. All rights reserved.

INDEX MINERALS

Some minerals, such as quartz, can form at many different temperatures and pressures. Other minerals, such as garnet, form only at certain temperatures and pressures. Therefore, rocks that contain minerals like garnet probably also formed at those temperatures and pressures. Geologists can use such minerals as index minerals.

Index minerals can indicate the temperature and pressure or depth at which a rock formed. These minerals help geologists learn the temperature and pressure at which a rock formed. Chlorite, muscovite, and garnet are index minerals for metamorphic rocks.

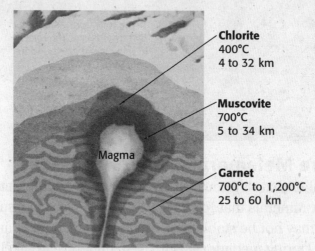

Chlorite
400°C
4 to 32 km

Muscovite
700°C
5 to 34 km

Magma

Garnet
700°C to 1,200°C
25 to 60 km

Geologists can use some minerals as index minerals. These minerals help geologists learn the temperature and pressure at which a rock formed. For example, a rock containing garnet most likely formed at a higher temperature and pressure than a rock containing chlorite.

How Do Geologists Classify Metamorphic Rocks?

Texture is an important feature that is used in classifying metamorphic rock. The texture of a metamorphic rock refers to the arrangement of the minerals in the rock. All metamorphic rocks have one of two textures—foliated or nonfoliated.

FOLIATED METAMORPHIC ROCK

In a **foliated** metamorphic rock, the minerals are arranged in stripes or bands. Most foliated rocks contain crystals of flat minerals, such as mica. These crystals are lined up with each other and form the bands in the rock. ☑

The figure on the next page shows how one kind of foliated rock, gneiss, can form. Gneiss may start out as the sedimentary rock shale. Heat and pressure can change shale to slate, phyllite, schist, or gneiss.

Critical Thinking

7. Infer Why can't geologists use minerals like quartz to determine the temperature and pressure that a rock formed at?

_____

_____

_____

TAKE A LOOK

8. Identify Which index mineral in the figure forms at the lowest temperature?

_____

_____

✓ **READING CHECK**

9. Define What is a foliated metamorphic rock?

_____

_____

Copyright © by Holt, Rinehart and Winston; a Division of Houghton Mifflin Harcourt Publishing Company. All rights reserved.

Sedimentary shale

The metamorphic rock slate can form when shale is placed under heat and pressure.

Slate

Phyllite can form when slate is put under heat and pressure.

Phyllite

Gneiss can form when schist is put under heat and pressure. The minerals in gneiss line up in bands, so gneiss is a foliated rock.

When phyllite is put under heat and pressure, schist can form.

Schist

Gneiss

TAKE A LOOK
10. Infer Which rock in the figure has been put under the most heat and pressure?

_____
_____

UNFOLIATED METAMORPHIC ROCK

In a **nonfoliated** metamorphic rock, the mineral crystals are not arranged in bands or stripes. Most nonfoliated rocks are made of only a few minerals. Metamorphism can cause the mineral crystals in a rock to get bigger. ☑

Quartzite is an example of a nonfoliated metamorphic rock. Quartzite can form from the sedimentary rock quartz sandstone. Quartz sandstone is made of grains of quartz sand that have been cemented together. The quartz crystals in these grains can grow larger during metamorphism. The quartz crystals in quartzite can be much larger than those in quartz sandstone.

✓ **READING CHECK**
11. Describe What can happen to the sizes of mineral crystals during metamorphism?

_____
_____

Type of Metamorphic Rock	Description	Example
Foliated		gneiss
Nonfoliated		quartzite

TAKE A LOOK
12. Define Fill in the blank spaces in the table.

Copyright © by Holt, Rinehart and Winston; a Division of Houghton Mifflin Harcourt Publishing Company. All rights reserved.

Section 4 Review

GLE 0707.Inq.3, GLE 0707.Inq.5, GLE 0707.7.2 ◢TN◣

SECTION VOCABULARY

foliated the texture of metamorphic rock in which the mineral grains are arranged in planes or bands	**nonfoliated** the texture of metamorphic rock in which the mineral grains are not arranged in planes or bands

1. Compare How are foliated metamorphic rocks different from nonfoliated metamorphic rocks?

_____

_____

_____

2. Define What is regional metamorphism?

_____

_____

3. Describe What is an index mineral? Give two examples of index minerals for metamorphic rocks.

_____

_____

_____

4. Explain How do index minerals help geologists?

_____

_____

5. Describe How does quartzite form?

_____

_____

_____

6. Apply Concepts A geologist finds two metamorphic rocks. One contains chlorite. The other contains garnet. Which rock probably formed at the greatest depth? Explain your answer.

_____

_____

_____

Copyright © by Holt, Rinehart and Winston; a Division of Houghton Mifflin Harcourt Publishing Company. All rights reserved.

CHAPTER 15 Plate Tectonics

SECTION 1 **Inside the Earth**

BEFORE YOU READ

After you read this section, you should be able to answer these questions:

• What are the layers inside Earth?

• How do scientists study Earth's interior?

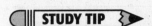
Tennessee Science Standards

GLE 0707.Inq.2
GLE 0707.Inq.5
GLE 0707.T/E.1
GLE 0707.7.3

What Is Earth Made Of?

Scientists divide the Earth into three layers based on composition: the crust, the mantle, and the core. These divisions are based on the compounds that make up each layer. A *compound* is a substance composed of two or more elements. The densest elements make up the core. Less-dense compounds make up the crust and mantle.

STUDY TIP

Summarize As you read, make a chart showing the features of Earth's layers. Include both the compositional layers and the physical layers.

THE CRUST

The thinnest, outermost layer of the Earth is the **crust**. There are two main kinds of crust: continental crust and oceanic crust. *Continental crust* forms the continents. It is thicker and less dense than oceanic crust. Continental crust can be up to 100 km thick. *Oceanic crust* is found beneath the oceans. It contains more iron than continental crust. Most oceanic crust is 5 km to 7 km thick. ☑

READING CHECK

1. Compare How is oceanic crust different from continental crust?

_____

_____

_____

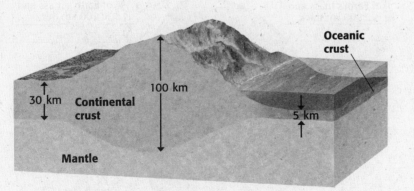

Oceanic crust is thinner and denser than continental crust.

Math Focus

2. Identify What fraction of the thickness of the thickest continental crust is the thickness of the oceanic crust? Give your answer as a reduced fraction.

_____

Copyright © by Holt, Rinehart and Winston; a Division of Houghton Mifflin Harcourt Publishing Company. All rights reserved.

THE MANTLE

The layer of the Earth between the crust and the core is the **mantle**. The mantle is much thicker than the crust. It contains most of the Earth's mass. The mantle contains more magnesium and less aluminum than the crust. This makes the mantle denser than the crust. ☑

No one has ever visited the mantle. The crust is too thick to drill through to reach the mantle. Therefore, scientists must use observations of Earth's surface to draw conclusions about the mantle. In some places, mantle rock pushes to the surface. This allows scientists to study the rock directly.

Another place scientists look for clues about the mantle is the ocean floor. Melted rock from the mantle flows out from active volcanoes on the ocean floor. These underwater volcanoes have given scientists many clues about the composition of the mantle. ☑

THE CORE

The layer beneath the mantle that extends to the center of the Earth is the **core**. Scientists think the core is made mostly of iron and smaller amounts of nickel. Scientists do not think that the core contains large amounts of oxygen, silicon, aluminum, or magnesium.

☑ **READING CHECK**

3. Explain Why is the mantle denser than the crust?

_____

_____

_____

☑ **READING CHECK**

4. Identify How can scientists learn about the mantle if they cannot study it directly?

_____

_____

TAKE A LOOK

5. List What are the three compositional layers of the Earth?

_____

_____

_____

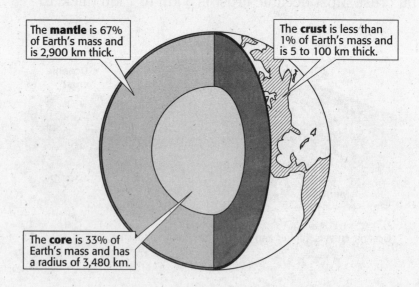

The **mantle** is 67% of Earth's mass and is 2,900 km thick.

The **crust** is less than 1% of Earth's mass and is 5 to 100 km thick.

The **core** is 33% of Earth's mass and has a radius of 3,480 km.

Copyright © by Holt, Rinehart and Winston; a Division of Houghton Mifflin Harcourt Publishing Company. All rights reserved.

SECTION 1 Inside the Earth *continued*

EARTH'S PHYSICAL STRUCTURE

Scientists also divide Earth into five layers based on physical properties. The outer layer is the **lithosphere**. It is a cool, stiff layer that includes all of the crust and a small part of the upper mantle. The lithosphere is divided into pieces. These pieces move slowly over Earth's surface. ☑

The **asthenosphere** is the layer beneath the lithosphere. It is a layer of hot, solid rock that flows very slowly. Beneath the asthenosphere is the **mesosphere**, which is the lower part of the mantle. The mesosphere flows more slowly than the asthensosphere.

There are two physical layers in Earth's core. The outer layer is the *outer core*. It is made of liquid iron and nickel. At the center of Earth is the *inner core*, which is a ball of solid iron and nickel. The inner core is solid because it is under very high pressure.

✓ **READING CHECK**
6. Define What is the lithosphere?

_____

_____

_____

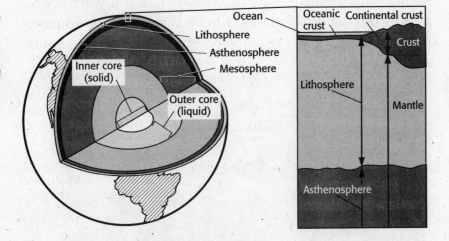

Critical Thinking
7. Infer What do you think is the reason that scientists divide the Earth into two different sets of layers?

_____

_____

_____

_____

What Are Tectonic Plates?

Pieces of the lithosphere that move around on top of the asthenosphere are called **tectonic plates**. Tectonic plates can contain different kinds of lithosphere. Some plates contain mostly oceanic lithosphere. Others contain mostly continental lithosphere. Some contain both continental and oceanic lithosphere. The figure on the top of the next page shows Earth's tectonic plates.

TAKE A LOOK
8. Describe What are the five layers of Earth, based on physical properties?

_____

_____

_____

_____

Copyright © by Holt, Rinehart and Winston; a Division of Houghton Mifflin Harcourt Publishing Company. All rights reserved.

SECTION 1 Inside the Earth *continued*

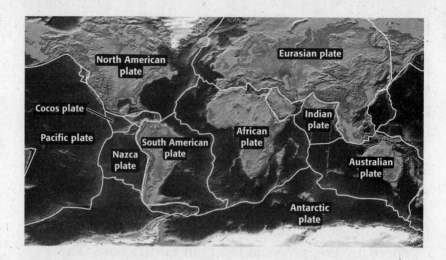

TAKE A LOOK

9. Identify Give the name of one plate that contains mostly oceanic lithosphere and of one plate that contains mostly continental lithosphere.

Oceanic: _____

Continental: _____

STRUCTURE OF A TECTONIC PLATE

The tectonic plates that make up the lithosphere are like pieces of a giant jigsaw puzzle. The figure below shows what a single plate might look like it if were separated from the other plates. Notice that the plate contains both continental and oceanic crust. It also contains some mantle material.

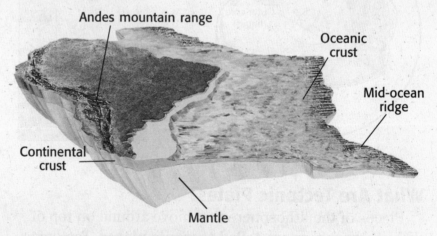

TAKE A LOOK

10. Compare Which type of crust is thicker, oceanic crust or continental crust?

_____

This figure shows what the South American plate might look like if it were lifted off the asthenosphere. Notice that the plate is thickest where it contains continental crust and thinnest where it contains oceanic crust.

Copyright © by Holt, Rinehart and Winston; a Division of Houghton Mifflin Harcourt Publishing Company. All rights reserved.

How Do Scientists Study Earth's Interior?

How do scientists know things about the deepest parts of the Earth? No one has ever been to these places. Scientists have never even drilled through the crust, which is only a thin layer on the surface of the Earth. So how do we know so much about the mantle and the core?

Much of what scientists know about Earth's layers comes from studying earthquakes. Earthquakes create vibrations called *seismic waves*. Seismic waves travel at different speeds through the different layers of Earth. Their speed depends on the density and composition of the material that they pass through. Therefore, scientists can learn about the layers inside the Earth by studying seismic waves. ☑

Scientists detect seismic waves using instruments called *seismometers*. Seismometers measure the times at which seismic waves arrive at different distances from an earthquake. Seismologists can use these distances and travel times to calculate the density and thickness of each physical layer of the Earth. The figure below shows how seismic waves travel through the Earth.

READING CHECK

11. Define What are seismic waves?

_____

_____

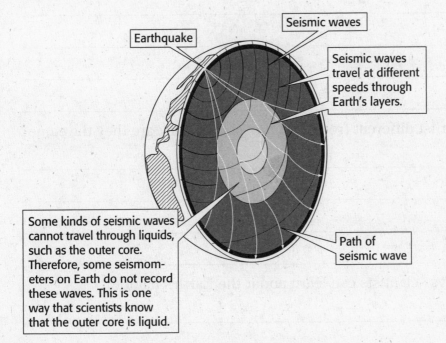

Earthquake

Seismic waves

Seismic waves travel at different speeds through Earth's layers.

Some kinds of seismic waves cannot travel through liquids, such as the outer core. Therefore, some seismometers on Earth do not record these waves. This is one way that scientists know that the outer core is liquid.

Path of seismic wave

TAKE A LOOK

12. Explain What is one way that scientists know the outer core is liquid?

_____

_____

_____

Copyright © by Holt, Rinehart and Winston; a Division of Houghton Mifflin Harcourt Publishing Company. All rights reserved.

Section 1 Review

GLE 0707.Inq.2, GLE 0707.Inq.5, GLE 0707.T/E.1, GLE 0707.7.3 **TN**

SECTION VOCABULARY

asthenosphere the soft layer of the mantle on which the tectonic plates move	**mantle** the layer of rock between the Earth's crust and core
core the central part of the Earth below the mantle	**mesosphere** the strong, lower part of the mantle between the asthenosphere and the outer core
crust the thin and solid outermost layer of the Earth above the mantle	**tectonic plates** a block of lithosphere that consists of the crust and the rigid, outermost part of the mantle
lithosphere the solid, outer layer of Earth that consists of the crust and the rigid upper part of the mantle	

1. Describe Complete the table below.

	Crust	**Mantle**	**Core**
Thickness or radius			3,430 km
Location	outer layer of the Earth		
Percent of Earth's mass			

2. Compare How is the inner core similar to the outer core? How are they different?

_____

_____

_____

_____

3. Compare How is the crust different from the lithosphere? How are they the same?

_____

_____

_____

_____

4. Identify Give three ways scientists can learn about the Earth's mantle.

_____

_____

_____

Copyright © by Holt, Rinehart and Winston; a Division of Houghton Mifflin Harcourt Publishing Company. All rights reserved.

Name _____ Class _____ Date _____

BEFORE YOU READ

After you read this section, you should be able to answer these questions:

- What is continental drift?
- How are magnetic reversals related to sea-floor spreading?

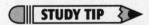 **Tennessee Science Standards**
GLE 0707.Inq.3
GLE 0707.Inq.5
GLE 0707.7.4

What Is Continental Drift?

Look at the map below. Can you see that South America and Africa seem to fit together, like the pieces of a jigsaw puzzle? In the early 1900s, a German scientist named Alfred Wegener made this same observation. Based on his observations, Wegener proposed the hypothesis of **continental drift**. According to this hypothesis, the continents once formed a single landmass. Then, they broke up and drifted to their current locations.

Continental drift can explain why the continents seem to fit together. For example, South America and Africa were once part of a single continent. They have since broken apart and moved to their current locations. ☑

Evidence for continental drift can also be found in fossils and rocks. For example, similar fossils have been found along the matching coastlines of South America and Africa. The organisms that formed these fossils could not have traveled across the Atlantic Ocean. Therefore, the two continents must once have been joined together.

STUDY TIP
Paired Summarizing Read this section silently. In pairs, take turns summarizing the material. Stop to discuss ideas that seem confusing.

READING CHECK
1. Explain Why do South America and Africa seem to fit together?

_____
_____
_____

Critical Thinking
2. Infer Which continent was once joined with Greenland? How do you know?

_____
_____
_____

TAKE A LOOK
3. Explain How do fossils indicate that the continents have moved with time?

_____
_____
_____

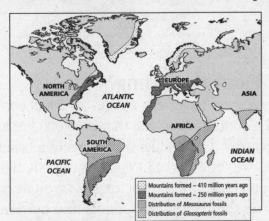

Similar fossils and rocks are found on widely separated continents. For example, *Glossopteris* and *Mesosaurus* fossils are found in Africa and in South America. These fossils and rocks indicate that, at one time, all of the continents were joined together.

Copyright © by Holt, Rinehart and Winston; a Division of Houghton Mifflin Harcourt Publishing Company. All rights reserved.

BREAKUP OF PANGAEA

About 245 million years ago, all of the continents were joined into a single *supercontinent*. This supercontinent was called *Pangaea*. The word *Pangaea* means "all Earth" in Greek. About 200 million years ago, Pangaea began breaking apart. It first separated into two large landmasses called Laurasia and Gondwana. The continents continued to break apart and slowly move to where they are today. ☑

As the continents moved, some of them collided. These collisions produced many of the landforms that we see today, such as mountain ranges and volcanoes.

READING CHECK

4. Identify When did Pangaea start to break apart?

_____

_____

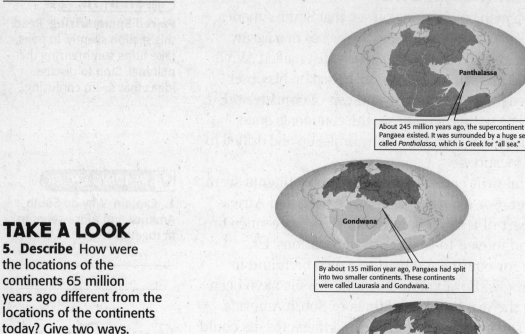

Panthalassa

About 245 million years ago, the supercontinent Pangaea existed. It was surrounded by a huge sea called *Panthalassa*, which is Greek for "all sea."

Gondwana

By about 135 million year ago, Pangaea had split into two smaller continents. These continents were called Laurasia and Gondwana.

Africa **South America** **Australia**

By about 65 million years ago, Laurasia and Gondwana had split into several smaller continents. These continents looked similar to the continents that exist today.

TAKE A LOOK

5. Describe How were the locations of the continents 65 million years ago different from the locations of the continents today? Give two ways.

_____

_____

_____

_____

What Is Sea-Floor Spreading?

Mid-ocean ridges are mountain chains on the ocean floor. They form a continuous chain that is 50,000 km long. The chain wraps around Earth like the seams of a baseball. Mid-ocean ridges are the sites of intense volcanic activity. ☑

At a mid-ocean ridge, melted rock rises through cracks in the sea floor. As the melted rock cools and hardens, it forms new crust. The newly formed crust pushes the older crust away from the mid-ocean ridge. This process is called **sea-floor spreading**.

READING CHECK

6. Define What is a mid-ocean ridge?

_____

_____

Copyright © by Holt, Rinehart and Winston; a Division of Houghton Mifflin Harcourt Publishing Company. All rights reserved.

SECTION 2 Restless Continents *continued*

SEA-FLOOR SPREADING AND MAGNETISM

In the 1960s, scientists studying the ocean floor discovered an interesting property of mid-ocean ridges. Using a tool that can record magnetism, they found magnetic patterns on the sea floor! The pattern on one side of a mid-ocean ridge was a mirror image of the pattern on the other side of the ridge. What caused the rocks to have these magnetic patterns?

Throughout Earth's history, the north and south magnetic poles have switched places many times. This process is called magnetic reversal. This process, together with sea-floor spreading, can explain the patterns of magnetism on the sea floor. ☑

Normal   **Reverse**

During times of normal polarity, such as today, a compass needle points toward the North Pole. During times of reverse polarity, a compass needle points toward the South Pole.

As ocean crust forms from melted rock, magnetic minerals form. These minerals act as compasses. As they form, they line up with Earth's magnetic north pole. When the melted rock cools, the minerals are stuck in place.

After Earth's magnetic field reverses, these minerals point to Earth's magnetic south pole. However, new rock that forms will have minerals that point to the magnetic north pole. Therefore, the ocean floor contains "stripes" of rock whose magnetic minerals point to the north or south magnetic poles.

This part of the sea floor formed when Earth's magnetic field was reversed.

This part of the sea floor formed when Earth's magnetic field was the same as it is today. As new sea floor formed, the rock was pushed away from the ridge.

Magma

■ Normal polarity lithosphere
▨ Reversed polarity lithosphere

Sea-floor spreading produces new oceanic lithosphere at mid-ocean ridges. The oldest oceanic crust is found far from the ridges, and the youngest crust is found very close to the ridges.

☑ **READING CHECK**

7. Define What is a magnetic reversal?

_____

_____

_____

TAKE A LOOK

8. Describe How are the "stripes" of magnetism on each side of the ridge related?

_____

_____

_____

Copyright © by Holt, Rinehart and Winston; a Division of Houghton Mifflin Harcourt Publishing Company. All rights reserved.

Section 2 Review

GLE 0707.Inq.3, GLE 0707.Inq.5, GLE 0707.7.4

SECTION VOCABULARY

continental drift the hypothesis that states that the continents once formed a single landmass, broke up, and drifted to their present locations	**sea-floor spreading** the process by which new oceanic lithosphere (sea floor) forms as magma rises to Earth's surface and solidifies at a mid-ocean ridge

1. Identify Give three pieces of evidence that support the idea of continental drift.

_____

_____

_____

2. Describe How does oceanic lithosphere form?

_____

_____

_____

3. Identify Does the oceanic lithosphere get older or younger as you move closer to the mid-ocean ridge?

_____

4. Explain How do the parallel magnetic "stripes" near mid-ocean ridges form?

_____

_____

_____

5. Apply Concepts The Earth is about 4.6 billion years old. However, the oldest sea floor is only about 180 million years old. What do you think is the reason for this? (Hint: Remember that new seafloor is constantly being created, but the Earth is not getting bigger with time.)

_____

_____

_____

Copyright © by Holt, Rinehart and Winston; a Division of Houghton Mifflin Harcourt Publishing Company. All rights reserved.

CHAPTER 15 Plate Tectonics
SECTION **3** **The Theory of Plate Tectonics**

BEFORE YOU READ

After you read this section, you should be able to answer these questions:

- What is the theory of plate tectonics?
- What are the three types of tectonic plate boundaries?

TN Tennessee Science
Standards
GLE 0707.Inq.3
GLE 0707.Inq.5
GLE 0707.7.3
GLE 0707.7.4

What Is the Theory of Plate Tectonics?

As scientists learned more about sea-floor spreading and magnetic reversals, they formed a theory to explain how continents move. The theory of **plate tectonics** states that Earth's lithosphere is broken into many pieces—tectonic plates—that move slowly over the asthenosphere.

Tectonic plates move very slowly—only a few centimeters per year. Scientists can detect this motion only by using special equipment, such as global positioning systems (GPS). This equipment is sensitive enough to pick up even small changes in a continent's location. ☑

What Happens Where Tectonic Plates Touch?

The places where tectonic plates meet are called *boundaries*. Some features, such as earthquakes and volcanoes, are more common at tectonic plate boundaries than at other places on Earth. Other features, such as mid-ocean ridges and ocean trenches, form only at plate boundaries.

There are three types of plate boundaries:

- divergent boundaries, where plates move apart;
- convergent boundaries, where plates move together; and
- transform boundaries, where plates slide past each other.

The features that form at a plate boundary depend on what kind of plate boundary it is.

DIVERGENT BOUNDARIES

A **divergent boundary** forms where plates are moving apart. Most divergent boundaries are found beneath the oceans. Mid-ocean ridges form at these divergent boundaries. Because the plates are pulling away from each other, cracks form in the lithosphere. Melted rock can rise through these cracks. When the melted rock cools and hardens, it becomes new lithosphere. ☑

STUDY TIP

Compare As you read, make a table showing the features of the three kinds of plate boundaries.

READING CHECK

1. Explain How do scientists detect tectonic plate motions?

_____

_____

_____

READING CHECK

2. Describe What features are found at most divergent boundaries?

_____

_____

_____

Copyright © by Holt, Rinehart and Winston; a Division of Houghton Mifflin Harcourt Publishing Company. All rights reserved.

CONVERGENT BOUNDARIES

A **convergent boundary** forms where plates are moving together. There are three different types of convergent boundaries:

- **Continent-Continent Boundaries** These form when continental lithosphere on one plate collides with continental lithosphere on another plate. Continent-continent convergent boundaries can produce very tall mountain ranges, such as the Himalayas.

- **Continent-Ocean Boundaries** These form when continental lithosphere on one plate collides with oceanic lithosphere on another plate. The denser oceanic lithosphere sinks underneath the continental lithosphere in a process called *subduction*. Subduction can cause a chain of mountains, such as the Andes, to form along the plate boundary.

- **Ocean-Ocean Boundaries** These form when oceanic lithosphere on one plate collides with oceanic lithosphere on another plate. One of the plates subducts beneath the other. A series of volcanic islands, called an *island arc*, can form along the plate boundary.

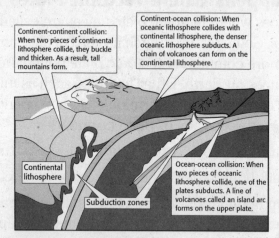

Continent-continent collision: When two pieces of continental lithosphere collide, they buckle and thicken. As a result, tall mountains form.

Continent-ocean collision: When oceanic lithosphere collides with continental lithosphere, the denser oceanic lithosphere subducts. A chain of volcanoes can form on the continental lithosphere.

Continental lithosphere

Ocean-ocean collision: When two pieces of oceanic lithosphere collide, one of the plates subducts. A line of volcanoes called an island arc forms on the upper plate.

Subduction zones

TRANSFORM BOUNDARIES

A **transform boundary** forms where plates slide past each other horizontally. Most transform boundaries are found near mid-ocean ridges. The ridges are broken into *segments*, or pieces. Transform boundaries separate the segments from one another.

One well-known transform boundary is the San Andreas fault system in California. It is located where the Pacific and North American plates slide past each other.

Critical Thinking

3. **Infer** Why do continent-continent convergent boundaries produce very tall mountain ranges?

_____

_____

_____

_____

_____

TENNESSEE STANDARDS CHECK

GLE 0707.7.4 Explain how earthquakes, mountain building, volcanoes, and sea floor spreading are associated with movements of the earth's <u>major</u> plates.

Word Help: major
of great importance or large scale

4. **Identify** Name two structures that can form at convergent plate boundaries.

_____

_____

Copyright © by Holt, Rinehart and Winston; a Division of Houghton Mifflin Harcourt Publishing Company. All rights reserved.

SECTION 3 The Theory of Plate Tectonics *continued*

Why Do Tectonic Plates Move?

Scientists do not know for sure what causes tectonic plates to move. They have three main hypotheses to explain plate movements: convection, slab pull, and ridge push.

Scientists used to think that convection in the mantle was the main force that caused plate motions. Remember that *convection* happens when matter carries heat from one place to another. Convection happens in the mantle as rock heats up and expands. As it expands, it becomes less dense and rises toward Earth's surface. ☑

As the hot material rises, cold, dense lithosphere sinks into the mantle at subduction zones. The rising hot material and the sinking cold material form *convection currents*. Until the 1990s, many scientists thought that these convection currents pulled the tectonic plates over Earth's surface.

Today, most scientists think that slab pull is the main force that causes plate motions. During subduction, oceanic lithosphere at the edge of a plate sinks into the mantle. The oceanic lithosphere sinks because it is colder and denser than the mantle. As the lithosphere at the edge of the plate sinks, it pulls the rest of the plate along with it. This process is called *slab pull*.

Another possible cause of plate motions is ridge push. At mid-ocean ridges, new oceanic lithosphere forms. This new lithosphere is warmer and less dense than the older lithosphere farther from the ridge. Therefore, it floats higher on the asthenosphere than the older lithosphere. As gravity pulls the new lithosphere down, the plate slides away from the mid-ocean ridge. This process is called *ridge push*.

☑ **READING CHECK**

5. Define What is convection?

_____

_____

_____

Critical Thinking

6. Compare How is slab pull different from ridge push?

_____

_____

_____

_____

_____

Driving Force	Description
Slab pull	Cold, sinking lithosphere at the edges of a tectonic plate pulls the rest of the plate across Earth's surface.
Ridge push	Gravity pulls newly formed lithosphere downward and away from the mid-ocean ridge. The rest of the plate moves because of this force.
Convection currents	Convection currents are produced when hot material in the mantle rises toward the surface and colder material sinks. The currents pull the plates over Earth's surface.

Copyright © by Holt, Rinehart and Winston; a Division of Houghton Mifflin Harcourt Publishing Company. All rights reserved.

Section 3 Review

GLE 0707.Inq.3, GLE 0707.Inq.5, GLE 0707.7.3, GLE 0707.7.4 TN

SECTION VOCABULARY

convergent boundary the boundary formed by the collision of two lithospheric plates	**plate tectonics** the theory that explains how large pieces of the Earth's outermost layer, called tectonic plates, move and change shape
divergent boundary the boundary between two tectonic plates that are moving away from each other	**transform boundary** the boundary between tectonic plates that are sliding past each other horizontally

1. Define Write your own definition for plate tectonics.

_____

_____

_____

2. Identify What are the three types of plate boundaries?

_____

_____

3. List Name three processes that may cause tectonic plates to move.

_____

_____

_____

4. Describe How fast do tectonic plates move?

_____

5. Identify Give two features that are found only at plate boundaries, and two features that are found most commonly at plate boundaries.

_____

_____

_____

_____

6. Explain Why does oceanic lithosphere sink beneath continental lithosphere at convergent boundaries?

_____

_____

_____

Copyright © by Holt, Rinehart and Winston; a Division of Houghton Mifflin Harcourt Publishing Company. All rights reserved.

CHAPTER 15 Plate Tectonics
SECTION 4 **Deforming the Earth's Crust**

TN **Tennessee Science Standards**
GLE 0707.Inq.2
GLE 0707.Inq.3
GLE 0707.Inq.5
GLE 0707.7.4

BEFORE YOU READ

After you read this section, you should be able to answer these questions:

• What happens when rock is placed under stress?

• What are three kinds of faults?

• How do mountains form?

What Is Deformation?

In the left-hand figure below, the girl is bending the spaghetti slowly and gently. The spaghetti bends, but it doesn't break. In the right-hand figure, the girl is bending the spaghetti quickly and with a lot of force. Some of the pieces of spaghetti have broken.

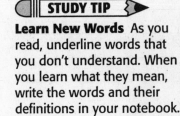

STUDY TIP

Learn New Words As you read, underline words that you don't understand. When you learn what they mean, write the words and their definitions in your notebook.

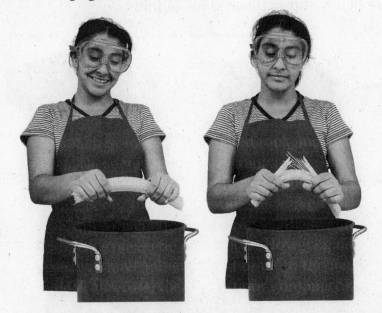

How can the same material bend in one situation but break in another? The answer is that the stress on the material is different in each case. *Stress* is the amount of force per unit area that is placed on an object. ☑

TAKE A LOOK
1. Describe Circle the picture in which the girl is putting the most force on the spaghetti.

READING CHECK
2. Define What is stress?

_____

_____

DEFORMATION

Like the spaghetti, rocks can bend or break under stress. When a rock is placed under stress, it *deforms*, or changes shape. When a small amount of stress is put on a rock slowly, the rock can bend. However, if the stress is very large or is applied quickly, the rock can break.

Copyright © by Holt, Rinehart and Winston; a Division of Houghton Mifflin Harcourt Publishing Company. All rights reserved.

What Happens When Rock Layers Bend?

Folding happens when rock layers bend under stress. Folding causes rock layers to look bent or buckled. The bends are called *folds*.

Most rock layers start out as horizontal layers. Therefore, when scientists see a fold, they know that deformation has happened. ☑

TYPES OF FOLDS

Three of the most common types of folds are synclines, anticlines, and monoclines. In a *syncline*, the oldest rocks are found on the outside of the fold. Most synclines are U-shaped. In an *anticline*, the youngest rocks are found on the outside of the fold. Most anticlines are ∩-shaped. In a *monocline*, rock layers are folded so that both ends of the fold are horizontal. The figure below shows these kinds of folds.

✓ READING CHECK

3. Explain How do folds indicate that deformation has happened?

_____

_____

_____

_____

_____

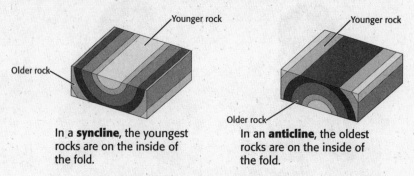

In a **syncline**, the youngest rocks are on the inside of the fold.

In an **anticline**, the oldest rocks are on the inside of the fold.

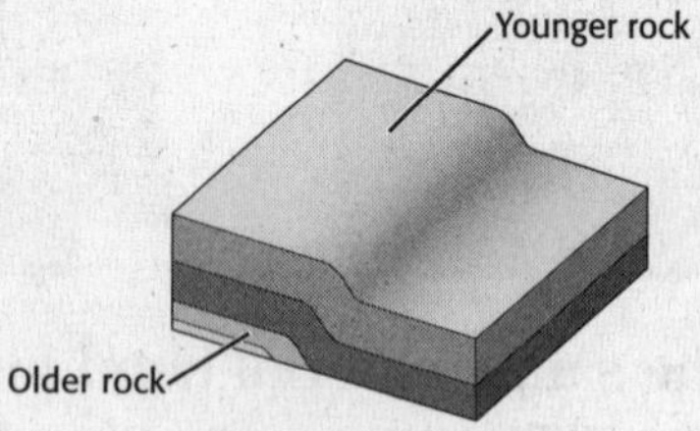

In a **monocline**, both sides of the fold are horizontal.

TAKE A LOOK

4. Identify Color the oldest rock layers in the figure blue. Color the youngest rock layers red.

What Happens When Rock Layers Break?

When rock is put under so much stress that it can no longer bend, it may break. The crack that forms when rocks break and move past each other is called a **fault**. The blocks of rock that are on either side of the fault are called *fault blocks*. When fault blocks move suddenly, they can cause earthquakes.

Copyright © by Holt, Rinehart and Winston; a Division of Houghton Mifflin Harcourt Publishing Company. All rights reserved.

HANGING WALL AND FOOTWALL

When a fault forms at an angle, one fault block is called the *hanging wall* and the other is called the *footwall*. The figure below shows the difference between the hanging wall and the footwall.

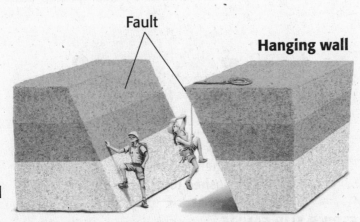

The footwall is the fault block that is below the fault. The hanging wall is the fault block that is above the fault.

Scientists classify faults by how the fault blocks have moved along the fault. There are three main kinds of faults: normal faults, reverse faults, and strike-slip faults.

NORMAL FAULTS

In a *normal fault*, the hanging wall moves down, or the footwall moves up, or both. Normal faults form when rock is under tension. **Tension** is stress that pulls rock apart. Therefore, normal faults are common along divergent boundaries, where Earth's crust stretches. ☑

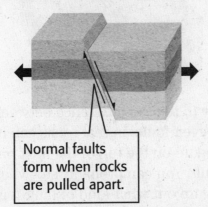

Normal faults form when rocks are pulled apart.

TAKE A LOOK
5. Compare How is the hanging wall different from the footwall?

_____

_____

✓ **READING CHECK**
6. Explain Why are normal faults common along divergent boundaries?

_____

_____

_____

Copyright © by Holt, Rinehart and Winston; a Division of Houghton Mifflin Harcourt Publishing Company. All rights reserved.

REVERSE FAULTS

In a *reverse fault*, the hanging wall moves up, or the footwall moves down, or both. Reverse faults form when rock is under compression. **Compression** is stress that pushes rock together. Therefore, reverse faults are common at convergent boundaries, where plates collide.

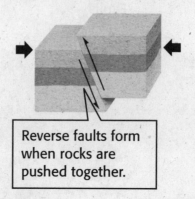

Reverse faults form when rocks are pushed together.

TAKE A LOOK
7. Identify Label the hanging walls and the footwalls on the normal and reverse faults.

STRIKE-SLIP FAULTS

In a *strike-slip fault*, the fault blocks move past each other horizontally. Strike-slip faults form when rock is under shear stress. *Shear stress* is stress that pushes different parts of the rock in different directions. Therefore, strike-slip faults are common along transform boundaries, where tectonic plates slide past each other.

Strike-slip faults form when rocks slide past each other horizontally.

TAKE A LOOK
8. Describe How do strike-slip faults form?

_____

_____

_____

It can be easy to tell the difference between faults in a diagram. However, faults in real rocks can be harder to tell apart. The figure on the top of the next page shows an example of a fault. You can probably see where the fault is. How can you figure out what kind of fault it is? One way is to look at the rock layers around the fault. The dark rock layer in the hanging wall is lower than the same layer in the footwall. Therefore, this is a normal fault.

Copyright © by Holt, Rinehart and Winston; a Division of Houghton Mifflin Harcourt Publishing Company. All rights reserved.

SECTION 4 Deforming the Earth's Crust *continued*

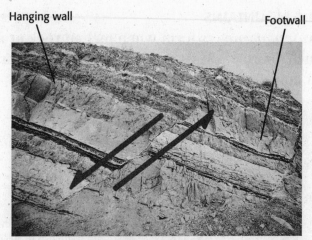

Hanging wall

Footwall

In these rocks, the hanging wall has moved down compared to the footwall. Therefore, this is a normal fault.

TAKE A LOOK
9. **Explain** How can you tell that this is a normal fault?

_____

_____

_____

How Do Mountains Form?

As tectonic plates move over Earth's surface, the edges of the plates grind against each other. This produces a lot of stress in Earth's lithosphere. Over very long periods of time, the movements of the plates can form mountains. Mountains can form in three main ways: through folding, faulting, or volcanism.

Critical Thinking
10. **Apply Concepts** Why does it take a very long time for most mountains to form?

_____

_____

_____

_____

FOLDED MOUNTAINS

Folded mountains form when rock layers are squeezed together and pushed upward. Folded mountains usually form at convergent boundaries, where continents collide. For example, the Appalachian Mountains formed hundreds of millions of years ago when North America collided with Europe and Africa.

FAULT-BLOCK MOUNTAINS

Fault-block mountains form when tension makes the lithosphere break into many normal faults. Along these faults, pieces of the lithosphere drop down compared with other pieces. This produces fault-block mountains. ☑

READING CHECK
11. **Identify** What kind of stress forms fault-block mountains?

_____

Fault-block mountains form when tension causes the crust to break into normal faults.

Copyright © by Holt, Rinehart and Winston; a Division of Houghton Mifflin Harcourt Publishing Company. All rights reserved.

SECTION 4 Deforming the Earth's Crust *continued*

VOLCANIC MOUNTAINS

Volcanic mountains form when melted rock erupts onto Earth's surface. Most major volcanic mountains are found at convergent boundaries.

The Andes mountains are examples of volcanic mountains. The Andes have formed where the Nazca plate is subducting beneath the South American plate.

Volcanic mountains can form on land or on the ocean floor. Volcanoes on the ocean floor can grow so tall that they rise above the surface of the ocean. These volcanoes form islands, such as the Hawaiian Islands.

Most of Earth's active volcanoes are concentrated around the edge of the Pacific Ocean. This area is known as the *Ring of Fire*.

Type of Mountain	Description
Folded	
Fault-block	
Volcanic	

How Can Rocks Move Vertically?

There are two types of vertical movements in the crust: uplift and subsidence. **Uplift** happens when parts of Earth's crust rise to higher elevations. Rocks that are uplifted may or may not be deformed. **Subsidence** happens when parts of the crust sink to lower elevations. Unlike some uplifted rocks, rocks that subside do not deform.

Say It

Investigate Find out more about a volcanic mountain chain, such as the Andes, the Hawaiian islands, or Japan. Share what you learn with a small group.

TAKE A LOOK

12. Identify What kind of convergent boundary have the Andes mountains formed on?

_____

TAKE A LOOK

13. Describe Fill in the table with the features of each kind of mountain. Include where the mountains form and what they are made of.

Copyright © by Holt, Rinehart and Winston; a Division of Houghton Mifflin Harcourt Publishing Company. All rights reserved.

SECTION 4 Deforming the Earth's Crust *continued*

CAUSES OF SUBSIDENCE AND UPLIFT

Temperature changes can cause uplift and subsidence. Hot rocks are less dense than cold rocks with the same composition. Therefore, as hot rocks cool, they may sink. If cold rocks are heated, they may rise. For example, the crust at mid-ocean ridges is hot. As it moves away from the ridge, it cools and subsides. Old, cold crust far from a ridge has a lower elevation than young, hot crust at the ridge.

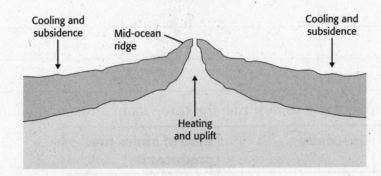

TAKE A LOOK
14. Explain Why does ocean crust far from a mid-ocean ridge subside?

_____

_____

Changes in the weight on the crust can also cause uplift or subsidence. For example, glaciers are huge, heavy bodies of ice. When they form on the crust, they can push the crust down and cause subsidence. If the glaciers melt, the weight on the crust decreases. The crust slowly rises back to its original elevation in a process called *rebound*.

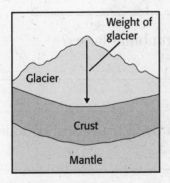 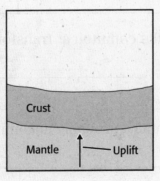

TAKE A LOOK
15. Identify What force caused the crust to subside in the left-hand figure?

_____

_____

Fault-block mountains are an example of a third way subsidence can happen. When the crust is under tension, rocks are stretched. They can break and form normal faults. The crust can sink along these faults, causing subsidence. This kind of subsidence is common in rift zones. A *rift zone* is a set of deep cracks that forms at a divergent boundary.

Copyright © by Holt, Rinehart and Winston; a Division of Houghton Mifflin Harcourt Publishing Company. All rights reserved.

Name _____ Class _____ Date _____

Section 4 Review

GLE 0707.Inq.2, GLE 0707.Inq.3, GLE 0707.Inq.5 TN
GLE 0707.7.4

SECTION VOCABULARY

compression stress that occurs when forces act to squeeze an object	**subsidence** the sinking of regions of the Earth's crust to lower elevations
fault a break in a body of rock along which one block slides relative to another	**tension** stress that occurs when forces act to stretch an object
folding the bending of rock layers due to stress	**uplift** the rising of regions of the Earth's crust to higher elevations

1. Compare How are folding and faulting similar? How are they different?

_____

_____

2. Describe Fill in the spaces in the table to describe the three main kinds of faults.

Kind of fault	Description	Kind of stress that produces it
Normal		
	Hanging wall moves up; footwall moves down.	
		shear stress

3. Explain Why are strike-slip faults common at transform boundaries?

_____

_____

4. Infer Why are fault-block mountains probably uncommon at transform boundaries?

_____

_____

5. Define What is the Ring of Fire?

_____

_____

Copyright © by Holt, Rinehart and Winston; a Division of Houghton Mifflin Harcourt Publishing Company. All rights reserved.

CHAPTER 16 Earthquakes

SECTION 1 **What Are Earthquakes?**

BEFORE YOU READ

After you read this section, you should be able to answer these questions:

• Where do most earthquakes happen?

• What makes an earthquake happen?

• What are seismic waves?

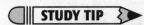

Tennessee Science Standards

GLE 707.Inq.2
GLE 707.Inq.5
GLE 707.7.4

What Is an Earthquake?

Have you ever been in an earthquake? An *earthquake* is a movement or shaking of the ground. Earthquakes happen when huge pieces of Earth's crust move suddenly and give off energy. This energy travels through the ground and makes it move. **Seismology** is the study of earthquakes. Scientists who study earthquakes are called *seismologists*.

STUDY TIP

Learn New Words As you read this section, circle words that you don't understand. When you learn what they mean, write the words and their definitions in your notebook.

Where Do Most Earthquakes Happen?

Most earthquakes happen at places where two tectonic plates touch. Tectonic plates are always moving. In some places, they move away from each other. In some places, they move toward each other. And in some places, they grind past each other.

The movements of the plates cause Earth's rocky crust to break. A place where the crust is broken is called a *fault*. Earthquakes happen when rock breaks and slides along a fault. ☑

Earthquakes and Plate Boundaries

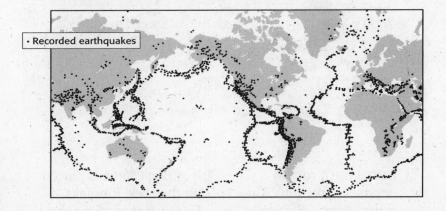

· Recorded earthquakes

READING CHECK

1. Define What is a fault?

is broken is called a fault.

TAKE A LOOK

2. Infer Use the earthquake locations to help you figure out where the tectonic plate boundaries are. Use a colored pen or marker to draw plate boundaries on the map.

Copyright © by Holt, Rinehart and Winston; a Division of Houghton Mifflin Harcourt Publishing Company. All rights reserved.

Why Do Earthquakes Happen?

When tectonic plates move, pressure builds up on the rock near the edges of the plates. When rock is put under pressure, it changes shape, or deforms. This is called **deformation**.

Some rock can bend and fold like clay. When the pressure is taken away, the rock stays folded. When rock stays folded after the pressure is gone, the change is called *plastic deformation*.

Folded Layers of Rock

TAKE A LOOK
3. Explain How do you know that the rock layers in the figure were once under a lot of pressure?

the pressure

TN TENNESSEE STANDARDS CHECK

GLE 0707.7.4 Explain how earthquakes, mountain buildings, volcanoes, and seafloor spreading are associated with movements of the Earth's <u>major</u> plates.

Word Help: <u>major</u> of great importance or large scale

4. Explain How does the movement of tectonic plates cause earthquakes?

the a move
back and
Forward

In some cases, rock acts more like a rubber band. It changes shape under pressure, but then it goes back to its original size and shape when the pressure goes away. This change is called *elastic deformation*.

Earthquakes happen when rock breaks under pressure. When the rock breaks, it snaps back to its original shape. This snap back is called **elastic rebound**. When the rock breaks and rebounds, it gives off energy. This energy creates faults and causes the ground to shake.

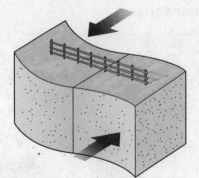

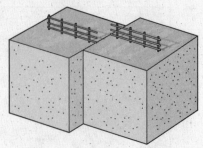

1. Forces push rock in opposite directions. The rock deforms elastically. It does not break.

2. If enough force is placed on the rock, it breaks. The rock slips along the fault. Energy is released.

Copyright © by Holt, Rinehart and Winston; a Division of Houghton Mifflin Harcourt Publishing Company. All rights reserved.

SECTION 1 What Are Earthquakes? *continued*

How Do Earthquakes Happen at Divergent Boundaries?

A *divergent boundary* is a place where two tectonic plates are moving away from each other. As the plates pull apart, the crust stretches. The crust breaks along faults. ☑

Most of the crust at divergent boundaries is thin and weak. Most earthquakes at divergent boundaries are small because only a little bit of pressure builds up before the rock breaks.

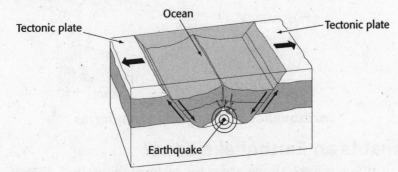

Earthquakes at Divergent Boundaries

✓ READING CHECK

5. Define What is a divergent boundary?

is a place where two tectonic plates

TAKE A LOOK
6. Identify Label the faults on the figure. Put a star where an earthquake is likely to happen.

How Do Earthquakes Happen at Convergent Boundaries?

A *convergent boundary* is a place where two tectonic plates collide. When two plates come together, the rock is put under a lot of pressure. The pressure grows and grows until the rock breaks.

The earthquakes that happen at convergent boundaries can be very strong because there is so much pressure. The strongest earthquakes ever recorded have all happened at convergent boundaries. ☑

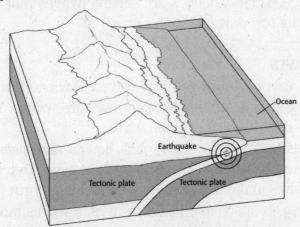

Earthquakes at Convergent Boundaries

✓ READING CHECK

7. Explain Why are many earthquakes at convergent boundaries very strong?

breaks

TAKE A LOOK
8. Identify Draw arrows on the figure to show the directions that the two tectonic plates are moving.

Copyright © by Holt, Rinehart and Winston; a Division of Houghton Mifflin Harcourt Publishing Company. All rights reserved.

How Do Earthquakes Happen at Transform Boundaries?

A *transform boundary* is a place where two tectonic plates slide past each other. As the plates move, pressure builds up on the rock. Eventually, the rock breaks and the plates slide past each other along a fault.

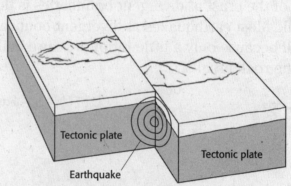

Tectonic plate

Tectonic plate

Earthquake

Earthquakes at Transform Boundaries

TAKE A LOOK
9. Identify Draw arrows showing the directions that the tectonic plates in the figure are moving.

What Is an Earthquake Zone?

A place where there are a lot of faults is called an *earthquake zone*. The San Andreas Fault Zone in California is an example of an earthquake zone. Most earthquake zones are near plate boundaries, but some are in the middle of tectonic plates.

Critical Thinking
10. Apply Concepts Why are most earthquake zones near plate boundaries?

_____

_____

How Does Earthquake Energy Travel?

When an earthquake occurs, a lot of energy is given off. This energy travels through the Earth in the form of waves called **seismic waves**.

There are two kinds of seismic waves. *Body waves* are seismic waves that travel through the inside of Earth to the surface. *Surface waves* are seismic waves that travel through the top part of Earth's crust. ☑

☑ **READING CHECK**
11. List What are the two kinds of seismic waves?

_____

_____

BODY WAVES

There are two kinds of body waves: P waves and S waves. **P waves** are also called pressure waves. They are the fastest kind of seismic wave.

P waves can move through solids, liquids, and gases. When a P wave travels through a rock, it squeezes and stretches the rock. P waves make the ground move back and forth.

S waves are also called shear waves. S waves move rock from side to side. They can travel only through solids. S waves travel more slowly than P waves.

Copyright © by Holt, Rinehart and Winston; a Division of Houghton Mifflin Harcourt Publishing Company. All rights reserved.

SECTION 1 What Are Earthquakes? *continued*

SURFACE WAVES

Surface waves travel along the top of Earth's crust. Only the very top part of the crust moves when a surface wave passes.

Surface waves travel much more slowly than body waves. When an earthquake happens, surface waves are the last waves to be felt. Surface waves cause a lot more damage to buildings and landforms than body waves do. ☑

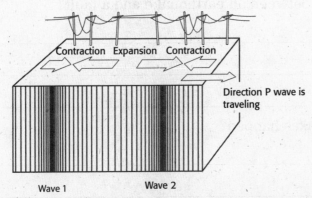

P waves are body waves that squeeze and stretch rock.

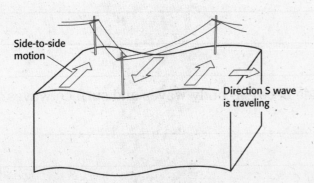

S waves are body waves that can move rock from side to side.

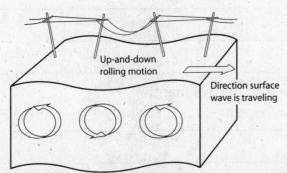

Surface waves can move the ground up and down in a circular motion.

Copyright © by Holt, Rinehart and Winston; a Division of Houghton Mifflin Harcourt Publishing Company. All rights reserved.

☑ **READING CHECK**

12. Compare Which kind of seismic wave travels the most slowly?

_____

_____

TAKE A LOOK

13. Compare How are the motions of P waves and S waves different?

_____

_____

_____

Critical Thinking

14. Infer What do you think is the reason surface waves usually cause the most damage?

_____

_____

_____

_____

Section 1 Review

GLE 0707.Inq.2, GLE 0707.Inq.5, GLE 0707.7.4 **TN**

SECTION VOCABULARY

deformation the bending, tilting, and breaking of the Earth's crust; the change in the shape of rock in response to stress	**S wave** a seismic wave that causes particles of rock to move in a side-to-side direction
elastic rebound the sudden return of elastically deformed rock to its undeformed shape	**seismic wave** a wave of energy that travels through the Earth, away from an earthquake in all directions
P wave a seismic wave that causes particles of rock to move in a back-and-forth direction	**seismology** the study of earthquakes

1. Compare What is the difference between an earthquake and a fault?

_____

_____

_____

2. Identify Where do most earthquakes happen?

_____

_____

3. Describe What causes earthquakes?

_____

_____

_____

4. Compare What is the main difference between body waves and surface waves?

_____

_____

_____

5. Apply Concepts Why are some earthquakes stronger than others?

_____

_____

_____

6. Infer Why do few earthquakes happen in Earth's mantle?

_____

_____

_____

Copyright © by Holt, Rinehart and Winston; a Division of Houghton Mifflin Harcourt Publishing Company. All rights reserved.

CHAPTER 16 Earthquakes

SECTION
2 **Earthquake Measurement**

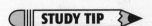

TN Tennessee Science Standards

GLE 0707.Inq.2
GLE 0707.Inq.3
GLE 0707.Inq.5

BEFORE YOU READ

After you read this section, you should be able to answer these questions:

• How do scientists know exactly where and when an earthquake happened?

• How are earthquakes measured?

How Do Scientists Study Earthquakes?

Scientists who study earthquakes use an important tool called a seismograph. A **seismograph** records vibrations that are caused by seismic waves. When the waves from an earthquake reach a seismograph, it records them as lines on a chart called a **seismogram**.

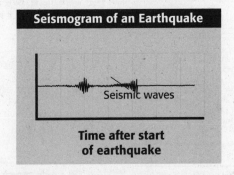

Seismogram of an Earthquake

Seismic waves

Time after start of earthquake

Remember that earthquakes happen when rock in Earth's crust breaks. The rock might break in one small area, but the earthquake can be felt many miles away.

The place inside the Earth where the rock first breaks is called the earthquake's **focus**. The place on Earth's surface that is right above the focus is called the **epicenter**. Seismologists can use seismograms to find the epicenter of an earthquake. ☑

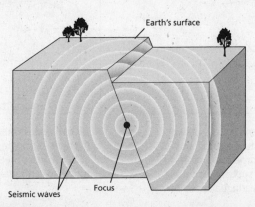

Earth's surface

Seismic waves

Focus

STUDY TIP

Ask Questions Read this section quietly to yourself. As you read, write down questions that you have. Discuss your questions in a small group.

✓ READING CHECK

1. Explain What is the difference between the epicenter and the focus of an earthquake?

_____

_____

_____

_____

TAKE A LOOK

2. Identify On the figure, mark the epicenter of the earthquake with a star.

Copyright © by Holt, Rinehart and Winston; a Division of Houghton Mifflin Harcourt Publishing Company. All rights reserved.

Name _____ Class _____ Date _____

How Do Seismologists Know When and Where an Earthquake Happened?

Seismograms help us learn when an earthquake happened. They can also help seismologists find the epicenter of an earthquake. The easiest way to do this is to use the S-P time method. This is how the S-P time method works:

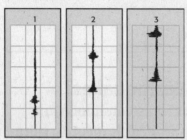

1. The seismologist uses seismograms of the earthquake made at three different places.

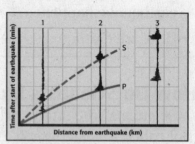

2. The seismologist lines up the P waves and S waves on each seismogram with the curves on a graph of time versus distance. The curves on the graph were made using information from earthquakes that happened in the past.

Math Focus

3. Read a Graph Look at the middle seismogram in step 3. What is the difference between the time the P waves arrived and the time the S waves arrived?

_____

4. Read a Graph Look at step 3. How far away from the epicenter is the farthest seismograph station?

_____

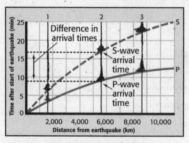

3. Then, the seismologist uses the graph to figure out the difference in arrival times of the P and S waves at each location. The seismologist can use the difference in arrival times to figure out when the earthquake happened. The seismologist can also determine how far away each station is from the epicenter of the earthquake.

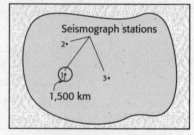

4. On a map, a circle is drawn around a seismograph station. The radius of the circle equals the distance from the seismograph to the epicenter. (This distance is taken from the time-distance graph.)

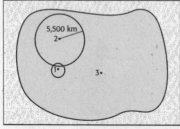

5. When a second circle is drawn around another seismograph station, the circle overlaps the first circle in two spots. One of these spots is the earthquake's epicenter.

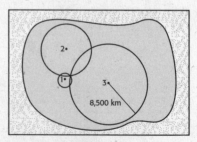

6. When a circle is drawn around the third seismograph station, all three circles meet in one spot—the earthquake's epicenter.

TAKE A LOOK

5. Identify On the map in step 6, draw a star at the earthquake's epicenter.

Copyright © by Holt, Rinehart and Winston; a Division of Houghton Mifflin Harcourt Publishing Company. All rights reserved.

What Is the Magnitude of an Earthquake?

Scientists study seismograms to find out how much the ground moved during an earthquake. They can use the seismograms to figure out how strong the earthquake was.

Have you ever heard someone say that an earthquake was 6.8 or 7.4 "on the Richter scale"? The Richter scale is used to describe the strength, or *magnitude*, of an earthquake. The higher the number, the stronger the earthquake.

Critical Thinking

6. Infer How do you think scientists use seismograms to determine the magnitude of an earthquake?

_____

_____

_____

_____

Richter magnitude	Effects
2.0	can be detected only by a seismograph
3.0	can be felt at the epicenter
4.0	can be felt by most people in the area
5.0	causes damage at the epicenter
6.0	can cause widespread damage
7.0	can cause great, widespread damage

The Richter scale can be used to compare the magnitudes of different earthquakes. When the Richter magnitude of an earthquake goes up by one unit, the amount of ground shaking caused by the earthquake goes up 10 times. For example, an earthquake with a magnitude of 5.0 is 10 times stronger than an earthquake with a magnitude of 4.0.

Math Focus

7. Calculate How many times stronger is a magnitude 6.0 earthquake than a magnitude 4.0 earthquake? Explain your answer.

_____

_____

_____

_____

What Is the Intensity of an Earthquake?

The *intensity* of an earthquake describes how much damage the earthquake caused and how much it was felt by people. Seismologists in the United States use the Modified Mercalli Intensity Scale to compare the intensity of different earthquakes.

The effects of an earthquake can be very different from place to place. An earthquake can have many different intensity numbers, even though it has only one magnitude.

Mercalli intensity (from I to XII)	Effects
I	shaking felt by only a few people
IV	shaking felt indoors by many, no damage
VIII	damage to some buildings
XII	total damage

Copyright © by Holt, Rinehart and Winston; a Division of Houghton Mifflin Harcourt Publishing Company. All rights reserved.

Section 2 Review

GLE 0707.Inq.2, GLE 0707.Inq.3, GLE 0707.Inq.5

SECTION VOCABULARY

epicenter the point on Earth's surface directly above an earthquake's starting point, or focus	**seismogram** a tracing of earthquake motion that is created by a seismograph
focus the point along a fault at which the first motion of an earthquake occurs	**seismograph** an instrument that records vibrations in the ground and determines the location and strength of an earthquake

1. Compare What is the difference between a seismograph and a seismogram?

_____

_____

_____

2. Apply Concepts Which city would more likely be the epicenter of an earthquake: San Francisco, California, or St. Paul, Minnesota? Explain your answer.

_____

_____

_____

3. Explain How does a seismologist use the graph of time versus distance for seismic waves to find the location of an earthquake's epicenter?

_____

_____

_____

4. Analyze Methods What can you learn from only one seismogram? What can't you learn?

_____

_____

_____

5. Infer How can an earthquake with a moderate magnitude have a high intensity?

_____

_____

_____

Copyright © by Holt, Rinehart and Winston; a Division of Houghton Mifflin Harcourt Publishing Company. All rights reserved.

CHAPTER 16 Earthquakes

SECTION 3
Earthquakes and Society

BEFORE YOU READ

After you read this section, you should be able to answer these questions:

• Can scientists predict when earthquakes will happen?

• Why do some buildings survive earthquakes better than others?

• How can you prepare for an earthquake?

TN **Tennessee Science Standards**

GLE 0707.Inq.2
GLE 0707.T/E.1
GLE 0707.T/E.2

What Is Earthquake Hazard?

Earthquake hazard tells how likely it is that a place will have a damaging earthquake in the future. Scientists look to the past to figure out earthquake-hazard levels. A place that has had a lot of strong earthquakes in the past has a high earthquake-hazard level. A place that has had few or no earthquakes has a much lower level.

STUDY TIP

Be Prepared As you read, underline important safety information that can help you to prepare for an earthquake.

Earthquake Hazard Map of the Continental United States

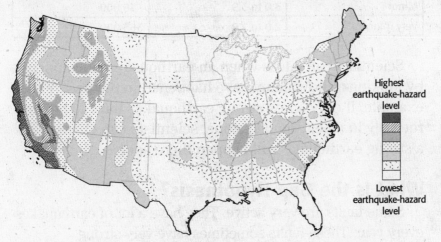

Highest earthquake-hazard level

Lowest earthquake-hazard level

TAKE A LOOK
1. **Identify** On the map, find the place where you live. What is its earthquake-hazard level?

Look at the map above. Notice that California has the highest earthquake-hazard level in the country. The San Andreas Fault Zone runs through most of California, and a lot of earthquakes happen there. Minnesota has a very low earthquake-hazard level. Very few strong earthquakes have been recorded in Minnesota.

Copyright © by Holt, Rinehart and Winston; a Division of Houghton Mifflin Harcourt Publishing Company. All rights reserved.

Can Scientists Predict Earthquakes?

You know that earthquakes have different magnitudes. You can probably guess that earthquakes don't happen on a set schedule. But what you may not know is that the strength of earthquakes is related to how often they happen.

Scientists can't predict earthquakes. However, by looking at how often earthquakes have happened in the past, they can estimate where and when an earthquake is likely to happen.

Look at the table below. It shows the number of earthquakes of different sizes that happen every year. There are many more weak earthquakes than strong earthquakes every year.

Math Focus
2. Calculate About how many times more light earthquakes than strong earthquakes happen every year?

_____

Description	Magnitude on the Richter scale	Average number per year
Great	8.0 and higher	1
Major	7.0 to 7.9	18
Strong	6.0 to 6.9	120
Moderate	5.0 to 5.9	800
Light	4.0 to 4.9	6,200
Minor	3.0 to 3.9	49,000
Very minor	2.0 to 2.9	365,000

Scientists can guess when an earthquake will happen by looking at how many have happened in the past. For example, if only a few strong earthquakes have happened recently in an earthquake zone, scientists can guess that a strong earthquake will happen there soon.

What Is the Gap Hypothesis?

Critical Thinking
3. Apply Concepts What do you think makes strong earthquakes more likely to happen in seismic gaps?

_____
_____
_____
_____
_____

Some faults are very active. They have a lot of earthquakes every year. These faults sometimes have very strong earthquakes. A part of an active fault that hasn't had a strong earthquake in a long time is called a **seismic gap**.

The **gap hypothesis** says that if an active fault hasn't had a strong earthquake in a long time, it is likely to have one soon. In other words, it says that strong earthquakes are more likely to happen in seismic gaps.

Copyright © by Holt, Rinehart and Winston; a Division of Houghton Mifflin Harcourt Publishing Company. All rights reserved.

How Do Earthquakes Affect Buildings?

Have you ever seen pictures of a city after a strong earthquake has hit? You may have noticed that some buildings don't have very much damage. Other buildings, however, are totally destroyed. Engineers can study the damage to learn how to make buildings that are stronger and safer.

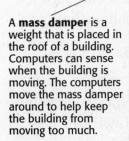

A **mass damper** is a weight that is placed in the roof of a building. Computers can sense when the building is moving. The computers move the mass damper around to help keep the building from moving too much.

Steel **cross braces** are found between the floors in a building. They help to keep the building from breaking when it moves from side to side.

Flexible pipes can help to prevent water lines and gas lines from breaking. The pipes can bend and twist without breaking.

An **active tendon system** is like a mass damper, except it is placed under the building.

Base isolators can absorb energy during an earthquake. They keep seismic waves from moving through the building. Base isolators are made of rubber, steel, and lead.

Critical Thinking

4. List Give three factors that can affect how much a building will be damaged by an earthquake.

_____

_____

_____

TAKE A LOOK

5. Compare How is a mass damper different from an active tendon system?

_____

_____

_____

Copyright © by Holt, Rinehart and Winston; a Division of Houghton Mifflin Harcourt Publishing Company. All rights reserved.

How Can You Prepare for an Earthquake?

If you live in a place where earthquakes happen often, you and your family should have an earthquake plan. You should practice your plan so you will be prepared if an earthquake happens. ☑

How Can You Make an Earthquake Plan?

There are several things to include in your earthquake plan.

SAFE HOME

Put heavy things near the floor so that they do not fall during an earthquake. Make sure things that can burn are kept away from electric wires and other things that can start a fire.

SAFE PLACES IN YOUR HOME

Make sure you know a safe place in each room in your home. Safe places are areas far from windows or heavy objects that could fall or break. ☑

PLAN TO MEET OTHERS

Talk to your family, friends, or neighbors and set up a place where you all will meet after an earthquake. If you all know where to meet one another, it will be easy to make sure that everyone is safe.

EARTHQUAKE KIT

Your earthquake kit should have things that you might need after an earthquake. Remember that you may not have electricity or running water after an earthquake.

READING CHECK

6. Explain Why is it important to make and practice an earthquake plan?

_____

_____

_____

_____

READING CHECK

7. Identify Think about your bedroom. Write down a safe place in your bedroom that you can go during an earthquake.

_____

_____

_____

TAKE A LOOK

8. List List four foods that would be useful to have in an earthquake kit.

_____

_____

_____

What Should Be in an Earthquake Kit	
• water	• food that won't go bad
• a fire extinguisher	• a flashlight with batteries
• a small radio that runs on batteries	• extra batteries for the radio and flashlight
• medicines	• a first-aid kit

Copyright © by Holt, Rinehart and Winston; a Division of Houghton Mifflin Harcourt Publishing Company. All rights reserved.

SECTION 3 Earthquakes and Society *continued*

What Should You Do During an Earthquake?

If you are inside when an earthquake happens, crouch or lie facedown under a table or a desk. Make sure you are far away from windows or heavy objects that might fall. Cover your head with your hands. ☑

If you are outside during an earthquake, lie face down on the ground. Make sure you are far from buildings, power lines, and trees. Cover your head with your hands.

If you are in a car or bus, you should ask the driver to stop. Everyone should stay inside the car or bus until the earthquake is over.

READING CHECK

9. List Look around your classroom. List two places that you could go in case of an earthquake.

_____

_____

_____

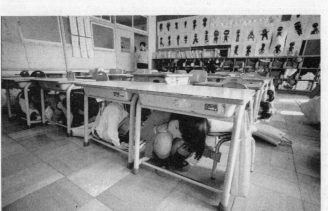

Say It

Share Experiences Have you ever been in an earthquake? In a small group, talk about what it was like.

What Should You Do After an Earthquake?

Being in an earthquake can be scary. After an earthquake happens, people are often confused about what happened. They may not know what to do or where to go.

After an earthquake, try to stay calm. Look around you. If you are near something dangerous, like a power line or broken glass, get away as quickly as you can. Never go into a building after an earthquake until your parent, a teacher, a police officer, or a firefighter tells you it is safe. ☑

Always remember that there could be aftershocks. Aftershocks are weaker earthquakes that can happen after a large earthquake. Even though they are weaker than the main earthquake, aftershocks can still be very strong and damaging.

Stick to your earthquake plan. Stay together with your family or friends so that they know you are safe.

READING CHECK

10. Identify Who should you ask if you want to know whether it is safe to go back into a building after an earthquake?

_____

_____

Copyright © by Holt, Rinehart and Winston; a Division of Houghton Mifflin Harcourt Publishing Company. All rights reserved.

Section 3 Review

GLE 0707.T/E.1, GLE 0707.T/E.2, GLE 0607.Inq.2 TN

SECTION VOCABULARY

gap hypothesis a hypothesis that is based on the idea that a major earthquake is more likely to occur along the part of an active fault where no earthquakes have occurred for a certain period of time	**seismic gap** an area along a fault where relatively few earthquakes have occurred recently but where strong earthquakes have occurred in the past

1. Identify Why are seismologists interested in seismic gaps?

_____

_____

_____

2. Describe Fill in the chart below to show what you should do during an earthquake.

If you are...	Then you should...
...inside a building	
	...lie face down on the ground with your hands on your head, far from power lines or fire hazards.
...in a car or bus	

3. Identify What do engineers do to learn how to make a building more likely to survive an earthquake?

_____

_____

4. Identify Relationships What is the relationship between the strength of an earthquake and how often it occurs?

_____

_____

5. Infer In most cases, you should stay inside a car or a bus in an earthquake. When might it be best to leave a car or a bus during an earthquake?

_____

_____

_____

Copyright © by Holt, Rinehart and Winston; a Division of Houghton Mifflin Harcourt Publishing Company. All rights reserved.

CHAPTER 17 Volcanoes

SECTION 1

Volcanic Eruptions

TN Tennessee Science
Standards
GLE 0707 Inq.2
GLE 0707 Inq.3
GLE 0707 Inq.5

BEFORE YOU READ

After you read this section, you should be able to answer these questions:

• What are two kinds of volcanic eruptions?

• How does the composition of magma affect eruptions?

• What are two ways that magma can erupt from a volcano?

What Is a Volcano?

When you think of a volcano, what comes into your mind? Most people think of a steep mountain with smoke coming out. In fact, a **volcano** is any place where gases and *magma*, or melted rock, come out of the ground. A volcano can be a tall mountain or a small hole in the ground.

THE PARTS OF A VOLCANO

If you could look inside an erupting volcano, it would look similar to the figure below. Below the volcano is a body of magma called a **magma chamber**. The magma from the magma chamber rises to the surface and erupts at the volcano. Magma escapes from the volcano through openings in the Earth's crust called **vents**. When magma flows onto the Earth's surface, it is called *lava*. ☑

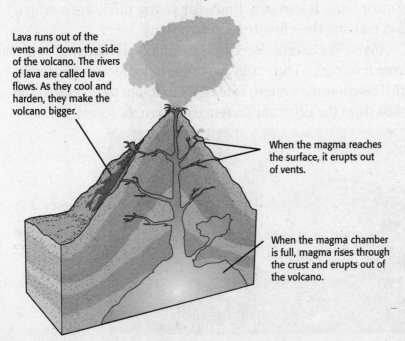

Lava runs out of the vents and down the side of the volcano. The rivers of lava are called lava flows. As they cool and harden, they make the volcano bigger.

When the magma reaches the surface, it erupts out of vents.

When the magma chamber is full, magma rises through the crust and erupts out of the volcano.

STUDY TIP

Compare After you read this section, make a chart that describes the features of each kind of lava and pyroclastic material.

READING CHECK

1. Define What is a magma chamber?

Below the volcano is a body of magma called a magma chamber.

TAKE A LOOK

2. Describe What makes volcanoes grow larger?

it make vacano is the lava

Copyright © by Holt, Rinehart and Winston; a Division of Houghton Mifflin Harcourt Publishing Company. All rights reserved.

SECTION 1 Volcanic Eruptions *continued*

What Happens When Volcanoes Erupt?

Many people think that all volcanic eruptions are alike. However, this is not the case. Scientists put volcanic eruptions into two groups: nonexplosive eruptions and explosive eruptions.

NONEXPLOSIVE ERUPTIONS

Nonexplosive volcanic eruptions are the most common type of eruption. These eruptions produce fairly calm flows of lava. The lava flows over the Earth's surface. Nonexplosive eruptions do not produce very much ash or dust, but they can release huge amounts of lava. For example, most of the rock of the ocean floor was produced by nonexplosive eruptions. ☑

EXPLOSIVE ERUPTIONS

Explosive eruptions are much less common than nonexplosive eruptions. However, explosive eruptions can be more destructive than nonexplosive eruptions. During an explosive eruption, clouds of hot ash, gas, and rock fragments shoot rapidly out of a volcano.

Most explosive eruptions do not produce lava flows. Instead of flowing calmly over the Earth's surface, magma sprays into the air in tiny droplets. The droplets harden to form particles called *ash*. The ash from an explosive eruption can reach the upper parts of the Earth's atmosphere. It can stay there for years, blocking sunlight and causing the climate to get cooler.

An explosive eruption can blast millions of tons of material from a volcano. The explosive eruption of Mount St. Helens in 1980 caused an entire side of a mountain to collapse. The blast from the eruption flattened 600 km² of forest.

READING CHECK

3. Identify What is the most common type of volcanic eruption?

lava rock of the Ocean

Critical Thinking

4. Compare How are nonexplosive eruptions different from explosive eruptions? Give two ways.

Most eruption to do not produce lava flows Instead.

Say It

Investigate Find out more information about the eruptions of Mount St. Helens. Share your findings with a small group.

The eruption of Mount St. Helens in 1980 was an explosive eruption. It was very destructive.

Copyright © by Holt, Rinehart and Winston; a Division of Houghton Mifflin Harcourt Publishing Company. All rights reserved.

Why Do Volcanoes Erupt?

By comparing magma from different eruptions, scientists have been able to figure out why volcanoes erupt in different ways. The main factor affecting an eruption is the composition of the magma. The amounts of water, silica, and gas in the magma determine the type of eruption. ☑

WATER CONTENT

If magma contains a lot of water, an explosive eruption is more likely. Beneath the surface, magma is under high pressure. The high pressure allows water to dissolve into the magma. If the magma rises quickly, the pressure suddenly decreases and the water turns to bubbles of gas. As the gases expand, they cause an explosion.

This is similar to what happens when you shake a can of soda and open it. When you shake the can, the gas dissolved in the soda forms bubbles. Pressure builds up inside the can. When you open the can, the pressure causes the soda to shoot out.

SILICA AND GAS CONTENT

The amount of silica in magma also affects how explosive an eruption is. *Silica* is a compound made of the elements silicon and oxygen. Magma that contains a lot of silica is very thick and stiff. It flows slowly and may harden inside a volcano's vents, blocking them. As more magma pushes up from below, the pressure increases. If enough pressure builds up, the volcano can explode. ☑

Silica-rich magma may be so stiff that water vapor and other gases cannot move out of the magma. Trapped bubbles of gas may expand until they explode. When they explode, the magma shatters and ash is blasted from the vent. Magma with less silica is thinner and runnier. Therefore, gases can move out of the magma easily, and explosive eruptions are less likely.

Material	How it affects eruptions
Water	
Silica	

READING CHECK

5. Identify What is the main factor that determines how a volcano erupts?

with bubble lave

READING CHECK

6. Describe How can magma that contains a lot of silica cause an explosive eruption?

Volcano via block then

TAKE A LOOK

7. Identify Relationships Fill in the blank spaces in the table.

Copyright © by Holt, Rinehart and Winston; a Division of Houghton Mifflin Harcourt Publishing Company. All rights reserved.

How Can Magma Erupt from a Volcano?

There are two main ways that magma can erupt from a volcano: as lava or as pyroclastic material. *Pyroclastic material* is hardened magma that is blasted into the air. Nonexplosive eruptions produce mostly lava. Explosive eruptions produce mostly pyroclastic material. ☑

Most eruptions produce either lava or pyroclastic material, but not both. However, a single volcano may erupt many times. It may produce lava during some eruptions and pyroclastic material during others.

✓ READING CHECK

8. Define What is pyroclastic material?

_____

_____

TYPES OF LAVA

Geologists classify lava by the shapes it forms when it cools. Some kinds of lava form smooth surfaces. Others form sharp, jagged edges as they cool. The figure below shows four kinds of lava flows.

Aa is lava that forms a thick, brittle crust as it cools. The crust is torn into sharp pieces as lava moves underneath it.

Pahoehoe is lava that forms a thin, flexible crust as it cools. The crust wrinkles as the lava moves underneath it.

Blocky lava is cool, stiff lava that does not travel very far from the volcano. Blocky lava usually oozes from a volcano and forms piles of rocks with sharp edges.

Pillow lava is lava that erupts under water. As it cools, it forms rounded lumps that look like pillows.

TAKE A LOOK

9. Compare How are aa and blocky lava similar?

_____

_____

TYPES OF PYROCLASTIC MATERIAL

Pyroclastic material forms when magma explodes from a volcano. The magma solidifies in the air. Pyroclastic material also forms when powerful eruptions shatter existing rock.

Geologists classify pyroclastic material by the size of its pieces. Pieces of pyroclastic material can be the size of houses or as small as dust particles. The figure on the top of the next page shows four kinds of pyroclastic materials.

Copyright © by Holt, Rinehart and Winston; a Division of Houghton Mifflin Harcourt Publishing Company. All rights reserved.

SECTION 1 Volcanic Eruptions *continued*

Volcanic bombs are large blobs of lava that harden in the air.

Lapilli are small bits of lava that harden before they hit the ground. Lapilli are usually about the size of pebbles."

Volcanic ash forms when gases trapped in magma or lava form bubbles. When the bubbles explode, they create millions of tiny pieces.

Volcanic blocks are large pieces of solid rock that come out of a volcano.

TAKE A LOOK
10. Describe How do lapilli form?

_____

_____

_____

PYROCLASTIC FLOWS

A *pyroclastic flow* is a dangerous type of volcanic flow. Pyroclastic flows form when ash and dust race down the side of a volcano like a river. Pyroclastic flows are very dangerous. They can be as hot as 700°C and can move at 200 km/h. A pyroclastic flow can bury or destroy everything in its path. A pyroclastic flow from the eruption of Mount Pinatubo is shown in the figure below.

Math Focus
11. Convert How fast can pyroclastic flows move? Give your answer in miles per hour.
1 km = 0.62 mi

_____

This pyroclastic flow formed during the 1991 eruption of Mount Pinatubo, in the Philippines.

Copyright © by Holt, Rinehart and Winston; a Division of Houghton Mifflin Harcourt Publishing Company. All rights reserved.

Section 1 Review

GLE 0707.Inq.2, GLE 0707.Inq.3, GLE 0707.Inq.5

SECTION VOCABULARY

magma chamber the body of molten rock that feeds a volcano **vent** an opening at the surface of the Earth through which volcanic material passes	**volcano** a vent or fissure in the Earth's surface through which magma and gases are expelled

1. Compare How is lava different from magma?

_____

2. Identify What are the two kinds of volcanic eruptions?

_____

_____

3. Explain How does the amount of water in magma affect how a volcano erupts?

_____

4. Explain Why is magma that contains little silica less likely to erupt explosively?

_____

5. Compare How is pahoehoe lava different from pillow lava? How are they similar?

_____

_____

_____

6. Describe How do volcanic bombs form?

_____

_____

7. Describe How does volcanic ash form?

_____

_____

8. Define What is a pyroclastic flow?

_____

_____

9. Infer Do pyroclastic flows form during explosive or nonexplosive eruptions?

_____

Copyright © by Holt, Rinehart and Winston; a Division of Houghton Mifflin Harcourt Publishing Company. All rights reserved.

CHAPTER 17 Volcanoes

SECTION 2 Effects of Volcanic Eruptions

TN Tennessee Science Standards
GLE 0707Inq.3
GLE 0707Inq.5

BEFORE YOU READ

After you read this section, you should be able to answer these questions:

• How can volcanoes affect climate?

• What are three kinds of volcanoes?

• What are three structures that volcanic eruptions can form?

How Can Volcanoes Affect Climate?

In 1815, a huge volcanic explosion happened on Mount Tambora in Indonesia. Historians estimate that the explosion killed 12,000 people. As many as 80,000 people died from hunger and disease following the explosion. However, the explosion did not affect only the people living in Indonesia. It also affected the climate worldwide.

Ash and dust from the explosion flew into the upper atmosphere. There, they spread across the Earth. They blocked sunlight from reaching the Earth's surface. As a result, global temperatures dropped. In 1816, there was a snowstorm in June! The colder temperatures caused food shortages in North America and Europe.

STUDY TIP

Compare After you read this section, make a chart comparing the three kinds of volcanoes. Describe how each type of volcano forms and what it looks like.

TAKE A LOOK

1. Identify What causes global temperatures to drop after a large explosive eruption?

_____

_____

_____

A large explosive eruption	→ produces →	a lot of ash and dust

which ↓

Global cooling	← causing ←	block sunlight from reaching Earth's surface

In 1991, an explosive eruption on Mount Pinatubo caused global temperatures to drop. Explosive eruptions may cause global temperatures to decrease by 0.5°C to 1°C. This may seem like a small change, but even small temperature changes can disrupt world climates.

How Can Volcanoes Affect the Earth's Surface?

In addition to affecting climate, volcanoes can have important effects on the Earth's surface. Volcanoes produce many unique *landforms*, or surface features.

The most well-known volcanic landforms are the volcanoes themselves. There are three main kinds of volcanoes: shield volcanoes, cinder cone volcanoes, and composite volcanoes.

Copyright © by Holt, Rinehart and Winston; a Division of Houghton Mifflin Harcourt Publishing Company. All rights reserved.

SECTION 2 Effects of Volcanic Eruptions *continued*

TENNESSEE STANDARDS CHECK

GLE 0707.7.4 Explain how earthquakes, mountain buildings, volcanoes, and seafloor spreading are associated with movements of the earth's <u>major</u> plates.

Word Help: <u>major</u>
of great importance of large scale

2. Describe How do shield volcanoes form?

_____

_____

_____

_____

SHIELD VOLCANOES

Shield volcanoes form when layers of lava from many nonexplosive eruptions build up. The lava that forms shield volcanoes is thin and runny. Therefore, it spreads out in thin layers over a wide area. This produces a volcano with a wide base and gently sloping sides.

Shield volcanoes can be very large. For example, Mauna Kea in Hawaii is a shield volcano. Measured from the base on the ocean floor, Mauna Kea is taller than Mount Everest!

Shield volcanoes form when many layers of lava build up over time.

CINDER CONE VOLCANOES

Cinder cone volcanoes are made of pyroclastic material. The pyroclastic material is produced from explosive eruptions. As it piles up, it forms a mountain with steep slopes. Cinder cones are small. Most of them erupt for only a short time. For example, Paricutín is a cinder cone volcano in Mexico. In 1943, Paricutín appeared in a cornfield. It erupted for only nine years. ☑

Most cinder cone volcanoes are found in clusters. They may be found on the sides of other volcanoes. They erode quickly because the pyroclastic material is loose and not stuck together.

☑ READING CHECK

3. Identify What are cinder cone volcanoes made of?

_____

_____

TAKE A LOOK

4. Identify Which type of volcanic eruption produces cinder cone volcanoes?

_____

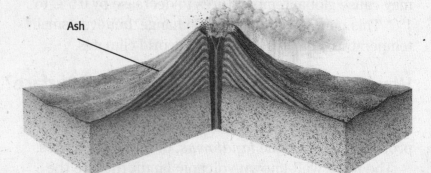

Cinder cone volcanoes form when ash from explosive eruptions piles up. Most cinder cones are small.

Copyright © by Holt, Rinehart and Winston; a Division of Houghton Mifflin Harcourt Publishing Company. All rights reserved.

COMPOSITE VOLCANOES

Composite volcanoes are the most common type of volcano. They form when a volcano erupts both explosively and nonexplosively. They have layers of lava flows and pyroclastic material. They usually have a broad base and sides that get steeper toward the top. Mount St. Helens is a composite volcano.

Critical Thinking

5. Infer The word stratum means "layer." Why are composite volcanoes sometimes also called stratovolcanoes?

_____

_____

Lava flows

Ash layers

Composite volcanoes form from layers of ash and lava. Most have steep sides.

TAKE A LOOK
6. Identify What two materials are composite volcanoes made of?

_____

_____

What Are Other Types of Volcanic Landforms?

In addition to volcanoes, other landforms are created by volcanic activity. The landforms include craters, calderas, and lava plateaus.

CRATERS

A **crater** is a funnel-shaped pit around the central vent at the top of a volcano. Lava and pyroclastic material can pile up around the vent. This produces a crater in the middle of the cone. ☑

CALDERAS

A **caldera** is a large *depression*, or pit, that forms when a magma chamber collapses. The ground over the magma chamber sinks, forming a caldera. Calderas can look similar to craters, but calderas are much larger.

READING CHECK
7. Define What is a crater?

_____

_____

_____

LAVA PLATEAUS

A **lava plateau** is a large area of land covered by a huge volume of lava. Lava plateaus are the largest volcanic landforms. They do not form at tall volcanoes. Instead, lava plateaus form when a large volume of lava erupts from a crack in the crust. Most of the lava on the Earth's surface is found in lava plateaus.

Copyright © by Holt, Rinehart and Winston; a Division of Houghton Mifflin Harcourt Publishing Company. All rights reserved.

Section 2 Review

GLE 0707.Inq.3, GLE 0707.Inq.5 **TN**

SECTION VOCABULARY

caldera a large, semicircular depression that forms when the magma chamber below a volcano partially empties and causes the ground above to sink	**crater** a funnel-shaped pit near the top of the central vent of a volcano
	lava plateau a wide, flat landform that results from repeated nonexplosive eruptions of lava that spread over a large area

1. Compare Explain how a crater is different from a caldera.

_____

_____

2. Describe How can volcanoes affect climate?

_____

_____

_____

3. Identify What are the three main types of volcanoes?

_____

_____

_____

4. Explain Why do shield volcanoes have wide bases?

_____

_____

5. Explain Why do cinder cone volcanoes erode quickly?

_____

6. Identify What is the largest kind of volcanic landform?

_____

7. Apply Concepts Does the lava that forms shield volcanoes probably have a lot of silica or water in it? Explain your answer.

_____

_____

_____

Copyright © by Holt, Rinehart and Winston; a Division of Houghton Mifflin Harcourt Publishing Company. All rights reserved.

CHAPTER 17 | Volcanoes
SECTION 3 Causes of Volcanic Eruptions

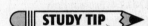 **Tennessee Science Standards**
GLE 0707.Inq.2
GLE 0707.Inq.3
GLE 0707.Inq.5
GLE 0707.7.3
GLE 0707.7.4

BEFORE YOU READ

After you read this section, you should be able to answer these questions:

- How does magma form?
- Where do volcanoes form?
- How can geologists predict volcanic eruptions?

How Does Magma Form?

Magma forms deep in the Earth's crust and in the upper parts of the mantle. In these areas, the temperature and pressure are very high. Changes in pressure and temperature can cause magma to form.

Part of the upper mantle is made of very hot, solid rock. The rock is so hot that it can flow, like soft chewing gum, even though it is solid. If rock of this temperature were found at the Earth's surface, it would be *molten*, or melted. The rock in the mantle does not melt because it is under high pressure. This pressure is produced by the weight of the rock above the mantle. ☑

In the figure below, the curved line shows the melting point of a rock. The *melting point* is the temperature at which the rock melts for a certain pressure.

STUDY TIP

Describe After you read this section, make flowcharts showing how magma forms at divergent boundaries and at convergent boundaries.

✓ READING CHECK

1. Explain Why doesn't the hot rock in the mantle melt?

_____

_____

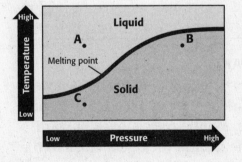

The curved line shows the melting point of the rock. Rock with the temperature and pressure of point A is liquid. Rock at the same temperature but higher pressure (B) is solid. Rock at the same pressure but lower temperature (C) is also solid.

Math Focus
2. Describe A rock starts out at point C. Then, its temperature increases. What will happen to the rock if its temperature continues to rise?

MAGMA FORMATION IN THE MANTLE

The temperature of the mantle is fairly constant. Magma usually forms because of a decrease in pressure. Therefore, a lot of magma forms at the boundary between separating tectonic plates, where pressure decreases. Magma is less dense than the solid rock it forms from. Therefore, it rises toward the surface and erupts.

Copyright © by Holt, Rinehart and Winston; a Division of Houghton Mifflin Harcourt Publishing Company. All rights reserved.

SECTION 3 Causes of Volcanic Eruptions *continued*

Where Do Volcanoes Form?

The locations of volcanoes give clues about how volcanoes form. The figure below shows the locations of some of the world's major active volcanoes. The map also shows the boundaries between tectonic plates. Most volcanoes are found at tectonic plate boundaries. For example, there are many volcanoes on the plate boundaries surrounding the Pacific Ocean. Therefore, the area is sometimes called the *Ring of Fire*.

Remember that tectonic plate boundaries are areas where plates collide, separate, or slide past one another. Most volcanoes are found where plates move together or apart. About 15% of active volcanoes on land form where plates separate, and about 80% form where plates collide. The remaining few volcanoes on land are found far from tectonic plate boundaries.

Volcanoes and Tectonic Plate Boundaries

- - - Plate boundary
• Active volcano

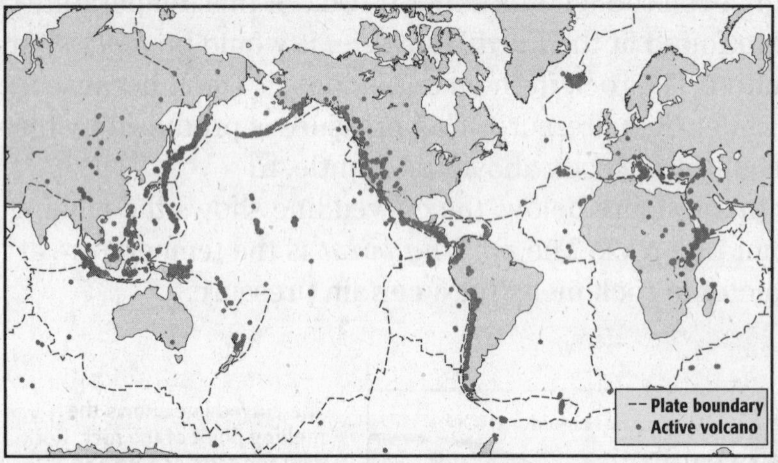

WHERE PLATES MOVE APART

At a *divergent boundary*, tectonic plates move away from each other. A set of deep cracks called a **rift zone** forms between the plates. Mantle rock moves upward to fill in the gap. When the mantle rock gets close to the surface, the pressure decreases. The decrease in pressure causes the mantle rock to melt, forming magma. The magma rises through the rift zones and erupts. ☑

Most divergent boundaries are on the ocean floor. Lava that flows from undersea rift zones produces volcanoes and mountain chains. These volcanoes and mountain chains are called *mid-ocean ridges*. The mid-ocean ridges circle the ocean floor.

TN TENNESSEE
STANDARDS CHECK

GLE 0707.7.4 Explain how earthquakes, mountain building, volcanoes, and seafloor spreading are associated with movements of the earth's <u>major</u> plates.

Word Help: <u>major</u>
of great importance or large scale

3. Describe Where are most volcanoes found?

_____

_____

✓ **READING CHECK**

4. Identify What causes magma to melt at divergent boundaries?

_____

_____

Copyright © by Holt, Rinehart and Winston; a Division of Houghton Mifflin Harcourt Publishing Company. All rights reserved.

SECTION 3 Causes of Volcanic Eruptions *continued*

How Magma Forms at a Divergent Boundary

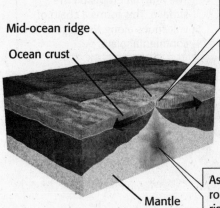

Mid-ocean ridge

Ocean crust

Mantle

At a divergent boundary, plates move apart. When the plates move apart, the pressure on the mantle below them decreases.

As the pressure decreases, mantle rock starts to melt. The magma rises toward the surface.

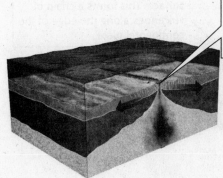

When the magma reaches the surface, it erupts onto the sea floor. When it cools and hardens, new ocean crust forms.

TAKE A LOOK
5. Explain How does new ocean crust form?

_____

_____

_____

WHERE PLATES MOVE TOGETHER

At a *convergent boundary*, the tectonic plates collide. When an oceanic plate collides with a continental plate, the oceanic plate slides under the continental plate. This is called *subduction*. The oceanic crust sinks into the mantle because it is more dense than the continental crust. ☑

As the ocean crust sinks, the temperature and pressure on it increase. Because the ocean crust forms below the ocean, the rock contains a lot of water. The heat and pressure on the ocean crust cause this water to be released.

The water mixes with the mantle rock above the oceanic plate. When the mantle rock mixes with water, it can melt at a lower temperature. The mantle rock begins to melt at the subduction zone. The magma rises to the surface and erupts as a volcano.

READING CHECK

6. Explain Why does oceanic crust sink below continental crust?

_____

_____

Copyright © by Holt, Rinehart and Winston; a Division of Houghton Mifflin Harcourt Publishing Company. All rights reserved.

SECTION 3 Causes of Volcanic Eruptions *continued*

How Magma Forms at a Convergent Boundary

The magma erupts on the surface. This forms a chain of volcanoes along the edge of the continental plate.

The magma erupts on the surface. This forms a chain of volcanoes along the edge of the continental plate.

TAKE A LOOK
7. Explain How does sub-duction produce magma?

_____

_____

_____

_____

 **READING CHECK**

8. Define What is a hot spot?

_____

_____

_____

IN THE MIDDLE OF PLATES

Although most volcanoes form at plate boundaries, not all volcanoes form there. Some volcanoes, such as the Hawaiian Islands, form at hot spots. **Hot spots** are places on the Earth's surface where volcanoes form far from plate boundaries. Most scientists think that hot spots form above hot columns of mantle rock called *mantle plumes*. Some scientists think that hot spots form where magma rises through cracks in the Earth's crust. ☑

Long chains of volcanoes are common at hot spots. One theory to explain this is that a mantle plume stays in one place while the plate moves over it. Another theory states that hot-spot volcanoes occur in long chains because they form along cracks in the Earth's crust. Scientists are not sure which of these theories is correct. It is possible that some hot spots form over plumes, but others form over cracks.

Copyright © by Holt, Rinehart and Winston; a Division of Houghton Mifflin Harcourt Publishing Company. All rights reserved.

How Can We Predict Volcanic Eruptions?

Scientists cannot always predict when and where a volcano will erupt. However, by studying ancient and modern volcanoes, scientists have been able to identify some signs that an eruption may happen.

One feature that scientists use to predict whether an eruption will happen is the state of the volcano. Geologists put volcanoes into three groups based on how active they are.

- *Extinct* volcanoes have not erupted in recorded history and probably will not erupt again.
- *Dormant* volcanoes are currently not erupting, but they may erupt again.
- *Active* volcanoes are currently erupting or show signs of erupting in the near future.

Critical Thinking

9. Compare What is the difference between dormant volcanoes and extinct volcanoes?

_____

_____

_____

SMALL QUAKES AND VOLCANIC GASES

Most active volcanoes produce small earthquakes as the magma within them moves upward. This happens because the magma pushes on the rocks as it rises. In many cases, the number and strength of these earthquakes increases before a volcanic eruption. Therefore, monitoring earthquakes is one of the best ways to predict an eruption.

Scientists also study the volume and composition of gases given off by the volcano. Just before an eruption, many volcanoes give off more gas. The composition of the gas may also change before an eruption. By monitoring the gases, scientists can predict when an eruption may happen.

Critical Thinking

10. Infer Why may a volcano that is about to erupt give off more gas?
(Hint: Why are some eruptions explosive?)

_____

_____

_____

_____

MEASURING SLOPE AND TEMPERATURE

As magma rises before an eruption, it can cause the Earth's surface to swell. The side of a volcano may even bulge. Scientists can use an instrument called a *tiltmeter* to measure the slope of the volcano's sides. Changes in the slope can indicate that an eruption is likely. ☑

One of the newest methods for predicting volcanic eruptions involves using satellite images. Satellites can record the surface temperatures at and around volcanoes. As magma rises, the surface temperature of the volcano may increase. Therefore, an increase in surface temperature can indicate that an eruption is likely.

✓ READING CHECK

11. Explain Why may the Earth's surface swell before an eruption?

_____

_____

Copyright © by Holt, Rinehart and Winston; a Division of Houghton Mifflin Harcourt Publishing Company. All rights reserved.

Section 3 Review

GLE 0707.Inq.2, GLE 0707.Inq.3, GLE 0707.Inq.5, TN
GLE 0707.7.3, GLE 0707.7.4

SECTION VOCABULARY

hot spot a volcanically active area of Earth's surface far from a tectonic plate boundary	**rift zone** an area of deep cracks that forms between two tectonic plates that are pulling away from each other

1. Identify Where do rift zones form?

_____

2. Apply Concepts The map below shows the locations of many volcanoes. On the map, circle three volcanoes that are probably found at hot spots.

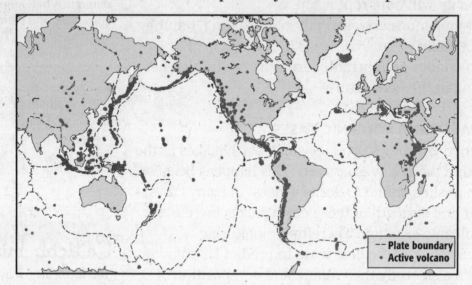

3. Identify What is the most common cause of magma formation in the mantle?

_____

4. Describe Why is sea floor spreading associated with Earth's plates?

_____

_____

_____

5. List Give four signs that a volcanic eruption is likely.

_____

_____

_____

Copyright © by Holt, Rinehart and Winston; a Division of Houghton Mifflin Harcourt Publishing Company. All rights reserved.

Name _____ Class _____ Date _____

TN Tennessee Science Standards
GLE 0707.Inq.5
GLE 0707.7.5

BEFORE YOU READ

After you read this section, you should be able to answer these questions:

• What is the difference between a renewable resource and a nonrenewable resource?

• How can you protect natural resources?

What Are Earth's Resources?

Earth provides what you need to survive. You breathe air from Earth's atmosphere. You drink water from Earth's rivers, lakes, and other water bodies. You eat food from Earth's living things.

A **natural resource** is any material from Earth that is used by people. Air, soil, fresh water, petroleum, rocks, minerals, forests, and wildlife are examples of natural resources. People use some natural resources, such as coal and wind, for energy. The energy in these resources comes from energy from the sun. The figure below shows some examples of natural resources. ☑

STUDY TIP

Summarize After you read this section, make a chart giving the definitions of renewable and nonrenewable resources. In the chart, include two examples of each kind of resource.

READING CHECK

1. Define In your own words, write a definition of *natural resource*.

_____

_____

_____

Examples of Natural Resources

Gasoline and plastic are both made from oil that is pumped out of Earth's crust.

These windmills produce electricity from the energy in wind. The energy in wind comes mainly from the sun.

This pile of lumber is made of wood, which comes from trees.

TAKE A LOOK

2. Identify Give two examples of natural resources that are not shown in the figure.

_____

_____

Copyright © by Holt, Rinehart and Winston; a Division of Houghton Mifflin Harcourt Publishing Company. All rights reserved.

What Types of Resources Exist on Earth?

Natural resources are grouped based on how fast they can be replaced. Some natural resources are nonrenewable. Others are renewable.

NONRENEWABLE RESOURCES

Some resources, such as coal, petroleum, and natural gas, take millions of years to form. A **nonrenewable resource** is a resource that is used much faster than it can be replaced. *Renew* means "to begin again." When nonrenewable resources are used up, people can no longer use them. ☑

RENEWABLE RESOURCES

Some natural resources, such as trees and fresh water, can grow or be replaced quickly. A **renewable resource** is a natural resource that can be replaced as quickly as people use it.

Many renewable resources are renewable only if people do not use them too quickly. For example, wood is usually considered a renewable resource. However, if people cut down trees faster than the trees can grow back, wood is no longer a renewable resource. Some renewable resources, such as the sun, will never be used up, no matter how fast people use them.

READING CHECK

3. Define What is a nonrenewable resource?

_____

_____

TAKE A LOOK

4. Explain Describe how some renewable resources can become nonrenewable resources.

_____

_____

_____

_____

Fresh water and trees are both renewable resources. However, they can be used up if people use them too quickly.

Copyright © by Holt, Rinehart and Winston; a Division of Houghton Mifflin Harcourt Publishing Company. All rights reserved.

How Can We Protect Natural Resources?

Whether the natural resources you use are renewable or nonrenewable, you should be careful how you use them. In order to *conserve* natural resources, you should try to use them only when you have to. For example, leaving the water running while you are brushing your teeth wastes clean water. Turning the water off while you brush your teeth saves water so that it can be used in the future.

The energy we use to heat our homes, drive our cars, and run our computers comes from natural resources. Most of these resources are nonrenewable. If we use too much energy now, we might use up these resources. Therefore, reducing the amount of energy you use can help to conserve natural resources. You can conserve energy by being careful to use it only when you need to. The table shows some ways you can conserve energy.

Instead of...	You can...
...leaving the lights on all the time	...turn them off when you're not in the room
...running the washing machine when it is only half full	...run it only when it is full
...using a car to travel everywhere	...walk, ride a bike, or use public transportation when you can

Recycling is another important way that you can help to conserve natural resources. **Recycling** means using things that have been thrown away to make new objects. Objects made from recycled materials use fewer natural resources than objects made from new materials. Recycling also helps to conserve energy. For example, it takes less energy to recycle an aluminum can than to make a new one. ☑

Newspaper, aluminum cans, some plastic containers, and many types of paper can be recycled. Check with your community's recycling center to see what kinds of materials you can recycle.

Conserving resources also means taking care of them even when you are not using them. For example, it is important to keep our drinking water clean. Polluted water can harm the living things, including humans, that need water in order to live.

Critical Thinking

5. Explain Why is it important to conserve all natural resources, even if they are renewable resources?

_____

_____

_____

_____

_____

_____

TAKE A LOOK
6. Brainstorm Fill in the blank spaces in the table with some other ways you can conserve natural resources.

✓ **READING CHECK**
7. Identify How does recycling conserve natural resources?

_____

_____

_____

_____

Copyright © by Holt, Rinehart and Winston; a Division of Houghton Mifflin Harcourt Publishing Company. All rights reserved.

Section 1 Review

GLE 0707.Inq.5, GLE 0707.7.5

SECTION VOCABULARY

natural resource any natural material that is used by humans, such as water, petroleum, minerals, forests, and animals	**recycling** the process of recovering valuable or useful materials from waste or scrap; the process of reusing some things
nonrenewable resource a resource that forms at a rate that is much slower than the rate at which the resource is consumed	**renewable resource** a natural resource that can be replaced at the same rate at which the resource is consumed

1. Identify What is the difference between a renewable resource and a nonrenewable resource?

_____

_____

_____

2. List Give four ways to conserve natural resources.

_____

_____

_____

_____

3. Explain Why is wood usually considered a renewable resource? When would it be considered a nonrenewable resource?

_____

_____

_____

4. Describe What does it mean to conserve natural resources?

_____

_____

_____

5. Explain Why are coal, oil, and natural gas considered nonrenewable resources, even though they come from living things that can reproduce?

_____

_____

Copyright © by Holt, Rinehart and Winston; a Division of Houghton Mifflin Harcourt Publishing Company. All rights reserved.

CHAPTER 18 Energy Resources

SECTION 2 Fossil Fuels

TN Tennessee Science Standards
GLE 0707.Inq.5
GLE 0707.T/E.1
GLE 0707.7.5
GLE 0707.7.6

BEFORE YOU READ

After you read this section, you should be able to answer these questions:

- What are the different kinds of fossil fuels?
- How do fossil fuels form?
- What are the problems with using fossil fuels?

What Are Fossil Fuels?

How do plants and animals that lived hundreds of millions of years ago affect your life today? Plants and animals that lived long ago provide much of the energy we use. If you turned on the lights or traveled to school in a car or bus, you probably used some of this energy.

Energy resources are natural resources that people use to produce energy, such as heat and electricity. Most of the energy we use comes from fossil fuels. A **fossil fuel** is an energy resource made from the remains of plants and tiny animals that lived long ago. The different kinds of fossil fuels are petroleum, coal, and natural gas. ☑

Fossil fuels are an important part of our everyday life. When fossil fuels burn, they release a lot of energy. Power plants use the energy to produce electricity. Cars use the energy to move.

However, there are also some problems with using fossil fuels. Fossil fuels are nonrenewable, which means that they cannot be replaced once they have been used. Also, when they burn, they release pollution.

STUDY TIP

Compare In your notebook, make a table to show the similarities and differences between different kinds of fossil fuels.

READING CHECK

1. Identify Where do we get most of the energy we use?

_____

_____

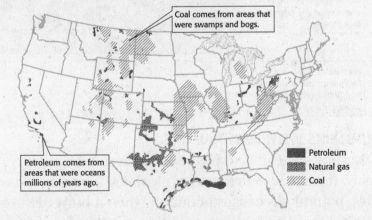

Coal comes from areas that were swamps and bogs.

Petroleum comes from areas that were oceans millions of years ago.

▓ Petroleum
▒ Natural gas
╱ Coal

TAKE A LOOK

2. Describe In general, where are the natural gas deposits in the United States?

_____

_____

Copyright © by Holt, Rinehart and Winston; a Division of Houghton Mifflin Harcourt Publishing Company. All rights reserved.

SECTION 2 Fossil Fuels *continued*

Say It

Discuss Have you ever seen or used methane, propane, or butane? In a small group, talk about what these different gases are used for.

☑ **READING CHECK**

3. Identify What is most natural gas used for?

_____

_____

TN TENNESSEE STANDARDS CHECK

GLE 0707.7.5 Differentiate between renewable and nonrenewable resources in terms of their use by man.

Word Help: resources anything that can be used to take care of a need

4. Explain Is natural gas a renewable or nonrenewable resource? Explain why.

_____

_____

_____

_____

_____

_____

What Is Natural Gas?

A *hydrocarbon* is a compound that contains the elements carbon, hydrogen, and oxygen. **Natural gas** is a mixture of hydrocarbons that are in the form of gases. Natural gas includes methane, propane, and butane, which can be separated from one another.

Most natural gas is used for heating. Your home may be heated by natural gas. Your kitchen stove may run on natural gas. Some natural gas is used for creating electrical energy, and some cars are able to run on natural gas, too. ☑

HOW NATURAL GAS FORMS

When tiny sea creatures die, their remains settle to the ocean floor and are buried in sediment. The sediment slowly becomes rock. Over millions of years, heat and pressure under the ground chemically change the remains. The carbon, hydrogen, and oxygen in them can become natural gas.

Natural gas is always forming. Some of the sea life that dies today will become natural gas millions of years from now.

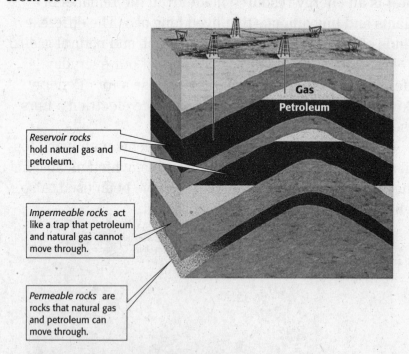

Reservoir rocks hold natural gas and petroleum.

Impermeable rocks act like a trap that petroleum and natural gas cannot move through.

Permeable rocks are rocks that natural gas and petroleum can move through.

PROBLEMS WITH NATURAL GAS

Natural gas can be dangerous because it burns very easily. It can cause fires and explosions. Like other fossil fuels, natural gas causes pollution when it burns. However, it does not cause as much pollution as other fossil fuels.

Copyright © by Holt, Rinehart and Winston; a Division of Houghton Mifflin Harcourt Publishing Company. All rights reserved.

SECTION 2 Fossil Fuels *continued*

What Is Petroleum?

Petroleum is a mixture of hydrocarbons that are in the form of liquids. It is also known as crude oil. At a *refinery*, petroleum is separated into many different products, including gasoline, jet fuel, kerosene, diesel fuel, and fuel oil. ☑

Petroleum products provide more than 40% of the world's energy, including fuel for airplanes, trains, boats, ships, and cars. Petroleum is so valuable that it is often called "black gold."

HOW PETROLEUM FORMS

Petroleum forms the same way natural gas does. Tiny sea creatures die and then get buried in sediment, which turns into rock. Some of their remains become petroleum, which is stored in permeable rock within Earth's crust.

PROBLEMS WITH PETROLEUM

Petroleum can be harmful to animals and their environment. For example, in June 2000, the carrier ship *Treasure* sank off the coast of South Africa. The ship spilled more than 400 tons of oil into the ocean. The oil covered penguins and other sea creatures, making it hard for them to swim, breathe, and eat.

Burning petroleum causes smog. **Smog** is a brownish haze that forms when sunlight reacts with pollution in the air. Smog can make it hard for people to breathe. Many cities in the world have problems with smog.

✓ **READING CHECK**

5. Describe What form is petroleum found in naturally?

_____

_____

Math Focus

6. Make a Graph Use the information below to show where the world's crude oil comes from.

Middle East: 66%

North America and South America: 15%

Europe and Asia: 12%

Africa: 7%

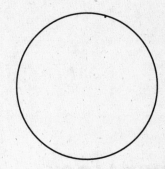

Copyright © by Holt, Rinehart and Winston; a Division of Houghton Mifflin Harcourt Publishing Company. All rights reserved.

What Is Coal?

Coal is a solid fossil fuel that is made of partly decayed plant material. Coal was once the main source of energy in the United States. Like other fossil fuels, coal releases heat when it is burned. Many people used to burn coal in stoves to heat their homes. Trains in the 1800s and 1900s were powered by coal-burning steam locomotives. Coal is now used in power plants to make electricity.

HOW COAL FORMS

When swamp plants die, they sink to the bottom of the swamp. If they do not decay completely, coal can start to form. Coal forms in several different stages.

TAKE A LOOK
7. Identify Name the three types of coal.

_____

_____

_____

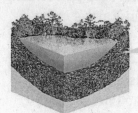

Stage 1: Formation of Peat
Dead swamp plants that have not decayed can turn into *peat*, a crumbly brown material made mostly of plant material and water. Dried peat is about 60% carbon. In some parts of the world, peat is dried and burned for fuel. Peat is not coal, but it can turn into coal.

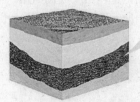

Stage 2: Formation of Lignite
If sediment buries the peat, pressure and temperature increase. The peat slowly changes into a type of coal called *lignite*. Lignite coal is harder than peat. Lignite is about 70% carbon.

Stage 3: Formation of Bituminous Coal
If more sediment is added, pressure and temperature force more water and gases out of the lignite. Lignite slowly changes into *bituminous* coal. Bituminous coal is about 80% carbon.

TAKE A LOOK
8. Identify Which kind of coal contains the most carbon?

_____

Stage 4: Formation of Anthracite
If more sediment accumulates, temperature and pressure continue to increase. Bituminous coal slowly changes into *anthracite*. Anthracite coal is the hardest type of coal. Anthracite coal is about 90% carbon.

Copyright © by Holt, Rinehart and Winston; a Division of Houghton Mifflin Harcourt Publishing Company. All rights reserved.

PROBLEMS WITH COAL

Mining coal can create environmental problems. When coal is mined from Earth's surface, people remove the layers of soil above the coal. This can harm the plants that need soil to grow and the animals that need soil for shelter. If the land is not restored after mining, wildlife habitats can be destroyed for years.

Coal that is on Earth's surface can cause pollution. Water that flows through coal can pick up poisonous metals. That water can then flow into streams and lakes and pollute water supplies.

When coal is burned without pollution controls, a gas called sulfur dioxide is released. Sulfur dioxide can combine with the water in the air to produce sulfuric acid. Sulfuric acid is one of the acids in acid precipitation. **Acid precipitation** is rain, sleet, or snow that contains a lot of acids, often because of air pollutants. Acid precipitation is also called "acid rain." Acid precipitation can harm wildlife, plants, and buildings.

In 1935, this statue had not been damaged by acid precipitation.

By 1994, acid precipitation had caused serious damage to the statue.

Critical Thinking

9. Infer What do you think is the reason that fossil fuels are still used today, even though they create many environmental problems?

_____

_____

_____

_____

_____

TAKE A LOOK

10. Define What is acid precipitation?

_____

_____

How Do We Obtain Fossil Fuels?

People remove fossil fuels from the Earth in different ways. The way that a fossil fuel is removed depends on the kind of fuel and where it is located. Remember that people remove coal from the Earth by mining it. People remove petroleum and natural gas by drilling into the rocks that contain the fuels. Then, the petroleum or natural gas is removed through a well. These wells may be on land or in the ocean.

Copyright © by Holt, Rinehart and Winston; a Division of Houghton Mifflin Harcourt Publishing Company. All rights reserved.

Section 2 Review

GLE 0707.Inq.5, GLE 0707.T/E.1, GLE 0707.7.5, GLE 0707.7.6 **TN**

SECTION VOCABULARY

acid precipitation precipitation such as rain, sleet, or snow, that contains a high concentration of acids, often because of pollution in the atmosphere

coal a fossil fuel that forms underground from partially decomposed plant material

fossil fuel a nonrenewable energy resource formed from the remains of organisms that lived long ago; examples include oil, coal, and natural gas

natural gas a mixture of gaseous hydrocarbons located under the surface of Earth, often near petroleum deposits; used as a fuel

petroleum a liquid mixture of complex hydrocarbon compounds; used widely as a fuel source

smog photochemical haze that forms when sunlight acts on industrial pollutants and burning fuels

1. Compare How are petroleum and natural gas different?

_____

_____

_____

2. Compare Fill in the table to compare the different kinds of fossil fuels.

Kind of fossil fuel	What it is	How it forms
Coal		
	a mixture of gases containing carbon, hydrogen, and oxygen	

3. Summarize What are some of the problems with using fossil fuels for energy?

_____

_____

_____

_____

_____

Copyright © by Holt, Rinehart and Winston; a Division of Houghton Mifflin Harcourt Publishing Company. All rights reserved.

CHAPTER 18 Energy Resources

SECTION 3 Alternative Resources

TN Tennessee Science Standards
GLE 0707.Inq.5
GLE 0707.7.5
GLE 0707.7.6

BEFORE YOU READ

After you read this section, you should be able to answer these questions:

• What are some kinds of alternative energy?

• What are some benefits of alternative energy?

• What are some problems with alternative energy?

What Is Alternative Energy?

What would your life be like if you couldn't turn on the lights, microwave your dinner, take a hot shower, or ride the bus to school? We get most of the energy we use for heating and electricity from fossil fuels. However, fossil fuels can be harmful to the environment and to living things. In addition, they are nonrenewable resources, so we cannot replace them when they are used up.

Many scientists are trying to find alternative energy sources. *Alternative energy sources* are sources of energy that are not fossil fuels. Some sources can be converted easily into usable energy. Others are not as easy to use.

What Is Nuclear Energy?

One kind of alternative energy source is nuclear energy. **Nuclear energy** is the energy that is released when atoms come together or break apart. Nuclear energy can be obtained in two main ways: fission and fusion. ☑

FISSION

Fission happens when an atom splits into two or more lighter atoms. Fission releases a large amount of energy. This energy can be used to generate electricity. All nuclear power that people use today is generated by fission.

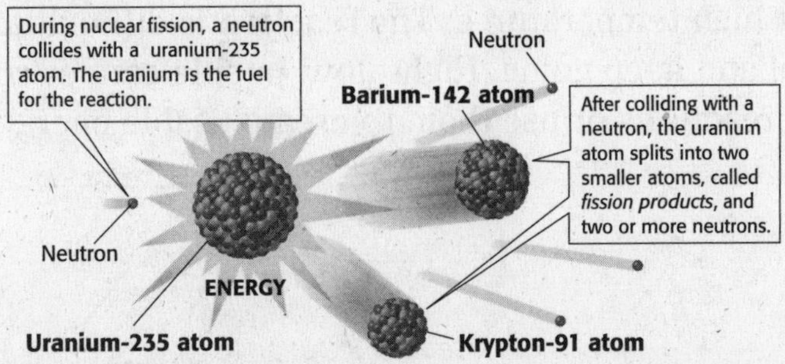

STUDY TIP

Compare and Contrast In your notebook, make a chart to show each kind of alternative energy source and its benefits and problems.

READING CHECK

1. List What are the two ways in which nuclear energy is produced?

_____

_____

_____

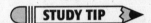

During nuclear fission, a neutron collides with a uranium-235 atom. The uranium is the fuel for the reaction.

Neutron

Barium-142 atom

After colliding with a neutron, the uranium atom splits into two smaller atoms, called *fission products*, and two or more neutrons.

Neutron

ENERGY

Uranium-235 atom

Krypton-91 atom

TAKE A LOOK

2. Identify What are the fission products in the figure?

_____

_____

_____

Copyright © by Holt, Rinehart and Winston; a Division of Houghton Mifflin Harcourt Publishing Company. All rights reserved.

FISSION'S BENEFITS AND PROBLEMS

One benefit of fission is that it does not cause air pollution. Mining uranium, the fuel for nuclear power, is less harmful to the environment than mining other energy sources, such as coal. ☑

However, nuclear fission power has several problems. The fission products created in nuclear power plants are poisonous. They must be stored for thousands of years. Nuclear fission plants can release harmful radiation into the environment. Also, nuclear power plants must release extra heat from the fission reaction. This extra heat cannot be used to make electricity. The extra heat can harm the environment.

FUSION

Fusion happens when two or more atoms join to form a heavier atom. This process occurs naturally in the sun. Fusion releases a lot of energy.

READING CHECK

3. Explain Why is nuclear energy called a "clean" energy source?

_____

_____

_____

TAKE A LOOK

4. Identify How many protons and how many neutrons are there in the helium-4 nucleus?

_____

_____

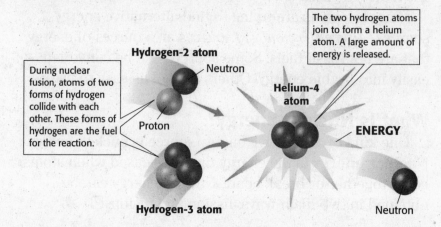

During nuclear fusion, atoms of two forms of hydrogen collide with each other. These forms of hydrogen are the fuel for the reaction.

Hydrogen-2 atom
Neutron
Proton

Hydrogen-3 atom

Helium-4 atom

The two hydrogen atoms join to form a helium atom. A large amount of energy is released.

ENERGY

Neutron

FUSION'S BENEFITS AND PROBLEMS

Fusion has two main benefits. First, fusion does not create a lot of dangerous wastes. Second, the fuels used in fusion are renewable.

The main problem with fusion is that it can take place only at high temperatures. The reaction is difficult to control and keep going. Right now, people cannot control fusion reactions or use them to create usable energy. ☑

READING CHECK

5. Describe What is the main problem with fusion?

_____

_____

_____

Copyright © by Holt, Rinehart and Winston; a Division of Houghton Mifflin Harcourt Publishing Company. All rights reserved.

What Is Wind Power?

Wind is air that is moving. Moving air contains energy. People can use windmills to turn the energy in wind into electricity. The electricity that is produced by windmills is called **wind power**. Large groups of windmills can make a lot of electricity.

Like all energy sources, wind power has benefits and problems. Since the wind can't be used up, wind energy is renewable. Wind power does not cause air pollution. However, in many areas, the wind isn't strong or regular enough to generate enough electricity for people to use.

This pickup truck shows how large the windmills are.

These windmills near Livermore, California, produce electricity.

What Are Fuel Cells?

What powers a car? You probably thought of gasoline. However, not all cars are powered by gasoline. Some cars are powered by fuel cells. *Fuel cells* are devices that change chemical energy into electrical power. The **chemical energy** is released when hydrogen and oxygen react to form water.

Fuel cells have been used in space travel since the 1960s. They have provided space crews with electrical energy and drinking water. Today, fuel cells are used to create electrical energy in some buildings and ships. ☑

The only waste product of fuel cells is water, so they do not create pollution. However, not very many cars today use fuel cells. The hydrogen and oxygen used in fuel cells can be expensive to make and to store. Many people hope that we will be able to use fuel cells to power cars in the future.

Copyright © by Holt, Rinehart and Winston; a Division of Houghton Mifflin Harcourt Publishing Company. All rights reserved.

Critical Thinking

6. Infer In most cases, people use a large number of windmills to create electricity. What do you think is the reason a lot of windmills are used, instead of just one or two?

_____

_____

_____

_____

TAKE A LOOK

7. Explain Based on what you see in the figure, what do you think is the reason windmills are not used in cities or other crowded areas?

_____

_____

✓ **READING CHECK**

8. Explain How could fuel cells give space crews electricity and water?

_____

_____

_____

SECTION 3 Alternative Resources *continued*

Say It

Share Experiences Have you ever used an object that was powered by sunlight? In a small group, talk about the different ways that sunlight can be used for energy.

What Is Solar Energy?

Most forms of energy originally come from the sun. For example, the fossil fuels we use today were made from plants. The plants got their energy from the sun. The heat and light that Earth gets from the sun is **solar energy**. This type of energy is a renewable resource.

People can use solar energy to create electricity. *Photovoltaic cells*, or solar cells, can change sunlight into electrical energy. Solar energy can also be used to heat buildings.

Solar energy does not produce pollution and is renewable. The energy from the sun is free. However, some climates don't have enough sunny days to be able to use solar energy all the time. Also, even though sunlight is free, solar cells are expensive to make.

What Is Hydroelectric Energy?

Water wheels have been used since ancient times to help people do work. Today, the energy of falling water is used to generate electrical energy. **Hydroelectric energy** is electrical energy produced from moving water.

Hydroelectric energy causes no air pollution and is considered renewable. Hydroelectric energy is generally not very expensive to produce. ☑

However, hydroelectric energy can be produced only in places that have a lot of fast-moving water. In addition, building a dam and a power plant to generate hydroelectric energy can be expensive. Dams can harm wildlife living in and around the river. Damming a river can cause flooding and erosion.

READING CHECK

9. Identify What are two benefits of hydroelectric energy?

_____

_____

This dam in California can create electricity because a lot of water moves through it every day.

Copyright © by Holt, Rinehart and Winston; a Division of Houghton Mifflin Harcourt Publishing Company. All rights reserved.

How Can Plants Be Used for Energy?

Plants store energy from the sun. Leaves, wood, and stems contain stored energy. Even the dung of plant-eating animals has a lot of stored energy. These sources of energy are called biomass. **Biomass** is organic matter that can be a source of energy.

Biomass is commonly burned in its solid form to release heat. However, biomass can also be changed into a liquid form. The sugar and starch in plants can be made into alcohol and used as fuel. Alcohol can be mixed with gasoline to make a fuel called **gasohol**.

Biomass is not very expensive. It is available almost everywhere. Since biomass grows quickly, it is considered a renewable resource. However, people must be careful not to use up biomass faster than it can grow back.

What Is Geothermal Energy?

Geothermal energy is energy produced by the heat within Earth. This heat makes solid rocks get very hot. If there is any water contained within the solid rock, the water gets hot, too. The hot water can be used to generate electricity and to heat buildings. ☑

Geothermal energy is considered renewable because the heat inside Earth will last for millions of years. Geothermal energy does not create air pollution or harm the environment. However, this kind of energy can be used only where hot rock is near the surface.

Critical Thinking

10. Infer What would happen if biomass were used at a faster rate than it was produced?

_____

_____

_____

☑ **READING CHECK**

11. List What are two uses for water that has been heated by hot rock?

_____

_____

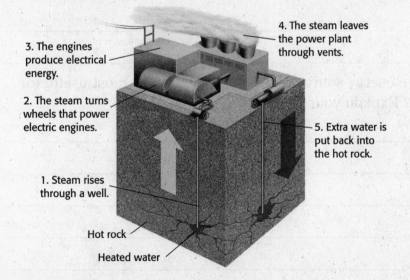

3. The engines produce electrical energy.

2. The steam turns wheels that power electric engines.

1. Steam rises through a well.

4. The steam leaves the power plant through vents.

5. Extra water is put back into the hot rock.

Hot rock

Heated water

TAKE A LOOK

12. Describe On the figure, draw arrows showing the path that the steam takes as it moves through the power plant.

Copyright © by Holt, Rinehart and Winston; a Division of Houghton Mifflin Harcourt Publishing Company. All rights reserved.

Section 3 Review

GLE 0707.Inq.5, GLE 0707.7.5, GLE 0707.7.6　**TN**

SECTION VOCABULARY

biomass organic matter that can be a source of energy	**hydroelectric energy** electrical energy produced by falling water
chemical energy the energy released when a chemical compound reacts to produce new compounds	**nuclear energy** the energy released by a fission or fusion reaction; the binding energy of the atomic nucleus
gasohol a mixture of gasoline and alcohol that is used as a fuel	**solar energy** the energy received by Earth from the sun in the form of radiation
geothermal energy the energy produced by heat within Earth	**wind power** the use of a windmill to drive an electric generator

1. Explain Why is solar energy considered a renewable resource?

_____

_____

_____

2. Identify When would biomass not be considered a renewable resource?

_____

_____

3. Apply Concepts Which place is *more likely* to be able to use geothermal energy: a city near a volcano or a city near a waterfall? Explain your answer.

_____

_____

_____

4. Identify Why is wind a useful energy source in some places, but not in others?

_____

_____

_____

5. Analyze Which alternative energy source do you think would be most useful for the place where you live? Explain your answer.

_____

_____

_____

_____

Copyright © by Holt, Rinehart and Winston; a Division of Houghton Mifflin Harcourt Publishing Company. All rights reserved.

Measuring Motion

TN Tennessee Science
Standards
GLE 0707.Inq.5
GLE 0707.T/E.1
GLE 0707.11.3

BEFORE YOU READ

After you read this section, you should be able to answer these questions:

• What is motion?

• How is motion shown by a graph?

• What are speed and velocity?

• What is acceleration?

What Is Motion?

Look around the room for a moment. What objects are in motion? Are students writing with pencils in their notebooks? Is the teacher writing on the board? Motion is all around you, even when you can't see it. Blood is circulating throughout your body. Earth orbits around the sun. Air particles shift in the wind.

When you watch an object move, you are watching it in relation to what is around it. Sometimes the objects around the object you are watching are at rest. An object that seems to stay in one place is called a *reference point*. When an object changes position over time in relation to a reference point, the object is in **motion**. ☑

You can use *standard reference directions* (such as north, south, east, west, right, and left) to describe an object's motion. You can also use features on Earth's surface, such as buildings or trees, as reference points. The figure below shows how a mountain can be used as a reference point to show the motion of a hot-air balloon.

STUDY TIP

Describe Study each graph carefully. In the margin next to the graph, write a sentence or two explaining what the graph shows.

READING CHECK

1. **Describe** What is the purpose of a reference point?

_____

_____

The hot-air balloon changed position relative to a reference point.

TAKE A LOOK

2. **Identify** What is the fixed reference point in the photos?

_____

_____

Copyright © by Holt, Rinehart and Winston; a Division of Houghton Mifflin Harcourt Publishing Company. All rights reserved.

How Can Motion Be Shown?

In the figure below, a sign-up sheet is being passed around a classroom. You can follow its path. The paper begins its journey at the reference point, the origin.

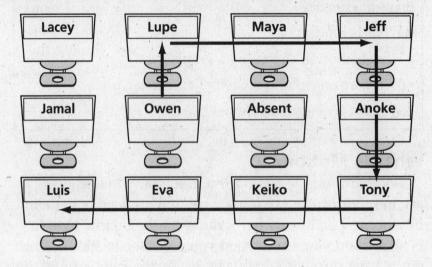

The path taken by a field trip sign-up sheet.

TAKE A LOOK

3. Identify What is the origin, or reference point, of the paper?

_____

_____

The figure below shows a graph of the position of the sign-up sheet as it is passed around the class. The paper moves in this order:

1. One positive unit on the y-axis
2. Two positive units on the x-axis
3. Two negative units on the y-axis
4. Three negative units on the x-axis

The graph provides a method of using standard reference directions to show motion.

TN TENNESSEE STANDARDS CHECK

GLE 0707.Inq.5 Communicate scientific understanding using descriptions, explanations, and models.

World Help: communicate to make known; to tell

4. Identify What is the shortest path that the paper could take to return to Owen's desk? The paper cannot move diagonally.

_____

_____

_____

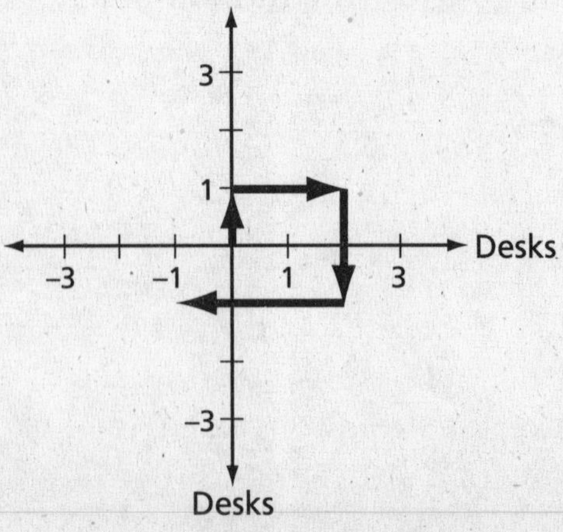

The position of the sign-up sheet as it moves through the classroom.

Copyright © by Holt, Rinehart and Winston; a Division of Houghton Mifflin Harcourt Publishing Company. All rights reserved.

SECTION 1 Measuring Motion *continued*

What Is Speed?

Speed is the rate at which an object moves. It is the distance traveled divided by the time taken to travel that distance. Most of the time, objects do not travel at a constant speed. For example, when running a race, you might begin slowly but then sprint across the finish line.

So, it is useful to calculate *average speed*. We use the following equation:

$$average\ speed = \frac{total\ distance}{total\ time}$$

Suppose that it takes you 2 s to walk 4 m down a hallway. You can use the equation above to find your average speed:

$$average\ speed = \frac{4\ m}{2\ s} = 2\ m/s$$

Your speed is 2 m/s. Units for speed include meters per second (m/s), kilometers per hour (km/h), feet per second (ft/s), and miles per hour (mi/h).

How Can You Show Speed on a Graph?

You can show speed on a graph by showing how the position of an object changes over time. The *x*-axis shows the time it takes to move from place to place. The *y*-axis shows distance from the reference point.

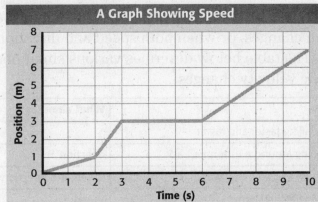

A graph of position versus time also shows the dog's speed during his walk. The more slanted the line, the faster the dog walked.

Suppose you watched a dog walk beside a fence. The graph above shows the total distance the dog walked in 10 s. The line is not straight because the dog did not walk the same distance in each second. The dog walked slowly for 2 s and then quickly for 1 s. From 3 s to 5 s, the dog did not move.

Copyright © by Holt, Rinehart and Winston; a Division of Houghton Mifflin Harcourt Publishing Company. All rights reserved.

Critical Thinking

5. Explain The average flight speed of a bald eagle is about 50 km/h. A scientist has measured an eagle flying 80 km/h. How is this possible?

_____

_____

_____

_____

Math Focus

6. Calculate Suppose you walk 10 m down a hallway in 2.5 s. What is your average speed? Show your work.

TAKE A LOOK

7. Apply Concepts Suppose the dog walks at a constant speed the whole way. On the graph, draw a line showing that the dog walks at a constant speed during the walk.

What Is Velocity?

Suppose that two birds leave the same tree at the same time. They both fly at 10 km/h for 5 min, then 5 km/h for 10 min. However, they don't end up in the same place. Why not?

The birds did not end up in the same place because they flew in different directions. Their speeds were the same, but because they flew in different directions, their velocities were different. **Velocity** is the speed of an object and its direction. ☑

The velocity of an object is constant as long as both speed and direction are constant. If a bus driving at 15 m/s south speeds up to 20 m/s south, its velocity changes. If the bus keeps moving at the same speed but changes direction from south to east, its velocity also changes. If the bus brakes to a stop, the velocity of the bus changes again.

The table below shows that velocity is a combination of both the speed of an object and its direction.

Speed	Direction	Velocity
15 m/s	south	15 m/s south
20 m/s	south	20 m/s south
20 m/s	east	20 m/s east
0 m/s	east	0 m/s east

Velocity changes when the speed changes, when the direction changes, or when both speed and direction change. The table below describes various situations in which the velocity changes.

Situation	What changes
Raindrop falling faster and faster	
Runner going around a turn on a track	direction
Car taking an exit off a highway	speed and direction
Train arriving at a station	speed
Baseball being caught by a catcher	speed
Baseball hit by a batter	
	speed and direction

READING CHECK

8. Analyze Someone tells you that the velocity of a car is 55 mi/h. Is this correct? Explain your answer.

_____

_____

 **Say It**

Share Experiences Have you ever experienced a change in velocity on an amusement park ride? In pairs, share an experience. Explain how the velocity changed—was it a change in speed, direction, or both?

TAKE A LOOK

9. Identify Fill in the empty boxes in the table.

Copyright © by Holt, Rinehart and Winston; a Division of Houghton Mifflin Harcourt Publishing Company. All rights reserved.

1 m/s 2 m/s 3 m/s 4 m/s 5 m/s

This cyclist moves faster and faster as he peddles his bike south.

TAKE A LOOK
10. Identify Is the cyclist accelerating? How do you know?

_____

_____

What Is Acceleration?

Acceleration is how quickly velocity changes. An object accelerates if its speed changes, its direction changes, or both its speed and direction change.

The units for acceleration are the units for velocity divided by a unit for time. The resulting unit is often meters per second per second (m/s/s or m/s^2).

Looking at the figure above, you can see that the speed increases by 1 m/s during each second. This means that the cyclist is accelerating at 1 m/s^2.

An increase in speed is referred to as *positive acceleration*. A decrease in speed is referred to as *negative acceleration* or *deceleration*. ☑

Acceleration can be shown on a graph of speed versus time. Suppose you are operating a remote control car. You push the lever on the remote to move the car forward. The graph below shows the car's acceleration as the car moves east. For the first 5 s, the car increases in speed. The car's acceleration is positive because the speed increases as time passes.

For the next 2 s, the speed of the car is constant. This means the car is no longer accelerating. Then the speed of the car begins to decrease. The car's acceleration is then negative because the speed decreases over time.

READING CHECK
11. Explain What happens to an object when it has negative acceleration?

_____

_____

Math Focus
12. Interpret Graphs Is the slope positive or negative when the car's speed increases? Is the slope positive or negative when the car's speed decreases? Some cars have a device called cruise control that keeps the car's speed from accelerating or decelerating beyond a set speed. How would this appear on a graph?

_____

_____

_____

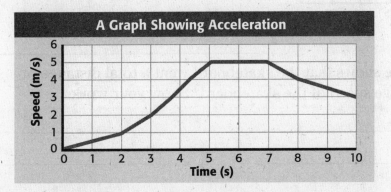

A Graph Showing Acceleration

The graph of speed versus time also shows that the acceleration of the car was positive and negative. Between 5 s and 7 s, it had no acceleration.

Copyright © by Holt, Rinehart and Winston; a Division of Houghton Mifflin Harcourt Publishing Company. All rights reserved.

Section 1 Review

GLE 0707.Inq.5, GLE 0707.11.3, GLE 0707.T/E.1

SECTION VOCABULARY

acceleration the rate at which the velocity changes over time; an object accelerates if its speed, direction, or both change	**motion** an object's change in position relative to a reference point
speed the distance traveled divided by the time interval during which the motion occurred	**velocity** the speed of an object in a particular direction

1. Identify What is the difference between speed and velocity?

_____

_____

_____

2. Complete a Graphic Organizer Fill in the graphic organizer for a car that starts from one stop sign and approaches the next stop sign. Use the following terms: constant *velocity*, *positive acceleration*, *deceleration*, and *at rest*.

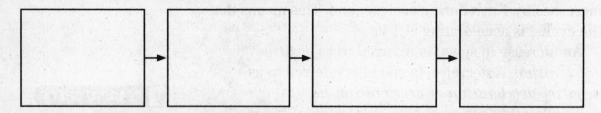

3. Interpret a Graph Describe the motion of the skateboard using the graph below. Write what the skateboard does from time = 0 s to time = 40 s.

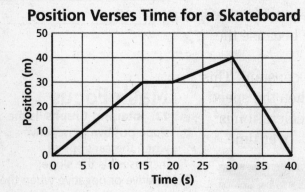

Position Verses Time for a Skateboard

_____

_____

_____

_____

_____

_____

4. Calculate The graph above shows that the skateboard went a total distance of 80 m. What was the average speed of the skateboard? Show your work.

_____

_____

_____

Copyright © by Holt, Rinehart and Winston; a Division of Houghton Mifflin Harcourt Publishing Company. All rights reserved.

CHAPTER 19 Matter in Motion

SECTION 2 **What Is a Force?**

BEFORE YOU READ

After you read this section, you should be able to answer these questions:

• What is a force?

• How do forces combine?

• What is a balanced force?

• What is an unbalanced force?

TN Tennessee Science Standards
GLE 0707.Inq.5

What Is a Force?

You probably hear people talk about force often. You may hear someone say, "That storm had a lot of force" or "Mrs. Larsen is the force behind the school dance." But what exactly is a force in science?

In science, a **force** is a push or a pull. All forces have two properties: direction and size. A **newton** (N) is the unit that describes the size of a force.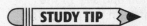

Forces act on the objects around us in ways that we can see. If you kick a ball, the ball receives a push from you. If you drag your backpack across the floor, the backpack is pulled by you.

Forces also act on objects around us in ways that we cannot see. For example, in the figure below, a student is sitting on a chair. What are the forces acting on the chair?

The student is pushing down on the chair, but the chair does not move. Why? The floor is balancing the force by pushing up on the chair. When the forces on an object are *balanced*, the object does not move.

STUDY TIP

Brainstorm As you read, think about objects you see every day. What kinds of forces are affecting them? How do the forces affect them?

 **READING CHECK**

1. List What two properties do all forces have?

_____

_____

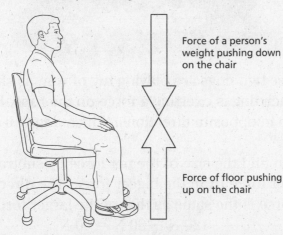

Force of a person's weight pushing down on the chair

Force of floor pushing up on the chair

A person sitting on a chair.

TAKE A LOOK

2. Explain Since the chair is not moving, what kind of forces are acting on it?

_____

_____

Copyright © by Holt, Rinehart and Winston; a Division of Houghton Mifflin Harcourt Publishing Company. All rights reserved.

How Do Forces Combine?

As you saw in the previous example, more than one force often acts on an object. When all of the forces acting on an object are added together, you determine the **net force** on the object. An object with a net force more than 0 N acting on it will change its state of motion.

TAKE A LOOK
3. Identify On the figure, draw an arrow showing the direction and size of the net force on the piano. The length of the arrow should represent the size of the force.

FORCES IN THE SAME DIRECTION

Suppose your music teacher asks you and a friend to move a piano, as shown in the figure above. You push the piano from one end and your friend pulls the piano from the other end. You and your friend are applying forces in the same direction. Adding the two forces gives you the size of the net force. The direction of the net force is the same as the direction of the forces.

$$125 \text{ N} + 120 \text{ N} = 245 \text{ N}$$
net force = 245 N to the right

FORCES IN DIFFERENT DIRECTIONS

Critical Thinking
4. Predict What would happen if both dogs pulled the rope with a force of 85 N?

_____

_____

_____

Suppose two dogs are playing tug of war, as shown above. Each dog is exerting a force on the rope. Here, the forces are in opposite directions. Which dog will win the tug of war?

You can find the size of the net force by subtracting the smaller force from the bigger force. The direction of the net force is the same as that of the larger force:

$$120 \text{ N} - 80 \text{ N} = 40 \text{ N}$$
net force = 40 N to the right

Copyright © by Holt, Rinehart and Winston; a Division of Houghton Mifflin Harcourt Publishing Company. All rights reserved.

SECTION 2 What Is a Force? *continued*

What Happens When Forces Are Balanced or Unbalanced?

Knowing the net force on an object lets you determine its effect on the motion of the object. Why? The net force tells you whether the forces on the object are balanced or unbalanced.

BALANCED FORCES

When the forces on an object produce a net force of 0 N, the forces are *balanced*. There is no change in the motion of the object. For example, a light hanging from the ceiling does not move. This is because the force of gravity pulls down on the light while the force of the cord pulls upward. ☑

The soccer ball moves because the players exert an unbalanced force on the ball each time they kick it.

UNBALANCED FORCES

When the net force on an object is not 0 N, the forces on the object are *unbalanced*. Unbalanced forces produce a change in motion of an object. Think about a soccer game. Players kick the ball to each other. When a player kicks the ball, the kick is an unbalanced force. It sends the ball in a new direction with a new speed. ☑

An object can continue to move when the unbalanced forces are removed. For example, when it is kicked, a soccer ball receives an unbalanced force. The ball continues to roll on the ground after the ball was kicked until an unbalanced force changes its motion.

☑ **READING CHECK**

5. Describe What happens to the motion of an object if the net force acting on it is 0 N?

_____

_____

☑ **READING CHECK**

6. Describe What will happen to an object that has an unbalanced force acting on it?

_____

_____

Copyright © by Holt, Rinehart and Winston; a Division of Houghton Mifflin Harcourt Publishing Company. All rights reserved.

Section 2 Review

GLE 0707.Inq.5 TN

SECTION VOCABULARY

force a push or a pull exerted on an object in order to change the motion of the object; force has size and direction	**newton** the SI unit for force (symbol, N)
net force the combination of all the forces acting on an object	

1. Explain If there are many forces acting on an object, how can the net force be 0?

_____

_____

_____

2. Apply Concepts Identify three forces acting on a bicycle when you ride it.

_____

_____

_____

3. Calculate Determine the net force on each of the objects shown below. Don't forget to give the direction of the force.

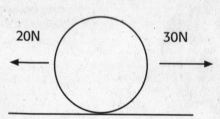

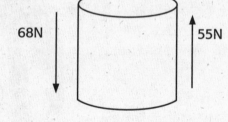

net force = _____   *net force =* _____

4. Explain How will the net force affect the motion of each object shown above?

_____

_____

_____

5. Describe What is the difference between balanced and unbalanced forces?

_____

_____

_____

_____

Copyright © by Holt, Rinehart and Winston; a Division of Houghton Mifflin Harcourt Publishing Company. All rights reserved.

CHAPTER 19 | Matter in Motion

SECTION **3** # Friction: A Force That Opposes Motion

TN Tennessee Science Standards
GLE 0707.Inq.2
GLE 0707.Inq.3
GLE 0707.Inq.5

BEFORE YOU READ

After you read this section, you should be able to answer these questions:

• What is friction?

• How does friction affect motion?

• What are the types of friction?

• How can friction be changed?

What Causes Friction?

Suppose you are playing soccer and you kick the ball far from you. You know that the ball will slow down and eventually stop. This means that the velocity of the ball will decrease to 0. You also know that an unbalanced force is needed to change the velocity of objects. So, what force is stopping the ball?

Friction is the force that opposes the motion between two surfaces that touch. Friction causes the ball to slow down and then stop. ☑

What causes friction? The surface of any object is rough. Even an object that feels smooth is covered with very tiny hills and valleys. When two surfaces touch, the hills and valleys of one surface stick to the hills and valleys of the other. This contact between surfaces causes friction.

If a force pushes two surfaces together even harder, the hills and valleys come closer together. This increases the friction between the surfaces.

STUDY TIP

Imagine As you read, think about the ways that friction affects your life. Make a list of things that might happen, or not happen, if friction did not exist.

READING CHECK

1. Describe What is friction?

_____

_____

_____

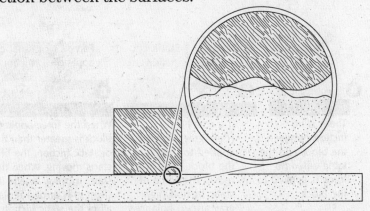

When the hills and valleys of one surface stick to the hills and valleys of another surface, friction is created.

TAKE A LOOK
2. Explain Why can't you see the hills and valleys without a close-up view of the objects?

_____

_____

Copyright © by Holt, Rinehart and Winston; a Division of Houghton Mifflin Harcourt Publishing Company. All rights reserved.

Critical Thinking

3. Apply Concepts Two identical balls begin rolling next to each other at the same velocity. One is on a smooth surface and one is on a rough surface. Which ball will stop first? Why

_____

_____

_____

What Affects the Amount of Friction?

Imagine that a ball is rolled over a carpeted floor and another ball is rolled over a wood floor. Which surface affects a ball more?

The smoothness of the surfaces of the objects affects how much friction exists. Friction is usually greater between materials that have rough surfaces than materials that have smooth surfaces. The carpet has greater friction than the wood floor, so the ball on the carpet stops first.

What Types of Friction Exist?

There are two types of friction: kinetic friction and static friction. *Kinetic friction* occurs when force is applied to an object and the object moves. When a cat slides along a countertop, the friction between the cat and the countertop is kinetic friction. The word kinetic means "moving."

The amount of kinetic friction between moving surfaces depends partly on how the surfaces move. In some cases, the surfaces slide past each other like pushing a box on the floor. In others, one surface rolls over another like a moving car on a road. There is usually less friction between surfaces that roll than between surfaces that slide.

Static friction occurs when force applied to an object does not cause the object to move. When you try to push a piece of furniture that will not move, the friction observed is static friction.

TAKE A LOOK
4. Describe When does static friction become kinetic friction?

_____

_____

_____

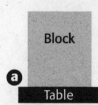

There is no friction between the block and the table when no force is applied to the block.

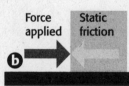

If a small force (dark gray arrow) is applied to the block, the block does not move. The force of static friction (light gray arrow) balances the force applied.

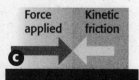

When the force applied to the block is greater than the force of static friction, the block starts moving. When the block starts moving, kinetic friction (light gray arrow) replaces all of the static friction and opposes the force applied.

Copyright © by Holt, Rinehart and Winston; a Division of Houghton Mifflin Harcourt Publishing Company. All rights reserved.

How Can Friction Be Decreased?

To reduce the amount of friction, you can apply a lubricant between two surfaces. A *lubricant* is a substance that reduces the friction between surfaces. Motor oil, wax, and grease are examples of lubricants.

You can also reduce friction by rolling, rather than sliding, an object. A refrigerator on rollers is much easier to move than one that just slides.

Another way of reducing friction is to smooth the surfaces that rub against each other. Skiers have their skis sanded down to make them smoother. This makes it easier for the skis to slide over the snow.

If you work on a bicycle, you may get dirty from the chain oil. This lubricant reduces friction between sections of the chain.

How Can Friction Be Increased?

Increasing the amount of friction between surfaces can be very important. For example, when the tires of a car grip the road better, the car stops and turns corners much better. Friction causes the tires to grip the road. Without friction, a car could not start moving or stop.

On icy roads, sand can be used to make the road surface rougher. Friction increases as surfaces are made rougher.

You can also increase friction by increasing the force between the two objects. Have you ever cleaned a dirty pan in the kitchen sink? You may have found that cleaning the pan with more force allows you to increase the amount of friction. This makes it easier to clean the pan. ☑

Critical Thinking

5. Infer How does a lubricant reduce the amount of friction?

_____

_____

_____

_____

TAKE A LOOK

6. Identify Why is it important to put oil on a bicycle chain?

_____

☑ **READING CHECK**

7. Identify Name two things that can be done to increase the friction between surfaces.

_____

_____

_____

Copyright © by Holt, Rinehart and Winston; a Division of Houghton Mifflin Harcourt Publishing Company. All rights reserved.

Section 3 Review

GLE 0707.Inq.2, GLE 0707.Inq.3, GLE 0707.Inq.5 **TN**

SECTION VOCABULARY

friction a force that opposes the motion between two surfaces that are in contact	

1. Describe What effect does friction have when you are trying to move an object at rest?

_____

2. Compare Explain the difference between static friction and kinetic friction. Give an example of each.

_____

_____

_____

3. Compare The figure on the left shows two surfaces up close. On the right, draw a sketch. Show what the surfaces of two objects that have less friction between them might look like.

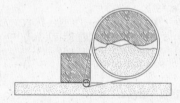

4. Analyze Name three common lubricants and describe why they are used.

_____

_____

5. Analyze In what direction does friction always act?

_____

6. Identify A car is driving on a flat road. When the driver hits the brakes, the car slows down and stops. What would happen if there were no friction between the tires and the road? Explain your answer.

_____

_____

Copyright © by Holt, Rinehart and Winston; a Division of Houghton Mifflin Harcourt Publishing Company. All rights reserved.

SECTION 4 Gravity: A Force of Attraction

BEFORE YOU READ

After you read this section, you should be able to answer these questions:

- What is gravity?
- How are weight and mass different?

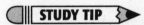 **TN** Tennessee Science Standards
GLE 0707.Inq.2
GLE 0707.Inq.3
GLE 0707.Inq.5

How Does Gravity Affect Matter?

Have you ever seen a video of astronauts on the moon? The astronauts bounce around like beach balls even though the space suits weighed 180 pounds on Earth. See the figure below. Why is it easier for a person to move on the moon than on Earth? The reason is that the moon has less gravity than Earth. **Gravity** is a force of attraction, or a pull, between objects. It is caused by their masses. ☑

All matter has mass. Gravity is a result of mass. Therefore all matter has gravity. This means that all objects attract all other objects in the universe! The force of gravity pulls objects toward each other. For example, gravity between the objects in the solar system holds the solar system together. Gravity holds you to Earth.

Small objects also have gravity. You have gravity. This book has gravity. Why don't you notice the book pulling on you or you pulling on the book? The reason is that the book's mass and your mass are both small. The force of gravity caused by small mass is not large enough to move either you or the book.

STUDY TIP

Discuss Ideas Take turns reading this section out loud with a partner. Stop to discuss ideas that seem confusing.

READING CHECK

1. **Describe** What is gravity?

_____
_____
_____

Critical Thinking

2. **Infer** Why can't you see two soccer balls attracting each other?

_____
_____
_____

Because the moon has less gravity than Earth does, walking on the moon's surface was a very bouncy experience for the Apollo astronauts.

Copyright © by Holt, Rinehart and Winston; a Division of Houghton Mifflin Harcourt Publishing Company. All rights reserved.

SECTION 4 Gravity: A Force of Attraction *continued*

What Is the Law of Universal Gravitation?

According to a story, Sir Isaac Newton, while sitting under an apple tree, watched an apple fall. This gave him a bright idea. Newton realized that an unbalanced force on the apple made it fall.

He then thought about the moon's orbit. Like many others, Newton had wondered what kept the planets in the sky. He realized that an unbalanced force on the moon kept it moving around Earth. Newton said that these forces are both gravity.

Newton's ideas are known as the *law of universal gravitation*. Newton said that all objects in the universe attract each other because of gravitational force. ☑

This law says that gravitational force depends on two things:

1. the masses of the objects
2. the distance between the objects

The word "universal" means that the law applies to all objects. Even seeds respond to gravity. They respond to gravity by sending roots down and green shoots up. But scientists do not understand how seeds can sense gravity. See the figure below. ☑

☑ **READING CHECK**

3. Describe What is the law of universal gravitation?

_____

_____

_____

_____

☑ **READING CHECK**

4. Identify What two things determine gravitational force?

_____

_____

TAKE A LOOK
5. Identify What is the unbalanced force that affects the motions of a falling apple and the moon?

_____

Sir Isaac Newton said that the same unbalanced force caused the motions of the apple and the moon.

Copyright © by Holt, Rinehart and Winston; a Division of Houghton Mifflin Harcourt Publishing Company. All rights reserved.

How Does Mass Affect Gravity?

Imagine an elephant and a cat. Because the elephant has a larger mass than the cat does, gravity between the elephant and Earth is larger. So, the cat is much easier to pick up than the elephant. The gravitational force between objects depends on the masses of the objects. See the figure below.

mass = 100 kg   mass = 100 kg

● Gravitational force is small between objects that have small masses.

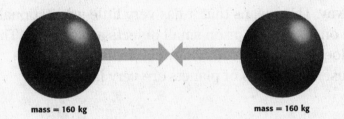

mass = 160 kg   mass = 160 kg

● If the mass of one or both objects increases, the gravitational force pulling them together increases.

The arrows indicate the gravitational force between two objects. The length of the arrows indicates the magnitude of the force.

TAKE A LOOK
6. Compare Is there more gravitational force between objects with small masses or objects with large masses?

_____

_____

_____

Mass also explains why an astronaut on the moon can jump around so easily. The moon has less mass than Earth does. This gives the moon a weaker pull on objects than the pull of Earth. The astronaut is not being pulled toward the moon as much as he is by Earth. So the astronaut can jump higher and more easily on the moon.

The universal law of gravitation can let us predict what happens to gravity when mass changes. According to the universal law of gravitation, suppose there is a 5 N force of gravity between two objects. If the mass of one object doubles and the other stays the same, the force of gravity also doubles.

Let's try a problem. The force due to gravity between two objects is 3 N. If the mass of one object triples and the other stays the same, what is the new force of gravity?

Solution: Since the mass of one object tripled and the other stayed the same, the force of gravity also triples. It is 9 N.

Math Focus
7. Infer Two objects of equal mass have a force of gravity of 6 N between them. Imagine the mass of one is cut in half and the other stays the same, what is the force due to gravity?

_____

Copyright © by Holt, Rinehart and Winston; a Division of Houghton Mifflin Harcourt Publishing Company. All rights reserved.

How Does Distance Affect Gravity?

The mass of the sun is 300,000 times bigger than that of Earth. However, if you jump up, you return to Earth every time you jump rather than flying toward the sun. If the sun has more mass, then why doesn't it have a larger gravitational pull on you?

This is because the gravitational force also depends on the distance between the objects. As the distance between two objects gets larger, the force of gravity gets much smaller. And as the distance between objects gets smaller, the force of gravity gets much bigger. This is shown in the figure below.

Although the sun has tremendous mass, it is also very far away. This means that it has very little gravitational force on your body or on small objects around you. The sun does have a large gravitational force on planets because the masses of planets are very large.

Critical Thinking

8. Analyze The sun is much more massive than Earth. Why is the force of gravity between you and the sun so much less than Earth's gravity and you?

_____

_____

● Gravitational force is large when the distance between two objects is small.

TAKE A LOOK

9. Describe Use the diagram to describe the effect of distance on gravitational force.

_____

_____

_____

_____

_____

● If the distance between two objects increases, the gravitational force pulling them together decreases rapidly.

The length of the arrows indicates the magnitude of the gravitational force between two objects.

Copyright © by Holt, Rinehart and Winston; a Division of Houghton Mifflin Harcourt Publishing Company. All rights reserved.

What Is the Difference Between Mass and Weight?

You have learned that gravity is a force of attraction between objects. **Weight** is a measure of the gravitational force on an object. The SI unit for weight is the newton (N).

Mass is a measure of the amount of matter in an object. This seems similar to weight, but it is not the same. An object's mass does not change when gravitational forces change, but its weight does. Mass is usually expressed in kilograms (kg) or grams (g). ☑

In the figure below, you can see the difference between mass and weight. Compare the astronaut's mass and weight on Earth to his mass and weight on the moon.

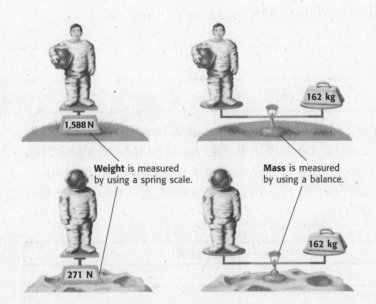

Weight is measured by using a spring scale.

Mass is measured by using a balance.

1,588 N

162 kg

271 N

162 kg

Gravity can cause objects to move because it is a type of force. But gravity also acts on objects that are not moving, or *static*. Earth's gravity pulls objects downward. However, not all objects move downward. Suppose a framed picture hangs from a wire. Gravity pulls the picture downward, but tension (the force in the wire) pulls the picture upward. The forces are balanced so that framed picture does not move.

☑ **READING CHECK**

10. Contrast How is mass different from weight?

_____

_____

_____

_____

TAKE A LOOK

11. Identify What is the weight of the astronaut on Earth? What is the weight of the astronaut on the moon?

_____

_____

Critical Thinking

12. Contrast What forces act on a framed picture on a shelf?

_____

_____

_____

Copyright © by Holt, Rinehart and Winston; a Division of Houghton Mifflin Harcourt Publishing Company. All rights reserved.

Name _____ Class _____ Date _____

Section 4 Review

GLE 0707.Inq.2, GLE 0707.Inq.3, GLE 0707.Inq.5 **TN**

SECTION VOCABULARY

gravity a force of attraction between objects that is due to their masses	**weight** a measure of the gravitational force exerted on an object; its value can change with the location of the object in the universe
mass a measure of the amount of matter in an object	

1. **Identify** What is gravity? What determines the gravitational force between objects?

_____

_____

2. **Describe** A spacecraft is moving toward Mars. Its rocket engines are turned off. As the spacecraft nears the planet, what will happen to the pull of Mars's gravity?

_____

_____

3. **Summarize** An astronaut travels from Earth to the moon. How does his mass change? How does his weight change? Explain.

_____

_____

_____

4. **Applying Concepts** An astronaut visits Planet X. Planet X has the same radius as Earth but has twice the mass of Earth. Fill in the table below to show the astronaut's mass and weight on Planet X. (Hint: Newton's law of universal gravitation says that when the mass of one object doubles, the force due to gravity also doubles.)

	Earth	Planet X
Mass of astronaut	80 kg	
Weight of astronaut	784 N	

5. **Select** Each of the spheres shown below is made of iron. Circle the pair of spheres that would have the greatest gravitational force between them. Below the spheres, explain the reason for your choice.

_____

Copyright © by Holt, Rinehart and Winston; a Division of Houghton Mifflin Harcourt Publishing Company. All rights reserved.

CHAPTER 20 Forces and Motion

SECTION 1 # Gravity and Motion

BEFORE YOU READ

After you read this section, you should be able to answer these questions:

• How does gravity affect objects?

• How does air resistance affect falling objects?

• What is free fall?

• Why does an object that is thrown horizontally follow a curved path?

TN **Tennessee Science Standards**
GLE 0707.Inq.2
GLE 0707.Inq.3
GLE 0707.Inq.5

How Does Gravity Affect Falling Objects?

In ancient Greece, a great thinker named Aristotle said that heavy objects fall faster than light objects. For almost 2,000 years, people thought this was true. Then, in the late 1500s, an Italian scientist named Galileo Galilei proved that heavy and light objects actually fall at the same rate.

It has been said that Galileo proved this by dropping two cannonballs from the top of a tower at the same time. The cannonballs were the same size, but one was much heavier than the other. The people watching saw both cannonballs hit the ground at the same time. ☑

Why don't heavy objects fall faster than light objects? Gravity pulls on heavy objects more than it pulls on light objects. However, heavy objects are harder to move than light objects. So, the extra force from gravity on the heavy object is balanced by how much harder it is to move.

Falling Balls

Time 1   ○   ⬤—Golf ball
             └ Table tennis ball

Time 2   ○

Time 3   ○

Time 4   ○

STUDY TIP

Practice After every page, stop reading and think about what you've read. Try to think of examples from everyday life. Don't go on to the next section until you think you understand.

✓ READING CHECK

1. Describe What did the people watching the cannonballs see that told them the cannonballs fell at the same rate?

_____

_____

TAKE A LOOK

2. Predict The golf ball is heavier than the table tennis ball. On the figure, draw three circles to show where the golf ball will be at times 2, 3, and 4.

Copyright © by Holt, Rinehart and Winston; a Division of Houghton Mifflin Harcourt Publishing Company. All rights reserved.

How Much Acceleration Does Gravity Cause?

Because of gravity, all objects accelerate, or speed up, toward Earth at a rate of 9.8 meters per second per second. This is written as 9.8 m/s/s or 9.8 m/s². So, for every second an object falls, its velocity (speed) increases by 9.8 m/s. This is shown in the figure below.

Math Focus

3. Calculate How fast is the ball moving at the end of the third second? Explain your answer.

_____

_____

_____

_____

_____

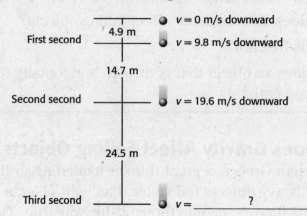

First second — 4.9 m — $v = 0$ m/s downward
$v = 9.8$ m/s downward

14.7 m

Second second — $v = 19.6$ m/s downward

24.5 m

Third second — $v = $ _____ ?

A falling object accelerates at a constant rate. The object falls faster and farther each second than it did the second before.

What Is the Velocity of a Falling Object?

Suppose you drop a rock from a cliff. How fast is it going when it reaches the bottom? If you have a stopwatch, you can calculate its final velocity.

If an object starts from rest and you know how long it falls, you can calculate its final velocity by using this equation:

$$v_{final} = g \times t$$

In the equation, v_{final} stands for final velocity in meters per second, g stands for the acceleration due to gravity (9.8 m/s²), and t stands for the time the object has been falling (in seconds).

If the rock took 4 s to hit the ground, how fast was it falling when it hit the ground?

Step 1: Write the equation.

$$v_{final} = g \times t$$

Step 2: Place values into the equation, and solve for the answer.

$$v_{final} = 9.8 \frac{m/s}{s} \times 4 \, s = 39.2 \text{ m/s}$$

The velocity of the rock was 39.2 m/s when it hit the ground.

Math Focus

4. Calculate A penny is dropped from the top of a tall stairwell. What is the velocity of the penny after it has fallen for 2 s? Show your work.

Copyright © by Holt, Rinehart and Winston; a Division of Houghton Mifflin Harcourt Publishing Company. All rights reserved.

How Can You Calculate How Long an Object Was Falling?

Suppose some workers are building a bridge. One of them drops a metal bolt from the top of the bridge. When the bolt hits the ground, it is moving 49 m/s. How long does it take the bolt to fall to the ground?

Step 1: Write the equation.

$$t = \frac{v_{final}}{g}$$

Step 2: Place values into the equation, and solve for the answer.

$$t = \frac{49 \text{ m/s}}{9.8 \frac{\text{m/s}}{\text{s}}} = 5 \text{ s}$$

The bolt fell for 5 s before it hit the ground.

How Does Air Resistance Affect Falling Objects?

Try dropping a pencil and a piece of paper from the same height. What happens? Does this simple experiment show what you just learned about falling objects? Now crumple the paper into a tight ball. Drop the crumpled paper and the pencil from the same height.

What happens? The flat paper falls more slowly than the crumpled paper because of air resistance. *Air resistance* is the force that opposes the motion of falling objects. ☑

How much air resistance will affect an object depends on the size, shape, and speed of the object. The flat paper has more surface area than the crumpled sheet. This causes the flat paper to fall more slowly.

How Air Resistance Affects Velocity

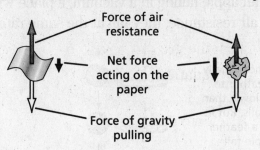

Force of air resistance

Net force acting on the paper

Force of gravity pulling

Math Focus

5. Calculate A rock falls from a cliff and hits the ground with a velocity of 98 m/s. How long does the rock fall? Show your work.

 **READING CHECK**

6. Identify Which has more air resistance, the flat paper or the crumpled paper?

_____

_____

TAKE A LOOK

7. Explain Why does the crumpled paper fall faster than the flat paper?

_____

_____

_____

Copyright © by Holt, Rinehart and Winston; a Division of Houghton Mifflin Harcourt Publishing Company. All rights reserved.

What Is Terminal Velocity?

As the speed of a falling body increases, air resistance also increases. The upward force of air resistance keeps increasing until it is equal to the downward force of gravity. At this point, the total force on the object is zero, so the object stops accelerating.

When the object stops accelerating, it does not stop moving. It falls without speeding up or slowing down. It falls at a constant velocity called the terminal velocity. **Terminal velocity** is the speed of an object when the force of air resistance equals the force of gravity. ☑

Air resistance causes the terminal velocity of hailstones to be between 5 m/s and 40 m/s. Without air resistance, they could reach the ground at a velocity of 350 m/s! Air resistance also slows sky divers to a safe landing velocity.

READING CHECK

8. Describe When does an object reach its terminal velocity?

_____

_____

TAKE A LOOK

9. Identify A sky diver is falling at terminal velocity. Draw and label an arrow showing the direction and size of the force due to gravity on the sky diver. Draw and label a second arrow showing the direction and size of the force of air resistance on the sky diver.

The parachute increases the air resistance of this sky diver and slows him to a safe terminal velocity.

What Is Free Fall?

Free fall is the motion of an object when gravity is the only force acting on the object. The figure below shows a feather and an apple falling in a vacuum, a place without any air. Without air resistance, they fall at the same rate.

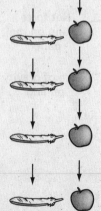

Air resistance usually causes a feather to fall more slowly than an apple falls. But in a vacuum, a feather and an apple fall with the same acceleration because both are in free fall.

TAKE A LOOK

10. Predict When air resistance acts on the apple and the feather, which falls faster?

_____

_____

Copyright © by Holt, Rinehart and Winston; a Division of Houghton Mifflin Harcourt Publishing Company. All rights reserved.

SECTION 1 Gravity and Motion *continued*

ORBITING OBJECTS ARE IN FREE FALL

Satellites and the space shuttle orbit Earth. You may have seen that astronauts inside the shuttle float unless they are belted to a seat. They seem weightless. In fact, they are not weightless, because they still have mass.

Weight is a measure of the pull of gravity on an object. Gravity acts between any two objects in the universe. Every object in the universe pulls on every other object. Every object with mass has weight. ☑

The force of gravity between two objects depends on the masses of the objects and how far apart they are. The more massive the objects are, the greater the force is. The closer the objects, the greater the force.

Your weight is determined mostly by the mass of Earth because it is so big and so close to you. If you were to move away from Earth, you would weigh less. However, you would always be attracted to Earth and to other objects, so you would always have weight.

Astronauts float in the shuttle because the shuttle is in free fall. That's right—the shuttle is always falling. Because the astronauts are in the shuttle, they are also falling. The astronauts and the shuttle are falling at the same rate. That is why the astronauts seem to float inside the shuttle. ☑

Isaac Newton first predicted this kind of free fall in the late 17th century. He reasoned that if a cannon were placed on a mountain and fired, the cannon ball would fall to Earth. Yet, if the cannon ball were shot with enough force, it would fall at the same rate that Earth's surface curves away. The cannon ball would never hit the ground, so it would orbit Earth. The figure below shows this "thought experiment."

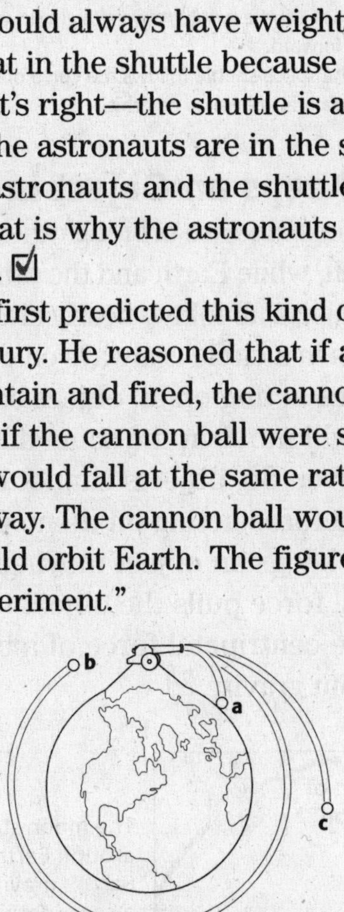

Newton's cannon is a "thought experiment." Newton reasoned that a cannon ball shot hard enough from a mountain top would orbit Earth.

✔ **READING CHECK**

11. Explain When will an object have no weight? Explain your answer.

_____

_____

✔ **READING CHECK**

12. Explain Why don't the astronauts in the orbiting shuttle fall to the floor?

_____

_____

_____

Critical Thinking

13. Infer Compared with cannon ball **b**, what do you think cannon ball **c** would do?

_____

_____

Copyright © by Holt, Rinehart and Winston; a Division of Houghton Mifflin Harcourt Publishing Company. All rights reserved.

What Motions Combine to Make an Object Orbit?

An object is in orbit (is orbiting) when it is going around another object in space. When the space shuttle orbits Earth, it is moving forward. Yet, the shuttle is also in free fall. The figure below shows how these two motions combine to cause orbiting.

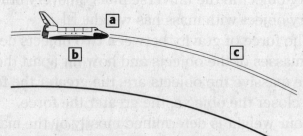

TAKE A LOOK

14. Identify On the figure, draw a line showing the path that the space shuttle would take if gravity were not acting on it.

a. The space shuttle moves forward at a constant speed. If there were no gravity, the space shuttle would continue to move in a straight line.

b. The space shuttle is in free fall because gravity pulls it toward Earth. The space shuttle would move straight down if it were not traveling forward.

c. The path of the space shuttle follows the curve of Earth's surface. This path is known as an orbit.

What Force Keeps an Object in Orbit?

Many objects in space are orbiting other objects. The moon orbits Earth, while Earth and the other planets orbit the sun. These objects all follow nearly circular paths. An object that travels in a circle is always changing direction.

If all the forces acting on an object balance each other out, the object will move in the same direction at the same speed forever. So, objects cannot orbit unless there is an unbalanced force acting on them. *Centripetal force* is the force that keeps an object moving in a circular path. Centripetal force pulls the object toward the center of the circle. The centripetal force of a body orbiting in space comes from gravity. ☑

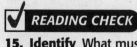

15. Identify What must be applied to an object to change its direction?

_____

_____

TAKE A LOOK

16. Identify Draw an arrow on the figure to show the direction that centripetal force acts on the moon.

Path of moon

The moon stays in orbit around Earth because Earth's gravity provides a centripetal force on the moon.

Copyright © by Holt, Rinehart and Winston; a Division of Houghton Mifflin Harcourt Publishing Company. All rights reserved.

SECTION 1 Gravity and Motion *continued*

What Is Projectile Motion?

Projectile motion is the curved path an object follows when it is thrown near the Earth's surface. The motion of a ball that has been thrown forward is an example of projectile motion.

Projectile motion is made up of two parts: horizontal motion and vertical motion. Horizontal motion is motion that is parallel to the ground. Vertical motion is motion that is perpendicular to the ground. The two motions do not affect each other. Instead, they combine to form the curved path we call projectile motion. ☑

When you throw a ball forward, your hand pushes the ball to make it move forward. This force gives the ball its horizontal motion. After the ball leaves your hand, no horizontal forces act on the ball (if we forget air resistance for now). So the ball's horizontal velocity does not change after it leaves your hand.

However, gravity affects the vertical part of projectile motion. Gravity pulls the ball straight down. All objects that are thrown accelerate downward because of gravity.

a After the ball leaves the pitcher's hand, the ball's _____ velocity is constant.

b The ball's vertical velocity increases because _____ causes it to accelerate downward.

c The two motions combine to form a _____ path.

READING CHECK

17. List What two motions combine to make projectile motion?

_____

_____

Critical Thinking

18. Infer If you are playing darts and you want to hit the bulls-eye, where should you aim?

_____

_____

TAKE A LOOK

19. List On the figure, fill in the three blanks with the correct words.

20. Apply Concepts If there were no air resistance, how fast would the ball's downward velocity be changing? Explain your answer.

_____

_____

_____

_____

Copyright © by Holt, Rinehart and Winston; a Division of Houghton Mifflin Harcourt Publishing Company. All rights reserved.

Section 1 Review

GLE 0707.Inq.2, GLE 0707.Inq.3, GLE 0707.Inq.5 **TN**

SECTION VOCABULARY

free fall the motion of a body when only the force of gravity is acting on the body **projectile motion** the curved path that an object follows when thrown, launched, or otherwise projected near the surface of Earth	**terminal velocity** the constant velocity of a falling object when the force of air resistance is equal in magnitude and opposite in direction to the force of gravity

1. **Explain** Is a parachutist in free fall? Why or why not?

_____

_____

_____

2. **Identify Cause and Effect** Complete the table below to show how forces affect objects.

Cause	Effect
Gravity acts on a falling object.	
	The falling object reaches terminal velocity.

3. **Calculate** A rock at rest falls off a cliff and hits the ground after 3.5 s. What is the rock's velocity just before it hits the ground? Show your work.

4. **Identify** What force must be applied to an object to keep it moving in a circular path?

_____

5. **Explain** Which part of projectile motion is affected by gravity? Explain how it is affected.

_____

_____

_____

_____

Copyright © by Holt, Rinehart and Winston; a Division of Houghton Mifflin Harcourt Publishing Company. All rights reserved.

CHAPTER 20 Forces and Motion

SECTION 2 Newton's Laws of Motion

BEFORE YOU READ

After you read this section, you should, be able to answer these questions:

• What is net force?

• What happens to objects that have no net force acting on them?

• How are mass, force, and acceleration related?

• How are force pairs related by Newton's third law of motion?

Tennessee Science Standards

GLE 0707.Inq.2
GLE 0707.Inq.3
GLE 0707.Inq.5
GLE 0707.T/E.1
GLE 0707.11.4

What Is a Net Force?

A *force* is a push or a pull. It is something that causes an object to change speed or direction. There are forces acting on all objects every second of the day. They are acting in all directions.

At first, this might not make sense. After all, there are many objects that are not moving. Are forces acting on an apple sitting on a desk? The answer is yes. Gravity is pulling the apple down. The desk is pushing the apple up.

In the figure below, the arrows represent the size and direction of the forces on the apple.

Forces Acting on an Apple

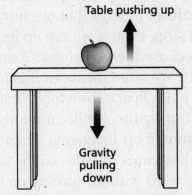

Table pushing up

Gravity pulling down

So, why doesn't the apple move? The apple is staying where it is because all the forces balance out. There are no unbalanced forces. That is, there is no net force on the apple. *Net force* is the total force acting on an object. If the net force on an object is zero, the object will not change speed or direction. ☑

STUDY TIP

Summarize in Pairs Read each of Newton's laws silently to yourself. After reading each law, talk about what you read with a partner. Together, try to figure out any ideas that you didn't understand.

TAKE A LOOK

1. Compare What is the size of the force pulling the apple down compared with the size of the force pushing it up?

_____

_____

READING CHECK

2. Explain What will not happen to an object if the net force acting on it is zero?

_____

_____

Copyright © by Holt, Rinehart and Winston; a Division of Houghton Mifflin Harcourt Publishing Company. All rights reserved.

What Is Newton's First Law of Motion?

Newton's first law of motion describes objects that have no unbalanced forces, or no net force, acting on them. It has two parts:

1. An object at rest will remain at rest.
2. An object moving at a constant velocity will continue to move at a constant velocity. ☑

PART 1: OBJECTS AT REST

An object that is not moving is said to be at rest. A golf ball on a tee is an example of an object at rest. An object at rest will not move unless an unbalanced force is applied to it. The golf ball will keep sitting on the tee until it is struck by a golf club.

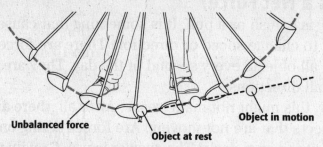

Unbalanced force   Object at rest   Object in motion

A golfball will remain at rest on a tee until it is acted on by the unbalanced force of a moving club.

PART 2: OBJECTS IN MOTION

The second part of Newton's first law can be hard to picture. On Earth, all objects that are moving eventually slow down and stop, even if we are no longer touching them. This is because there is always a net force acting on these objects. We will talk about this force later.

However, in outer space, Newton's first law can be seen easily. During the Apollo missions to the moon, the spacecraft turned off its engine when it was in space. It then drifted thousands of miles to the moon. It could keep moving forward without turning on its engines because there was no unbalanced force to slow it down.

Spacecraft traveling to the moon at constant velocity

READING CHECK

3. Identify What will happen to an object at rest if no unbalanced forces act on it?

_____

_____

4. Identify What will happen to an object moving at constant velocity?

_____

_____

TAKE A LOOK

5. Predict What would happen to the distances between the moving-ball images if the unbalanced force were greater?

_____

_____

TAKE A LOOK

6. Apply Concepts What must the spacecraft do to land softly on the moon?

_____

_____

Copyright © by Holt, Rinehart and Winston; a Division of Houghton Mifflin Harcourt Publishing Company. All rights reserved.

How Does Friction Affect Newton's First Law?

On Earth, friction makes observing Newton's first law difficult. If there were no friction, a ball would roll forever until something got in its way. Instead, it stops quickly because of friction.

Friction is a force that is produced whenever two surfaces touch each other. Friction always works against motion. ☑

Friction makes a rolling ball slow down and stop. It also makes a car slow down when its driver lets up on the gas pedal.

What Is Inertia?

Newton's first law is often called the *law of inertia*. **Inertia** is the ability of an object to resist any change in motion. In order to change an object's motion, a force has to overcome the object's inertia. So, in order to move an object that is not moving, you have to apply a force to it. Likewise, in order to change the motion of an object that is moving, you have to apply a force to it. The greater the object's inertia, the harder it is to change its motion.

How Are Mass and Inertia Related?

An object that has a small mass has less inertia than an object with a large mass. Imagine a golf ball and a bowling ball. Which one is easier to move?

The golf ball has much less mass than the bowling ball. The golf ball also has much less inertia. This means that a golf ball will be much easier to move than a bowling ball. ☑

Inertia makes it harder to accelerate a car than to accelerate a bicycle. Inertia also makes it easier to stop a moving bicycle than a car moving at the same speed.

☑ **READING CHECK**

7. Describe How does friction affect the forward motion of an object?

_____

_____

TN TENNESSEE STANDARDS CHECK

GLE 0707.11.4 Investigate how Newton's Laws of Motion explain an object's movement.

8. Explain When a moving car stops suddenly, why does a bag of groceries on the passenger seat fly forward into the dashboard?

_____

_____

_____

_____

☑ **READING CHECK**

9. Explain Why is a golf ball easier to throw than a bowling ball?

_____

_____

Copyright © by Holt, Rinehart and Winston; a Division of Houghton Mifflin Harcourt Publishing Company. All rights reserved.

SECTION 2 Newton's Laws of Motion *continued*

What Is Newton's Second Law of Motion?

Newton's second law of motion describes how an object moves when an unbalanced force acts on it. The second law has two parts:

1. The acceleration of an object depends on the mass of the object. If two objects are pushed or pulled by the same force, the object with the smaller mass will accelerate more. ☑

2. The acceleration of an object depends on the force applied to the object. If two objects have the same mass, the one you push harder will accelerate more.

READING CHECK

10. Apply Concepts Which object will accelerate more if the same force is applied to both: a pickup truck or a tractor-trailer truck?

TAKE A LOOK
11. Compare On the figure, draw arrows showing the size and direction of the force that the person is applying to the cart in each picture.

Acceleration

Acceleration

Acceleration

If the force applied to the carts is the same, the acceleration of the empty cart is greater than the acceleration of the loaded cart.

Acceleration increases when a larger force is exerted.

How Is Newton's Second Law Written as an Equation?

Newton's second law can be written as an equation. The equation shows how acceleration, mass, and net force are related to each other:

$$a = \frac{F}{m}, \text{ or } F = m \times a$$

In the equation, a is acceleration (in meters per second squared), m is mass (in kilograms), and F is net force (in newtons, N). One newton is equal to one kilogram multiplied by one meter per second squared. ☑

Newton's second law explains why all objects fall to Earth with the same acceleration. In the figure on the top of the next page, you can see how the larger force of gravity on the watermelon is balanced by its large mass.

READING CHECK

12. Identify What are the units of force?

Copyright © by Holt, Rinehart and Winston; a Division of Houghton Mifflin Harcourt Publishing Company. All rights reserved.
Interactive Reader and Study Guide   340   Forces and Motion

SECTION 2 Newton's Laws of Motion *continued*

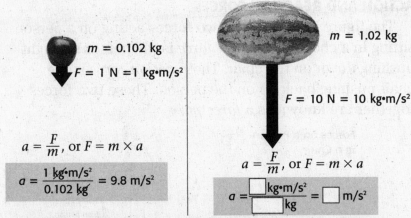

$m = 0.102$ kg

$F = 1$ N $= 1$ kg•m/s²

$m = 1.02$ kg

$F = 10$ N $= 10$ kg•m/s²

$a = \dfrac{F}{m}$, or $F = m \times a$

$a = \dfrac{1 \text{ kg•m/s}^2}{0.102 \text{ kg}} = 9.8 \text{ m/s}^2$

$a = \dfrac{F}{m}$, or $F = m \times a$

$a = \dfrac{\boxed{} \text{ kg•m/s}^2}{\boxed{} \text{ kg}} = \boxed{} \text{ m/s}^2$

The apple has less mass than the watermelon does. So, less force is needed to give the apple the same acceleration that the watermelon has.

Math Focus

13. Calculate In the figure, fill in the boxes with the correct numbers to calculate the acceleration of the watermelon.

How Can You Solve Problems Using Newton's Second Law?

You can use the equation $F = ma$ to calculate how much force you need to make a certain object accelerate a certain amount. Or you can use the equation $a = \dfrac{F}{m}$ to calculate how much an object will accelerate if a certain net force acts on it.

For example, what is the acceleration of a 3 kg mass if a force of 14.4 N is used to move the mass?

Step 1: Write the equation that you will use.

$$a = \frac{F}{m}$$

Step 2: Replace the letters in the equation with the values from the problem.

$$a = \frac{14.4 \text{ N}}{3 \text{ kg}} = \frac{14.4 \text{ kg•m/s}^2}{3 \text{ kg}} = 4.8 \text{ m/s}^2$$

Math Focus

14. Calculate What force is needed to accelerate a 1,250 kg car at a rate of 40 m/s²? Show your work in the space below.

What Is Newton's Third Law of Motion?

All forces act in pairs. Whenever one object exerts a force on a second object, the second object exerts a force on the first object. The forces are always equal in size and opposite in direction.

For example, when you sit on a chair, the force of your weight pushes down on the chair. At the same time, the chair pushes up on you with a force equal to your weight.

Copyright © by Holt, Rinehart and Winston; a Division of Houghton Mifflin Harcourt Publishing Company. All rights reserved.

ACTION AND REACTION FORCES

The figure below shows two forces acting on a person sitting in a chair. The *action force* is the person's weight pushing down on the chair. The *reaction force* is the chair pushing back up on the person. These two forces together are known as a *force pair*.

Forces on a Person in a Chair

Chair pushes up on person.

Person's weight pushes down on chair.

TAKE A LOOK
15. Identify On the figure, label the action force and the reaction force.

Action and reaction forces are also present when there is motion. The figures below show some more examples of action and reaction forces.

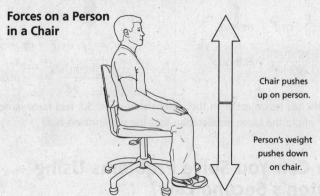

TAKE A LOOK
16. Describe How big is the reaction force compared with the action force in each picture?

_____

_____

Critical Thinking

17. Apply Concepts If the ball exerts a force on the bat, why doesn't the bat move backward?

_____

_____

_____

_____

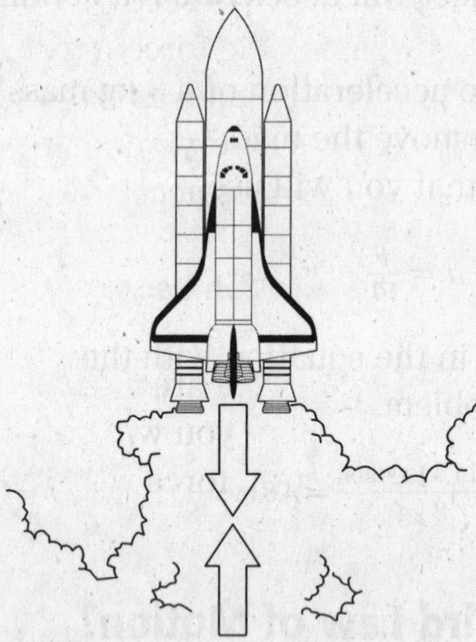

The space shuttle's thrusters push gases downward. The gases push the space shuttle upward with equal force.

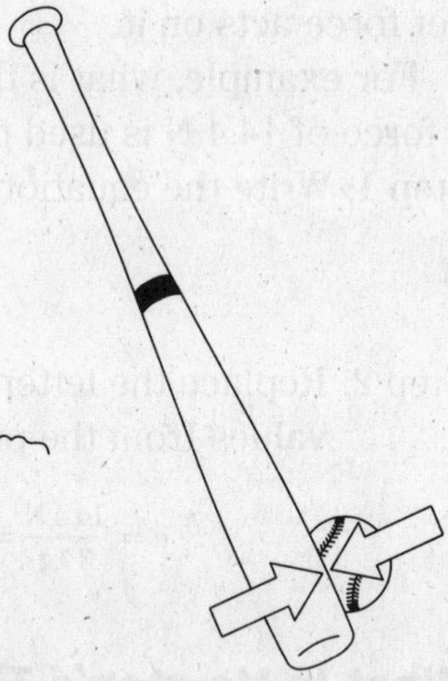

The bat exerts a force on the ball and sends the ball flying. The ball exerts an equal force on the bat.

The action force and reaction force always act on different objects. For example, when you sit in a chair, the action force (your weight) acts on the chair. However, the reaction force (the chair pushing up on you) acts on you.

Copyright © by Holt, Rinehart and Winston; a Division of Houghton Mifflin Harcourt Publishing Company. All rights reserved.

HARD-TO-SEE REACTION FORCES

In the figure below, a ball is falling toward the Earth's surface. The action force is the Earth's gravity pulling down on the ball. What is the reaction force?

The force of gravity between Earth and a falling object is a force pair.

Believe it or not, the reaction force is the ball pulling up on Earth. Have you ever felt this reaction force when you have dropped a ball? Of course not. However, both forces are present. So, why don't you see or feel Earth rise?

To answer this question, recall Newton's second law. Acceleration depends on the mass and the force on an object. The force acting on Earth is the same as the force acting on the ball. However, Earth has a very, very large mass. Because it has such a large mass, its acceleration is too small to see or feel. ☑

You can easily see the ball's acceleration because its mass is small compared with Earth's mass. Most of the objects that fall toward Earth's surface are much less massive than Earth. This means that you will probably never feel the effects of the reaction force when an object falls to Earth.

TAKE A LOOK
18. Identify On the figure, draw and label arrows showing the size and direction of the action force and the reaction force for the ball falling to the Earth.

✓ **READING CHECK**

19. Explain Why can't you feel the effect of the reaction force when an object falls to Earth?

_____

_____

_____

Copyright © by Holt, Rinehart and Winston; a Division of Houghton Mifflin Harcourt Publishing Company. All rights reserved.

Name ______________________ Class ______________ Date ____________

Section 2 Review

GLE 0707.Inq.2, GLE 0707.Inq.3, GLE 0707.Inq.5, GLE 0707.T/E.1, GLE 0707.11.4

SECTION VOCABULARY

inertia the tendency of an object to resist being moved or, if the object is moving, to resist a change in speed or direction until an outside force acts on the object	

1. Explain How are inertia and mass related?

____

2. Use Graphics The hockey puck shown below is moving on the ice at a constant velocity. The arrow represents the constant velocity. In the box, draw the arrow that represents the velocity of the puck if no unbalanced forces act on the puck.

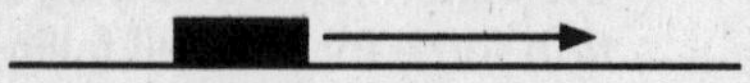

3. Use Graphics The ball shown below has two forces acting on it. The arrows represent the size and direction of the forces. In the box, draw the arrow that represents the net force on the ball.

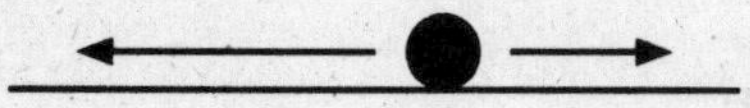

4. Identify Describe two things you can do to increase the acceleration of an object.

____

____

5. Identify Identify the action and reaction forces when you kick a soccer ball.

____

____

____

6. Calculate What force is needed to accelerate a 40 kg person at a rate of 4.5 m/s^2? Show your work.

____

____

____

Copyright © by Holt, Rinehart and Winston; a Division of Houghton Mifflin Harcourt Publishing Company. All rights reserved.

CHAPTER 20 Forces and Motion
SECTION
3 **Momentum**

 **Tennessee Science Standards**
GLE 0707.Inq.2
GLE 0707.Inq.3
GLE 0707.Inq.5
GLE 0707.11.4

BEFORE YOU READ

After you read this section, you should be able to answer these questions:

• What is momentum?

• How is momentum calculated?

• What is the law of conservation of momentum?

What Is Momentum?

Picture a compact car and a large truck moving at the same velocity. The drivers of both vehicles put on the brakes at the same time. Which vehicle will stop first? You most likely know that it will be the car. But why? The answer is momentum.

The momentum of an object depends on the object's mass and velocity. **Momentum** is the product of the mass and velocity of an object. In the figure below, a car and a truck are shown moving at the same velocity. Because the truck has a larger mass, it has a larger momentum. A greater force will be needed to stop the truck. ☑

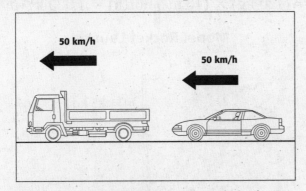
A truck and car traveling with the same velocity do not have the same momentum.

Object	Momentum
A train moving at 30 km/h	
A bird sitting on a branch high in a tree	
A truck moving at 30 km/h	
A rock sitting on a beach	

STUDY TIP

Visualize As you read, try to picture in your head the events that are described. If you have trouble imagining, draw a sketch to illustrate the event.

READING CHECK

1. Identify What is the momentum of an object?

_____

_____

TAKE A LOOK

2. Predict How could the momentum of the car be increased?

_____

_____

Critical Thinking

3. Apply Concepts Fill in the chart to the left to show which object has the most momentum, which object has a smaller amount of momentum, and which objects have no momentum.

Copyright © by Holt, Rinehart and Winston; a Division of Houghton Mifflin Harcourt Publishing Company. All rights reserved.

How Can You Calculate Momentum?

If you know what an object's mass is and how fast it is going, you can calculate its momentum. The equation for momentum is

$$p = m \times v$$

In this equation, p is momentum (in kilograms multiplied by meters per second), m is the mass of the object (in kilograms), and v is the velocity of the object (in meters per second). ☑

Like velocity, momentum has direction. The direction of an object's momentum is always the same as the direction of the object's velocity.

Use the following procedure to solve momentum problems:

Step 1: Write the momentum equation.
Step 2: Replace the letters in the equation with the values from the problem.

Let's try a problem. A 120 kg ostrich is running with a velocity of 16 m/s north. What is the momentum of the ostrich?

Step 1: The equation is $p = m \times v$.
Step 2: m is 120 kg and v is 16 m/s north. So,

$p = (120 \text{ kg}) \times (16 \text{ m/s north}) = 1{,}920 \text{ kg}\bullet\text{m/s north}$

Model Rocket Launch

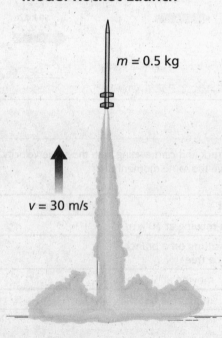

$m = 0.5$ kg

$v = 30$ m/s

READING CHECK

4. Identify What do you need to know in order to calculate an object's momentum?

_____

_____

Math Focus

5. Calculate A 6 kg bowling ball is moving at 10 m/s down the alley toward the pins. What is the momentum of the bowling ball? Show your work.

_____

TAKE A LOOK

6. Describe Use a metric ruler to draw an arrow next to the rocket to show its momentum. The size of the rocket's momentum is 15 kg•m/s. The scale for the arrow should be 1 cm = 10 kg•m/s.

Copyright © by Holt, Rinehart and Winston; a Division of Houghton Mifflin Harcourt Publishing Company. All rights reserved.

SECTION 3 Momentum *continued*

What Is the Law of Conservation of Momentum?

When a moving object hits an object at rest, some or all of the momentum of the first object is transferred to the second object. This means that the object at rest gains all or some of the moving object's momentum. During a collision, the total momentum of the two objects remains the same. Total momentum doesn't change. This is called the *law of conservation of momentum*.

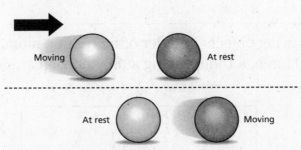

Momentum

Moving   At rest

At rest   Moving

The momentum before a collision is equal to the momentum after the collision.

TAKE A LOOK
7. Identify Draw an arrow showing the size and direction of the darker ball's momentum after its collision with the lighter ball.

The law of conservation of momentum is true for any colliding objects as long as there are no outside forces. For example, if someone holds down the darker ball in the collision shown above, it will not move. In that case, the momentum of the lighter ball would be transferred to the person holding the ball. The person is exerting an outside force. ☑

The law of conservation of momentum is true for objects that either stick together or bounce off each other during a collision. In both cases, the velocities of the objects will change so that their total momentum stays the same.

☑ READING CHECK
8. Identify When is the law of conservation of momentum not true for two objects that are interacting?

_____

_____

When football players tackle another player, they stick together. The velocity of each player changes after the collision because of conservation of momentum.

Although the bowling ball and bowling pins bounce off each other and move in different directions after a collision, momentum is neither gained nor lost.

📣 Say It
Explain Words In a group, discuss how the everyday use of the word *momentum* differs from its use in science.

Copyright © by Holt, Rinehart and Winston; a Division of Houghton Mifflin Harcourt Publishing Company. All rights reserved.

Section 3 Review

GLE 0707.Inq.2, GLE 0707.Inq.3, GLE 0707.Inq.5, GLE 0707.11.4

SECTION VOCABULARY

momentum a quantity defined as the product of the mass and velocity of an object

1. Explain A car and a train are moving at the same velocity. Do the two objects have the same momentum? Explain your answer.

_____

_____

_____

2. Show Relationships Put the following objects in order of increasing momentum: a parked car, a train moving at 50 km/h, a train moving at 80 km/h, a car moving at 50 km/h.

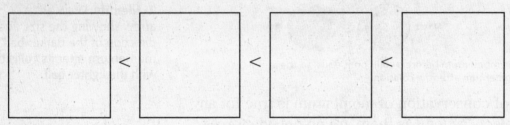

3. Calculate A 2.5 kg puppy is running with a velocity of 4.8 m/s south. What is the momentum of the puppy? Show your work.

4. Explain What is the law of conservation of momentum?

_____

_____

5. Calculate A ball has a momentum of 1 kg•m/s north. It hits another ball of equal mass that is at rest. If the first ball stops, what is the momentum of the other ball after the collision? (Assume there are no outside forces.) Explain your answer.

_____

_____

_____

Copyright © by Holt, Rinehart and Winston; a Division of Houghton Mifflin Harcourt Publishing Company. All rights reserved.

CHAPTER 21 | Work and Machines

SECTION
1 **Work and Power**

TN🠆 Tennessee Science
Standards
GLE 0707.Inq.3
GLE 0707.Inq.5
GLE 0707.11.2

BEFORE YOU READ

After you read this section, you should be able to answer these questions:

• What is work?

• How do we measure work?

• What is power and how is it calculated?

What Is Work?

You may think of work as a large homework assignment. You have to read a whole chapter by tomorrow. That sounds like a lot of work, but in science, work has a different meaning. **Work** is done when a force causes an object to move in the direction of the force. You might have to do a lot of thinking, but you are not using a force to move anything. You are doing work when you to turn the pages of the book or move your pen when writing. ☑

The student in the figure below is bowling. She is doing work. She applies a force to the bowling ball and the ball moves. When she lets go of the ball she stops doing work. The ball keeps rolling but she is not putting any more force on the ball.

🖉 **STUDY TIP**

As you read this section, write the questions in Before You Read in your science notebook and answer each one.

✓ **READING CHECK**

1. **Describe** When is work done on an object?

_____
_____
_____

You might be surprised to find out that bowling is work!

The direction of the force

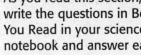

The direction of the bowling ball

TN🠆 **TENNESSEE STANDARDS CHECK**

GLE 0707.Inq.5 Communicate scientific understanding using descriptions, explanations, and models.

Word Help: communicate to make known; to tell

2. **Identify** How is energy transmitted to a bowling ball to make it move down the alley?

_____
_____

How Is Energy Transferred When Work Is Done?

The bowler in the figure above has done work on the bowling ball. Since the ball is moving, it now has *kinetic* energy. The bowler has transferred energy to the ball.

Copyright © by Holt, Rinehart and Winston; a Division of Houghton Mifflin Harcourt Publishing Company. All rights reserved.

SECTION 1 Work and Power *continued*

When Is Work Done On An Object?

Applying a force does not always mean that work is done. If you push a car but the car does not move, no work is done on the car. Pushing the car may have made you tired. If the car has not moved, no work is done on the car. When the car moves, work is done. If you apply a force to an object and it moves in that direction, then work is done.

How Are Work and Force Different?

You can apply a force to an object, but not do work on the object. Suppose you are carrying a heavy suitcase through an airport. The direction of the force you apply to hold the suitcase is up. The suitcase moves in the direction you are walking. The direction the suitcase is moving is not the same as the direction of the applied force. So when you carry the suitcase, no work is done on the suitcase. Work is done when you lift the suitcase off the ground.

Work is done on an object if two things happen.

1. An object moves when a force is applied.

2. The object moves in the direction of the force. ☑

In the figure below, you can see how a force can cause work to be done.

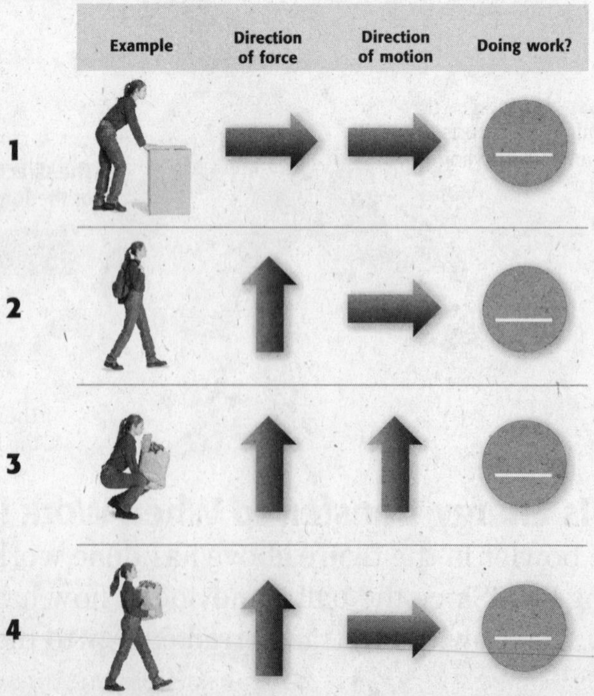

Copyright © by Holt, Rinehart and Winston; a Division of Houghton Mifflin Harcourt Publishing Company. All rights reserved.

✓ **READING CHECK**

3. Describe What two things must happen to do work on an object?

_____

_____

_____

_____

_____

TAKE A LOOK

4. Identify In the figure, there are four examples. For each example, decide if work is being done and write your answers in the figure.

SECTION 1 Work and Power *continued*

How Is Work Calculated?

An equation can be written to calculate the work (W) it takes to move an object. The equation shows how work, force, and distance are related to each other:

$$W = F \times d$$

In the equation, F is the force applied to an object (in newtons). d is the distance the object moves in the direction of the force (in meters). The unit of work is the newton-meter (N×m). This is also called a **joule** (J). When work is done on an object, energy is transferred to the object. The joule is a unit of energy.

Let's try a problem. How much work is done if you push a chair that weighs 60 N across a room for 5 m?

Step 1: Write the equation.

$$W = F \times d$$

Step 2: Place values into the equation, and solve for the answer.

$$W = 60 \text{ N} \times 5 \text{ m} = 300 \text{ J}$$

The work done on the chair is 300 joules.

Math Focus
5. Calculate How much work is done pushing a car 20 m with a force of 300 N? Show your work.

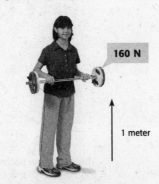

W = 80 N × 1 m = 80 J
The force to lift an object is the same as the force of gravity on the object. In other words, the object's weight is the force.

W = 160 N × 1 m = 160 J
The amount of work increases when the weight of an object increases. More force is needed to lift the object.

Math Focus
6. Calculate In the last figure, a barbell weighing 80 N is lifted 2 m off of the ground. How much work is done? Show your work.

W = _____
The amount of work also increases when the distance increases.

Copyright © by Holt, Rinehart and Winston; a Division of Houghton Mifflin Harcourt Publishing Company. All rights reserved.

SECTION 1 Work and Power *continued*

Two Paths, Same Work?

A car is pushed to the top of a hill using two different paths. The first path is a long road that has a low, gradual slope. The second path is a steep cliff. Pushing the car up the long road doesn't need as much force as pulling it up the steep cliff. But, believe it or not, the same amount of work is done either way. It is clear that you would need a different amount of force for each path.

Look at the figure below. Pushing the car up the long road uses a smaller force over a larger distance. Pulling it up the steep cliff uses a larger force over a smaller distance. The car ends up at the same place either way, so according to the Law of Conservation of Energy, the work done on the car is the same. When work is calculated for both paths, you get the same amount of work for each path. ☑

READING CHECK

7. Explain Suppose it takes the same amount of work for two people to move an object. One person applies less force in moving the object than the other. How can they both do the same amount of work?

_____

_____

_____

_____

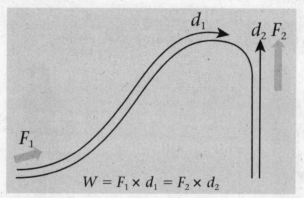

There are two paths to move the car to the top of the hill. d_1 follows the shape of the hill. d_2 is straight from the bottom to the top of the hill. The same work is done in both paths. The distance and force for each of the paths are different.

Let's do a calculation. Suppose path 1 needs a force of 200 N to push the car up the hill for 30 m. Path 2 needs a force of 600 N to pull the car up the hill 10 m. Show that the work is the same for both paths.

Step 1: Write the equation.

$$W_1 = F_1 \times d_1 \text{ and } W_2 = F_2 \times d_2$$

Step 2: Place values into the equation, and solve for the answer.

$$W_1 = 200 \text{ N} \times 30 \text{ m} \times 600 \text{ N and } W_2 =$$
$$600 \text{ N} \times 10 \text{ m} = 600 \text{ J}$$

The amount of work done is the same for both paths.

Copyright © by Holt, Rinehart and Winston; a Division of Houghton Mifflin Harcourt Publishing Company. All rights reserved.

SECTION 1 Work and Power *continued*

What Is Power?

The word *power* has a different meaning in science than how we often use the word. **Power** is how fast energy moves from one object to another.

Power measures how fast work is done. The *power output* of something is another way to say how much work can be done quickly. For example, a more powerful weightlifter can lift a barbell more quickly than a less powerful weightlifter.

How Is Power Calculated?

To calculate power (*P*), divide the work (*W*) by the time (*t*) it takes to do the work. This is shown in the following equation: $P = \dfrac{W}{t}$

Power is written in the units joules per second (J/s). This is called a **watt**. One *watt* (W) is the same as 1 J/s.

Let's do a problem. A stage manager at a play raises the curtain by doing 5,976 J of work on the curtain in 12 s. What is the power output of the stage manager?

Step 1: Write the equation.
$$P = \frac{W}{t}$$

Step 2: Place values into the equation, and solve for the answer.
$$P = \frac{5{,}976 \text{ J}}{12 \text{ s}} = 498 \text{ W}$$

Critical Thinking

8. Apply Concepts An escalator and a elevator can transport a person from one floor to the next. The escalator does it in 15 s and the elevator takes 10 s.

Which does more work on the person? Which has the greater power output?

_____

_____

_____

_____

Math Focus
9. Calculate A light bulb is on for 12 s, and during that time it uses 1,200 J of electrical energy. What is the wattage (power output) of the light bulb?

Wood can be sanded by hand or with an electric sander. The electric sander does the same amount of work faster.

TAKE A LOOK
10. Explain Why does the electric sander have a higher power output than sanding by hand?

_____

_____

Copyright © by Holt, Rinehart and Winston; a Division of Houghton Mifflin Harcourt Publishing Company. All rights reserved.

Section 1 Review

GLE 0707.11.2, GLE 0707.Inq.3, GLE 0707.Inq.5

SECTION VOCABULARY

joule the unit used to express energy; equivalent to the amount of work done by a force of 1 N acting through a distance of 1 m in the direction of the force (symbol, J) **power** the rate at which work is done or energy is transformed	**watt** the unit used to express power; equivalent to a joule per second (symbol, W) **work** the transfer of energy to an object by using a force that causes the object to move in the direction of the force

1. Explain Is work always done on an object when a force is applied to the object? Why or why not?

_____

_____

2. Analyze Work is done on a ball when a pitcher throws it. Is the pitcher still doing work on the ball as it flies through the air? Explain your answer.

_____

_____

3. Calculate A force of 10 N is used to push a shopping cart 10 m. How much work is done? Show how you got your answer.

_____

_____

_____

4. Compare How is the term work different than the term power?

_____

_____

_____

5. Identify How can you increase your power output by changing the amount of work that you do? How can you increase your power output by changing the time it takes you to do the work?

_____

_____

6. Calculate You did 120 J of work in 3 s. How much power did you use? Show how you got your answer.

_____

Copyright © by Holt, Rinehart and Winston; a Division of Houghton Mifflin Harcourt Publishing Company. All rights reserved.

CHAPTER 21 | Work and Machines

SECTION
2 **What Is a Machine?**

BEFORE YOU READ

After you read this section, you should be able to answer these questions:

• What is a machine?

• How does a machine make work easier?

• What is mechanical advantage?

• What is mechanical efficiency?

Tennessee Science Standards
GLE 0707.Inq.2
GLE 0707.Inq.3
GLE 0707.Inq.5
GLE 0707.11.2

What Is a Machine?

Imagine changing a flat tire without a jack to lift the car or a tire iron to remove the bolts. Would it be easy? No, you would need several people just to lift the car! Sometimes you need the help of machines to do work. A **machine** is something that makes work easier. It does this by lowering the size or direction of the force you apply. ☑

When you hear the word machine, what kind of objects do you think of? Not all machines are hard to use. You use many simple machines every day. Think about some of these machines. The following table lists some jobs you use a machine to do.

Work	Machine you could use
Removing the snow in your driveway	
Getting you to school in the morning	
Painting a room	
Picking up the leaves from your front yard	
Drying your hair	

STUDY TIP

Brainstorm Think of ways that machines make your work easier and write them down in your science notebook.

READING CHECK

1. Describe How does a machine make work easier?

_____

_____

_____

TAKE A LOOK
2. Identify Complete the table by filling in the last column.

Two Examples of Everyday Machines

Wheelchair

Scissors

Copyright © by Holt, Rinehart and Winston; a Division of Houghton Mifflin Harcourt Publishing Company. All rights reserved.

How Do Machines Make Work Easier?

The type of machine and how it is designed depend on the task that needs to be done. Engineers design simple or complex machines based on how much work is to be input and output by those machines. You can use a simple machine, such as a screwdriver, to remove the lid from a paint can. An example of this is shown in the figure below. The screwdriver is a type of *lever*. The tip of the screwdriver is put under the lid and you push down on the screwdriver. The tip of the screwdriver lifts the lid as you push down. In other words, you do work on the screwdriver, and the screwdriver does work on the lid.

WORK IN, WORK OUT

When you use a machine, you do work and the machine does work. The work you do on a machine is called the **work input**. The force you apply to the machine to do the work is the *input force*. The work done by the machine on another object is called the **work output**. The force the machine applies to do this work is the *output force*. ☑

✓ **READING CHECK**

3. Identify What is work input? What is work output?

_____

_____

_____

_____

TAKE A LOOK

4. Identify What is the force you put on a screwdriver called?

_____

What is the force the screwdriver puts on the lid called?

_____

You do work on the machine and the machine does work on something else.

Output force

Input force

HOW MACHINES HELP

Machines do not decrease the amount of work you do. Remember that work equals the force applied times the distance ($W = F \times d$). Machines lower the force that is needed to do the same work by increasing the distance the force is applied. This means less force is needed to do the same work.

In the figure above, you apply a force to the screwdriver and the screwdriver applies a force to the lid. The force the screwdriver puts on the lid is greater than the force you apply. Since you apply this force over a greater distance, your work is easier. ☑

✓ **READING CHECK**

5. Identify To make work easier, what force is lowered?

_____

Copyright © by Holt, Rinehart and Winston; a Division of Houghton Mifflin Harcourt Publishing Company. All rights reserved.

SECTION 2 What Is a Machine? *continued*

SAME WORK, DIFFERENT FORCE

Machines make work easier by lowering the size or direction (or both) of the input force. A machine doesn't change the amount of work done. A ramp can be used as a simple machine shown in the figure below. In this example a ramp makes work easier because the box is pushed with less force over a longer distance.

Input Force and Distance

The boy lifts the box. The input force is the same as the weight of the box.

The girl uses a ramp to lift the box. The input force is the less than the weight of the box. She applies this force for a longer distance.

TAKE A LOOK
6. Describe Notice that the box is lifted the same distance by the boy and the girl. Which does more work on the box?

_____

_____

Look at the boy lifting a box in the figure above. Suppose the box weighs 450 N and is lifted 1 m. How much work is done to move the box?

Step 1: Write the equation.
$$W = F \times d$$

Step 2: Place values into the equation, and solve.
$$W = 450 \text{ N} \times 1 \text{ m} = 450 \text{ N} \times \text{m, or } 450 \text{ J}$$

Look at the girl using a ramp. Suppose the force to push the box is 150 N. It is pushed 3 m. How much work is done to move the box?

Step 1: Write the equation.
$$W = F \times d$$

Step 2: Place values into the equation, and solve.
$$W = 150 \text{ N} \times 3 \text{ m} = 450 \text{ N} \times \text{m, or } 450 \text{ J}$$

Work done to move box is 450 J.

The same amount of work is done with or without the ramp. The boy uses more force and a shorter distance to lift the box. The girl uses less force and a longer distance to move the box. They each use a different force and a different distance to do the same work.

Math Focus
7. Calculate How much work is done when a 50 N force is applied to a 0.30 m screwdriver to lift a paint can lid?

Copyright © by Holt, Rinehart and Winston; a Division of Houghton Mifflin Harcourt Publishing Company. All rights reserved.

SECTION 2 What Is a Machine? *continued*

FORCE AND DISTANCE CHANGE TOGETHER

When a machine changes the size of the output force, the distance must change. When the output force increases, the distance the object moves must decrease. This is shown in the figure of the nutcracker below. The handle is squeezed with a smaller force than the output force that breaks the nut. So, the output force is applied over a smaller distance.

Machines Change the Size and/or Direction of a Force

Input force

Output force

A nutcracker increases the force but applies it over a shorter distance.

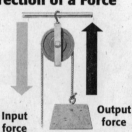

Input force

Output force

A simple pulley changes the direction of the input force, but the size of the output force is the same as the input force.

What Is Mechanical Advantage?

Some machines can increase the size of the force more than others. A machine's **mechanical advantage** tells you how much the force increases. The mechanical advantage compares the input force with the output force. Part of the design cycle for producing machines is to determine how much mechanical advantage they will provide.

CALCULATING MECHANICAL ADVANTAGE

A machine's mechanical advantage can be calculated by using the following equation:

$$mechanical\ advantage\ (MA) = \frac{output\ force}{input\ force}$$

Look at this example. You push a box weighing 500 N up a ramp (output force) by applying 50 N of force. What is the mechanical advantage of the ramp?

Step 1: Write the equation.

$$mechanical\ advantage\ (MA) = \frac{output\ force}{input\ force}$$

Step 2: Place values into the equation, and solve.

$$MA = \frac{500\ N}{50\ N} = 10$$

The mechanical advantage of the ramp is 10.

Math Focus

8. Explain Why is the input arrow shorter than the output arrow in the photo of the nutcracker? Why are the arrows the same length in the figure of the pulley?

_____

_____

_____

_____

Math Focus

9. Calculate What is the mechanical advantage of a nutcracker if the input force is 65 N and the output force is 130 N?

Copyright © by Holt, Rinehart and Winston; a Division of Houghton Mifflin Harcourt Publishing Company. All rights reserved.

What Is a Machine's Mechanical Efficiency?

No machine changes all of the input work into output work. Some of the work done by the machine is lost to *friction*. Friction is always present when two objects touch. The work done by the machine plus the work lost to friction is equal to the work input. This is known as the *Law of Conservation of Energy*.

The **mechanical efficiency** of a machine compares a machine's work output with the work input. A machine is said to be efficient if it doesn't lose much work to friction.

CALCULATING MECHANICAL EFFICIENCY

A machine's mechanical efficiency is calculated using the following equation:

$$mechanical\ advantage\ (MA) = \frac{work\ output}{work\ input} \times 100$$

The 100 in the equation means that mechanical efficiency is written as a percentage. It tells you the percentage of work input that gets done as work output.

Let's try a problem. You do 100 J of work on a machine and the work output is 40 J. What is the mechanical efficiency of the machine?

Step 1: Write the equation.

$$mechanical\ effeciency\ (ME) = \frac{work\ output}{work\ input} \times 100$$

Step 2: Place values into the equation, and solve.

$$ME = \frac{40\ J}{100\ J} \times 100 = 40\%$$

Process Chart

You apply an input force to a machine.

↓

The machine changes the size and/or direction of the force.

↓

The machine applies an _____ _____ on the object.

↓

The mechanical efficiency and/or advantage of the machine can be determined.

Math Focus

10. Calculate What is the mechanical efficiency of a simple pulley if the input work is 100 N and the output work is 90 N?

Math Focus

11. Identify Fill in the missing words on the process chart.

_____

Copyright © by Holt, Rinehart and Winston; a Division of Houghton Mifflin Harcourt Publishing Company. All rights reserved.

Section 2 Review

GLE 0707.Inq.2, GLE 0707.Inq.3, GLE 0707.Inq.5, GLE 0707.11.2

SECTION VOCABULARY

machine a device that helps do work by either overcoming a force or changing the direction of the applied force	**work input** the work done on a machine; the product of the input force and the distance through which the force is exerted
mechanical advantage a number that tells how many times a machine multiplies force	**work output** the work done by a machine; the product of the output force and the distance through which the force is exerted
mechanical efficiency a quantity, usually expressed as a percentage, that measures the ratio of work output to work input in a machine	

1. Explain Why is it easier to move a heavy box up a ramp than it is to lift the box off the ground?

_____

_____

2. Identify What are the two ways that a machine can make work easier?

_____

3. Compare What is the difference between work input and work output?

_____

_____

4. Calculate You apply an input force of 20 N to a hammer that applies an output force of 120 N to a nail. What is the mechanical advantage of the hammer? Show your work.

5. Explain Why is a machine's work output always less than the work input?

_____

6. Calculate What is the mechanical efficiency of a machine with a work input of 75 J and a work output of 25 J? Show your work.

Copyright © by Holt, Rinehart and Winston; a Division of Houghton Mifflin Harcourt Publishing Company. All rights reserved.

CHAPTER 21 Work and Machines
SECTION 3 Types of Machines

TN **Tennessee Science Standards**
GLE 0707.Inq.2
GLE 0707.Inq.5
GLE 0707.T/E.1
GLE 0707.11.1
GLE 0707.11.2

BEFORE YOU READ

After you read this section, you should be able to answer these questions:

• What are the six simple machines?

• What is a compound machine?

What Are the Six Types of Simple Machines?

All machines are made from one or more of the six simple machines. They are the lever, the pulley, the wheel and axle, the inclined plane, the wedge, and the screw. They each work differently to change the size or direction of the input force.

STUDY TIP

As you read through the section, study the figures of the types of machines. Make a list of the six simple machines and a sentence describing how each works.

What Is a Lever?

A commonly used simple machine is the **lever**. A *lever* has a bar that rotates at a fixed point, called a *fulcrum*. The force that is applied to the lever is the *input force*. The object that is being lifted by the lever is called the *load*. A lever is used to apply a force to move a load. There are three classes of levers. They all have a different location for the fulcrum, the load, and the input force on the bar. ☑

READING CHECK

1. Describe How does a lever do work?

_____

_____

_____

FIRST-CLASS LEVERS

In first-class levers, the fulcrum is between the input force and the load as shown in the figure below. The direction of the input force always changes in this type of lever. They can also be used to increase either the force or the distance of the work.

Examples of First-Class Levers

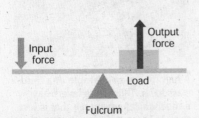

The fulcrum can be located closer to the load than to the input force. This lever has a mechanical advantage that is greater than 1. The output force is larger than the input force.

The fulcrum can be located exactly in the middle. This lever has a mechanical advantage that is equal to 1. The output force is the same as the input force.

Critical Thinking

2. Predict Suppose the fulcrum in the figure to the far left is located closer to the input force. How will this change the mechanical advantage of the lever? Explain.

_____

_____

_____

_____

Copyright © by Holt, Rinehart and Winston; a Division of Houghton Mifflin Harcourt Publishing Company. All rights reserved.

Critical Thinking

3. Explain How does a second-class lever differ from a first-class lever?

_____

_____

_____

_____

SECOND-CLASS LEVERS

In second-class levers, the load is between the fulcrum and the input force as shown in the figure below. They do not change the direction of the input force. Second-class levers are often used to increase the force of the work. You apply less force to the lever than the force it puts on the load. This happens because the force is applied over a larger distance.

Examples of Second-Class Levers

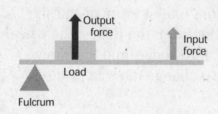

In a second-class lever, the output force, or load, is between the input force and the fulcrum.

A wheelbarrow is an example of a second-class lever. Second-class levers have a mechanical advantage that is greater than 1.

THIRD-CLASS LEVERS

In third-class levers, the input force is between the fulcrum and the load as shown in the figure below. The direction of the input force does not change and the input force does not increase. This means the output force is always less than the input force. Third-class levers do increase the distance that the output force works.

Critical Thinking

4. Explain Why can't a third-class lever have a mechanical advantage of 1 or more?

_____

_____

_____

Examples of Third-Class Levers

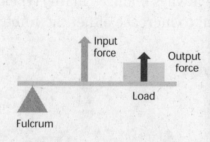

In a third-class lever, the input force is between the fulcrum and the load.

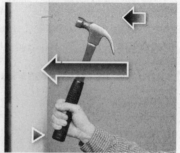

A hammer is an example of a third-class lever. Third-class levers have a mechanical advantage that is less than 1. The output force is less than the input force. Third-class levers increase the distance that the output force acts on.

Copyright © by Holt, Rinehart and Winston; a Division of Houghton Mifflin Harcourt Publishing Company. All rights reserved.

What Is a Pulley?

When you open window blinds by pulling on a cord, you are using a pulley. A **pulley** is a simple machine with a grooved wheel that holds a rope or a cable. An input force is applied to one end of the cable. The object being lifted is called the load. The load is attached to the other end. The different types of pulleys are shown in the figure at the bottom of the page. ☑

FIXED PULLEYS

A *fixed pulley* is connected to something that does not move, such as a ceiling. To use a fixed pulley, you pull down on the rope to lift the load. The direction of the force changes. Since the size of the output force is the same as the input force, the mechanical advantage (*MA*) is 1. An elevator is an example of a fixed pulley. ☑

MOVABLE PULLEYS

Moveable pulleys are connected directly to the object that is being moved, which is the load. The direction does not change, but the size of the force does. The mechanical advantage (*MA*) of a movable pulley is 2. This means that less force is needed to move a heavier load. Large construction cranes often use movable pulleys.

BLOCK AND TACKLES

If you use a fixed pulley and a movable pulley together, you form a pulley system. This is a *block and tackle*. The mechanical advantage (*MA*) of a block and tackle is equal to the number of sections of rope in the system.

READING CHECK

5. **Describe** What is a pulley?

_____
_____
_____

READING CHECK

6. **Describe** Why can't a fixed pulley have a mechanical advantage greater than 1?

_____
_____
_____

Types of Pulleys

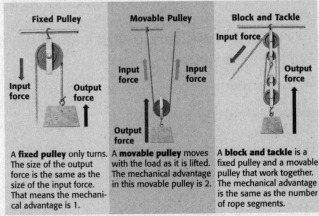

A **fixed pulley** only turns. The size of the output force is the same as the size of the input force. That means the mechanical advantage is 1.

A **movable pulley** moves with the load as it is lifted. The mechanical advantage in this movable pulley is 2.

A **block and tackle** is a fixed pulley and a movable pulley that work together. The mechanical advantage is the same as the number of rope segments.

TAKE A LOOK

7. **Identify** The section of rope labeled Input force for the block and tackle is not counted as a rope segment. There are four rope segments in this block and tackle. What is the mechanical advantage of the block and tackle?

_____

Copyright © by Holt, Rinehart and Winston; a Division of Houghton Mifflin Harcourt Publishing Company. All rights reserved.

Critical Thinking

8. Explain If the input force remains constant and the wheel is made smaller, what happens to the output force?

_____

_____

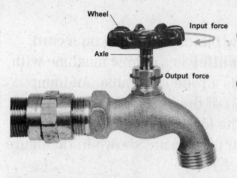

Wheel

Input force

Axle

Output force

a When a small input force is applied to the wheel, it turns in a circular distance.

b When the wheel turns, so does the axle. The axle is smaller than the wheel. Since the axle turns a smaller distance, the output force is larger than the input force.

What Is a Wheel and Axle?

Did you know that a faucet is a machine? The faucet in the figure above is an example of a **wheel and axle**. It is a simple machine that is made up of two round objects that move together. The larger object is the *wheel* and the smaller object is the *axle*. Some examples of a wheel and axle are doorknobs, wrenches, and steering wheels.

MECHANICAL ADVANTAGE OF A WHEEL AND AXLE

The mechanical advantage (*MA*) of a wheel and axle can be calculated. To do this you need to know the *radius* of both the wheel and the axle. Remember, the radius is the distance from the center to the edge of the round object. The equation to find the mechanical advantage (*MA*) of a wheel and axle is:

$$mechanical\ advantage\ (MA) = \frac{radius\ of\ wheel}{radius\ of\ axle}$$

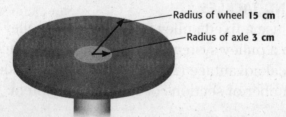

Radius of wheel **15 cm**

Radius of axle **3 cm**

Math Focus

9. Calculate A car has a wheel and axle. If the radius of the axle is 7.5 cm and the radius of the wheel is 75 cm, what is the mechanical advantage? Show your work.

The mechanical advantage of a wheel and axle is the wheel radius divided by the axle radius.

Let's calculate the mechanical advantage of the wheel and axle in the figure above.

Step 1: Write the equation.

$$mechanical\ advantage\ (MA) = \frac{radius\ of\ wheel}{radius\ of\ axle}$$

Step 2: Place values into the equation, and solve.

$$MA = \frac{15\ cm}{3\ cm} = 5$$

The mechanical advantage of this wheel and axle is 5.

Copyright © by Holt, Rinehart and Winston; a Division of Houghton Mifflin Harcourt Publishing Company. All rights reserved.

What Is an Inclined Plane?

The Egyptians built the Great Pyramid thousands of years ago using the **inclined plane**. An *inclined plane* is a simple machine that is a flat, slanted surface. A ramp is an example of an inclined plane.

Using an inclined plane to move a heavy object into a truck is easier than lifting the object. The input force is smaller than the object's weight. The same work is done, but it happens over a longer distance. ☑

Ramp Length   Ramp Height

You do work to push a piano up a ramp. This is the same amount of work you would do to lift it straight up. An inclined plane lets you apply a smaller force over a greater distance.

✔ READING CHECK

10. Explain How does an incline plane make lifting an object easier?

_____

_____

_____

_____

MECHANICAL ADVANTAGE OF INCLINED PLANES

The mechanical advantage (*MA*) of an inclined plane can also be calculated. The length of the inclined plane and the height the object that is lifted must be known. The equation to find the mechanical advantage (*MA*) of an inclined plane is:

$$\text{mechanical advantage } (MA) = \frac{\text{length of inclined plane}}{\text{height load raised}}$$

We can calculate the mechanical advantage (*MA*) of the inclined plane shown in the figure above.

Step 1: Write the equation.

$$\text{mechanical advantage } (MA) = \frac{\text{length of inclined plane}}{\text{height load raised}}$$

Step 2: Place values into the equation, and solve for the answer.

$$MA = \frac{3 \text{ m}}{0.6 \text{m}} = 5$$

The mechanical advantage (*MA*) is 5.

If the length of the inclined plane is much greater than the height, the mechanical advantage is large. That means an inclined plane with a gradual slope needs less force to move objects than a steep-sloped one.

Math Focus
11. Determine An inclined plane is 10 m and lifts a piano 2.5 m. What is the mechanical advantage of the inclined plane? Show your work.

Copyright © by Holt, Rinehart and Winston; a Division of Houghton Mifflin Harcourt Publishing Company. All rights reserved.

What Is a Wedge?

A knife is often used to cut because it is a **wedge**. A *wedge* is made of two inclined planes that move. Like an inclined plane, a wedge needs a small input force over a large distance. The output force of the wedge is much greater than the input force. Some useful wedges are doorstops, plows, ax heads, and chisels.

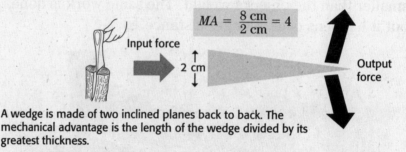

$$MA = \frac{8 \text{ cm}}{2 \text{ cm}} = 4$$

Input force

2 cm

Output force

A wedge is made of two inclined planes back to back. The mechanical advantage is the length of the wedge divided by its greatest thickness.

TAKE A LOOK
12. Predict What would happen to the mechanical advantage of the wedge if it were longer in length?

_____

MECHANICAL ADVANTAGE OF WEDGES

The mechanical advantage of a wedge can be found by dividing the length of the wedge by its greatest thickness. The equation to find a wedge's mechanical advantage is:

$$\text{mechanical advantage } (MA) = \frac{\text{length of wedge}}{\text{largest thickness of wedge}}$$

A wedge has a greater mechanical advantage if it is long and thin. When you sharpen a knife you are making the wedge thinner. This needs a smaller input force.

What Is a Screw?

A **screw** is an inclined plane that is wrapped around a cylinder. To turn a screw, a small force over a long distance is needed. The screw applies a large output force over a short distance. Screws are often used as fasteners.

Threads

If you could unwind a screw, you would have a very long inclined plane.

Say It

Demonstrate Take a long pencil and a piece of paper cut so it looks like an inclined plane. Roll the paper around the pencil so it looks like threads on a screw. Show the class how the paper looks like threads on a screw. Then unwind the paper showing that it looks like an inclined plane.

MECHANICAL ADVANTAGE OF SCREWS

To find the mechanical advantage of a screw you need to first unwind the inclined plane. Then, if you compare the length of the inclined plane with its height you can calculate the mechanical advantage. This is the same as calculating the mechanical advantage of an inclined plane. The longer the spiral on a screw and the closer the threads, the greater the screw's mechanical advantage.

Copyright © by Holt, Rinehart and Winston; a Division of Houghton Mifflin Harcourt Publishing Company. All rights reserved.

SECTION 3 Types of Machines *continued*

What Is a Compound Machine?

There are machines all around you. Many machines do not look like the six simple machines that you have read about. That is because most machines are **compound machines**. These are machines that are made of two or more simple machines. One purpose of engineering is to find different ways to combine simple machines to produce a particular compound machine that performs a specific task well. A block and tackle is one example of a compound machine. It is made of two or more pulleys. ☑

A common example of a compound machine is a can opener. A can opener may look simple, but it is made of three simple machines. They are the second-class lever, the wheel and axle, and the wedge. When you squeeze the handle, you are using a second-class lever. The blade is a wedge that cuts the can. When you turn the knob to open the can, you are using a wheel and axle.

A can opener is a compound machine. The handle is a second-class lever, the knob is a wheel and axle, and a wedge is used to open the can.

MECHANICAL EFFICIENCY OF COMPOUND MACHINES

The *mechanical efficiency* of most compound machines is low. Remember that mechanical efficiency tells you what percentage of work input gets done as work output. This is different than the mechanical advantage. The efficiency of compound machines is low because they usually have many moving parts. This means that there are more parts that contact each other and more friction. Recall that friction lowers output work. ☑

Cars and airplanes are compound machines that are made of many simple machines. It is important to lower the amount of friction in these compound machines. Friction can often damage machines. Grease is usually added to cars because it lowers the friction between the moving parts.

Copyright © by Holt, Rinehart and Winston; a Division of Houghton Mifflin Harcourt Publishing Company. All rights reserved.

READING CHECK

13. Describe What is a compound machine?

_____

_____

_____

TAKE A LOOK

14. Describe Describe the process of using a can opener. Tell the order in which each simple machine is used and what it does to open the can.

_____

_____

_____

_____

_____

READING CHECK

15. Identify Why do most compound machines have low mechanical efficiency?

_____

_____

Section 3 Review

GLE 0707.Inq.2, GLE 0707.Inq.5, GLE 0707.T/E.1, ▬ TN
GLE 0707.11.1, GLE 0707.11.2

SECTION VOCABULARY

compound machine a machine made of more than one simple machine	**screw** a simple machine that consists of an inclined plane wrapped around a cylinder
inclined plane a simple machine that is a straight, slanted surface, which facilitates the raising of loads; a ramp	**wedge** a simple machine that is made up of two inclined planes and that moves; often used for cutting
lever a simple machine that consists of a bar that pivots at a fixed point called a fulcrum	**wheel and axle** a simple machine consisting of two circular objects of different sizes; the wheel is the larger of the two circular objects
pulley a simple machine that consists of a wheel over which a rope, chain, or wire passes	

1. Compare Use a Venn Diagram to compare a first-class lever and a second-class lever.

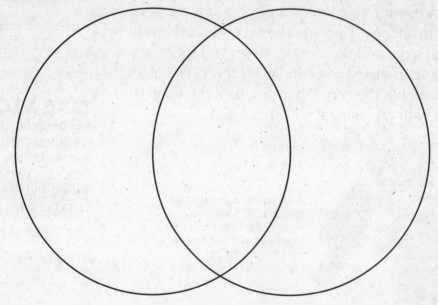

2. Calculate A screwdriver is used to put a screw into a piece of wood. The radius of the handle is 1.8 cm and the radius of shaft is 0.6 cm. What is the mechanical advantage of using the screwdriver? Show your work.

3. List Identify one everyday application for each of the six simple machines.

_____

_____

4. Analyze When there is a lot of friction in a machine, what is lowered and causes mechanical efficiency to be lowered?

_____

Copyright © by Holt, Rinehart and Winston; a Division of Houghton Mifflin Harcourt Publishing Company. All rights reserved.

CHAPTER 22 The Energy of Waves

SECTION 1 The Nature of Waves

TN Tennessee Science Standards
GLE 0707.Inq.5
GLE 0707.11.5
GLE 0707.11.6

BEFORE YOU READ

After you read this section, you should be able to answer these questions:

• What is a wave, and how does it transmit energy?

• How do waves move?

• What are the different types of waves?

What Is Wave Energy?

A **wave** is any disturbance that transmits energy through matter or empty space. Energy can be carried away from its source by a wave. However, the material through which the wave moves is not transmitted. For example, a ripple caused by a rock thrown into a pond does not move water out of the pond.

A wave travels through a material or substance called a **medium**. A medium may be a solid, a liquid, or a gas. The plural of medium is *media*.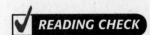

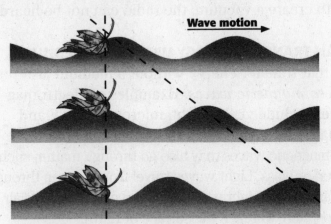

Wave motion →

A wave travels through the medium, but the medium does not travel. In a pond, lake or ocean, the medium through which a wave travels is the water. The waves in a pond travel towards the shore. However, the water and the leaf floating on the surface only travel up and down.

How Can Waves Do Work?

As a wave travels, it does work on everything in its path. The waves traveling through a pond do work on the water. Anything floating on the surface of the water moves up and down. The fact that any object on the water moves indicates that the waves are transferring energy. Waves can transfer energy through a medium or without a medium.

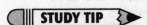
STUDY TIP

As you read the section, make a table of the types of waves. Have columns for the type of wave, what it moves through, its direction of motion, and how it transmits energy.

✓ READING CHECK

1. Identify What does a wave move through?

_____

✓ READING CHECK

2. Describe What indicates that a water wave transfers energy to a floating object?

_____

Copyright © by Holt, Rinehart and Winston; a Division of Houghton Mifflin Harcourt Publishing Company. All rights reserved.

SECTION 1 The Nature of Waves *continued*

WAVES CAN TRANSFER ENERGY THROUGH A MEDIUM

When a particle *vibrates* (moves back and forth), it can pass its energy to the particle next to it. The second particle will vibrate like the first particle and may pass the energy on to another particle. In this way, energy is transmitted through a medium.

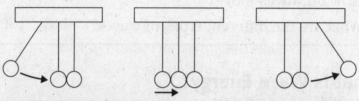

A particle can pass energy to the particle next to it. The particle receiving the energy will vibrate like the first particle. This is shown by the Newton's pendulum above. When the moving steel ball collides with another steel ball, its energy is given to that ball. Notice that the first ball stops, but its energy is passed on to a third ball.

TAKE A LOOK
3. Describe How did the last ball in the figure on the right gain energy?

_____

_____

_____

Waves that require a medium are called *mechanical waves*. Mechanical waves include sound waves, ocean waves, and earthquake waves. For example, consider a radio inside a jar. If all of the air from inside the jar is removed to create a vacuum, the radio can not be heard.

WAVES CAN TRANSFER ENERGY WITHOUT A MEDIUM

Waves that transfer energy without a medium are called *electromagnetic waves*. Examples of electromagnetic waves include visible light, microwaves, TV and radio signals, and X-rays used by dentists and doctors.

Electromagnetic waves may also go through matter, such as air, water, or glass. Light waves travel from the sun through space toward Earth. Light waves then travel through the air in the atmosphere to reach the surface of Earth.

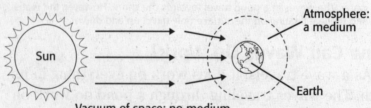

To reach the Earth, light travels from the sun, through the vacuum of space. The light then travels through the particles of the atmosphere before reaching the surface of the earth.

TN **TENNESSEE STANDARDS CHECK**

GLE 0707.11.6 Investigate the types and fundamental properties of waves.

4. Identify What kinds of waves need a medium to transfer their energy? What kinds of waves don't need a medium to transfer their energy?

_____

_____

_____

_____

_____

Copyright © by Holt, Rinehart and Winston; a Division of Houghton Mifflin Harcourt Publishing Company. All rights reserved.

What Are the Different Types of Waves?

Waves transfer energy through vibrations. However, the way particles in a wave vibrate depends on the type of wave. Waves are classified based on the direction in which wave particles vibrate compared with the direction in which waves move. There are two main types of waves, *transverse waves* and *longitudinal waves*. ☑

TRANSVERSE WAVES

Waves in which the particles vibrate in an up-and-down motion are called **transverse waves**. Particles in a transverse wave move at right angles relative to the direction of the wave. See the figure below. The highest point in a transverse wave is a *crest*. The lowest point in a transverse wave is a *trough*. ☑

READING CHECK

5. Identify What are the two main types of waves?

_____

_____

READING CHECK

6. Identify What is the direction of a transverse wave relative to its direction of motion?

_____

_____

Motion of a Transverse Wave

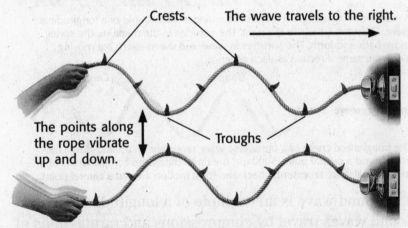

Crests

The wave travels to the right.

The points along the rope vibrate up and down.

Troughs

A wave traveling down a length of rope is an example of a transverse wave. The wave travels to the right. The particles in the medium, the rope, travel up-and-down. The particles in the wave and the medium are moving at right angles to each other.

All electromagnetic waves are transverse waves. Remember, electromagnetic waves can travel through space or through a medium. Electromagnetic waves are transverse waves because the wave vibrations are at right angles to the direction the wave is traveling. ☑

READING CHECK

7. Describe Why is an electromagnetic wave identified as a transverse wave?

_____

_____

_____

_____

Copyright © by Holt, Rinehart and Winston; a Division of Houghton Mifflin Harcourt Publishing Company. All rights reserved.

LONGITUDINAL WAVES

Waves in which the particles of the medium vibrate back and forth along the path of the wave are called **longitudinal waves**. For example, pushing together two ends of a spring causes the coils to crowd together. When you let go, a longitudinal wave is created in the spring that travels along the length of the spring. ☑

In a longitudinal wave, a *compression* is the location of the crowded particles. A *rarefaction* is where the particles are spread apart.

Compressions and rarefactions are similar to crests and troughs in a transverse wave. See the figure below.

Comparing Longitudinal and Transverse Waves

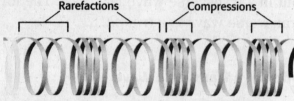

Longitudinal wave

A wave traveling along the length of a spring is an example of a longitudinal wave. The wave travels to the right. The particles in the medium, the spring, move back-and-forth. The particles in wave and the medium are moving along the same direction as each other.

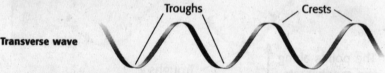

Transverse wave

The troughs and crests of a transverse wave represent an up-and-down motion around a central point. Similarly, the rarefactions and compressions of a longitudinal wave represent a back-and-forth motion around a central point.

A sound wave is an example of a longitudinal wave. Sound waves travel by compressions and rarefactions of air particles.

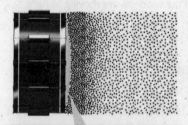

When the drumhead moves out after being hit, a compression is created in the air particles.

When the drumhead moves back in, a rarefaction is created.

Sound energy is carried away from a drum by a longitudinal wave through the air.

Copyright © by Holt, Rinehart and Winston; a Division of Houghton Mifflin Harcourt Publishing Company. All rights reserved.

☑ READING CHECK

8. Describe How do the particles of a longitudinal wave vibrate?

_____

_____

_____

Critical Thinking

9. Describe How could you produce a transverse wave in a spring?

_____

_____

TAKE A LOOK

10. Identify Circle the compression part of the wave in the second figure.

SECTION 1 The Nature of Waves *continued*

SURFACE WAVE

When waves move at or near the surface between two media a *surface wave* may form. For example, this occurs when an ocean wave comes into shallow water at the shore. Surface waves travel in both transverse and longitudinal motion. A particle in a surface wave will appear to move in a circular motion.

Ocean waves are surface waves. A floating bottle shows the circular motion of particles in a surface wave.

Summary of Wave Types and Their Motion Through Space

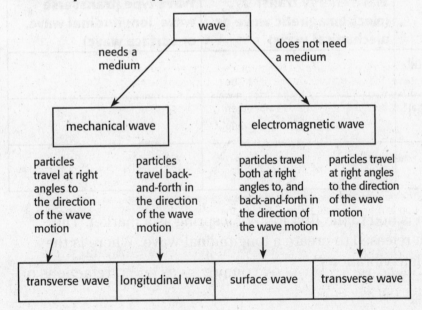

Critical Thinking

11. Identify An ocean wave forms a surface wave as it comes into shallow water. What are the two media involved in forming the surface wave?

_____

_____

TAKE A LOOK
12. Identify Which type of wave must have a medium in order to travel?

_____

Copyright © by Holt, Rinehart and Winston; a Division of Houghton Mifflin Harcourt Publishing Company. All rights reserved.

Section 1 Review

GLE 0707.Inq.5, GLE 0707.11.5, GLE 0707.11.6 TN

SECTION VOCABULARY

medium a physical environment in which phenomena occur	**transverse wave** a wave in which the particles of the medium move perpendicularly to the direction the wave is traveling
longitudinal wave a wave in which the particles of the medium vibrate parallel to the direction of wave motion	**wave** a periodic disturbance in a solid, liquid, or gas as energy is transmitted through a medium

1. Describe How does energy travel through a wave in a medium?

_____

_____

_____

2. Identify Determine the method of energy transfer, and the wave type for each wave source.

Wave source	Wave energy transfer (electromagnetic wave or mechanical wave)	Wave type (transverse wave, longitudinal wave, or surface wave)
Light emitted from a light bulb		
Sound coming from a violin		
Rock dropped in a pond		

3. Apply Concepts A ribbon is tied to the first loop of a spring as a marker. The spring is pulled and then released to create a longitudinal wave. Where is the ribbon after three complete vibrations?

_____

4. Recall Label each wave part as a crest, trough, compression, or rarefaction according to its description.

Wave Part	Description
	particles are crowded toward each other
	particles are at their highest point
	particles are at their lowest point
	particles are spread away from each other

Copyright © by Holt, Rinehart and Winston; a Division of Houghton Mifflin Harcourt Publishing Company. All rights reserved.

CHAPTER 22 The Energy of Waves

SECTION
2 **Properties of Waves**

BEFORE YOU READ

After you read this section, you should be able to answer these questions:

- What are ways to describe a wave?
- What determines the energy of a wave?

TN **Tennessee Science Standards**
GLE 0707.Inq.2
GLE 0707.Inq.3
GLE 0707.Inq.5
GLE 0707.11.5
GLE 0707.11.6

What Is the Amplitude of a Wave?

An earthquake in the ocean can make a transverse wave called a *tsunami*. The waves can get very tall as they reach land. The *amplitude* of a transverse wave is related to the height of the wave. **Amplitude** is the maximum distance the particles of the wave vibrate away from their rest position. The rest position, shown in the figure below, is the location of the particles of the medium before the wave gets there. ☑

STUDY TIP

With a partner, discuss the properties of a wave and how they affect the energy of a wave.

READING CHECK

1. Describe What is the amplitude of a wave?

_____

_____

_____

_____

The amplitude of a transverse wave is measured from the rest position to the crest or the trough of the wave.

RELATIONSHIP BETWEEN AMPLITUDE AND ENERGY

Taller waves have larger amplitudes, and shorter waves have smaller amplitudes. It takes more energy to create a wave with a large amplitude. Therefore, it carries more energy. ☑

READING CHECK

2. Explain Why do waves of large amplitude carry more energy?

_____

_____

_____

A wave with a larger amplitude carries more energy.

A wave with a smaller amplitude carries less energy.

Copyright © by Holt, Rinehart and Winston; a Division of Houghton Mifflin Harcourt Publishing Company. All rights reserved.

What Is the Wavelength of a Wave?

The **wavelength** is the distance between any point on a wave and the identical point on the next wave. A wavelength is the distance between two neighboring crests or two compressions. The distance between two troughs or two rarefactions next to each other is also a wavelength. See the figure below.

Measuring Wavelengths

Longitudinal wave

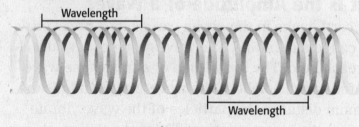

Transverse wave

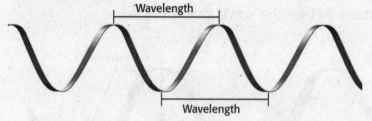

A wavelength can be measured between any two identical points on neighboring waves.

Math Focus

3. Infer What is the distance between a crest and a trough next to a crest? Hint: use the figure for help.

_____

_____

RELATIONSHIP BETWEEN WAVELENGTH AND ENERGY

Suppose you have two waves of the same amplitude. The wave with a shorter wavelength carries more energy than the wave with a longer wavelength.

A wave with a short wavelength carries more energy.

A wave with a long wavelength carries less energy.

TN TENNESSEE STANDARDS CHECK

GLE 0707.11.6 Investigate the types and fundamental properties of waves.

4. Identify There are two groups of waves with the same amplitude. One contains waves of short wavelength and the other has waves of long wavelength. Which waves have the most energy?

_____

_____

_____

Copyright © by Holt, Rinehart and Winston; a Division of Houghton Mifflin Harcourt Publishing Company. All rights reserved.

What Is the Frequency of a Wave?

The **frequency** of a wave is the number of waves produced in a given amount of time. Frequency is often expressed in *hertz* (Hz). One hertz equals one wave per second (1 Hz = 1/s).

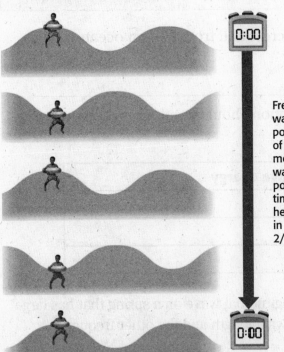

Frequency is the number of waves that pass a certain point in a given amount of time. Frequency can be measured by counting the waves that pass a certain point in a certain amount of time. In the image shown here, two waves passed by in 10 s. The frequency is 2/10 s = 0.2 1/s = 0.2 Hz.

Math Focus

5. Determine If a source of waves produces 30 waves per second, what is the frequency in hertz?

_____

Critical Thinking

6. Applying Concepts What would the time of the lower clock read if the frequency of the wave were 0.1 Hz?

_____

RELATIONSHIP BETWEEN FREQUENCY AND ENERGY

When the amplitudes of two waves are equal, the wave with the higher frequency has more energy. Lower-frequency waves carry less energy. ☑

What Is Wave Speed?

Wave speed is the speed at which a wave travels through a medium. Wave speed is symbolized by *v*. The speed of a wave is a property of the medium. Changing the medium of a wave changes its speed. For example, light travels faster in air than in water.

The equation for calculating speed is $v = \lambda \times f$. λ is the Greek letter *lambda* and means the wavelength. *f* is the frequency of the wave.

Let's do a problem. What is the speed of a wave whose wavelength is 5 m and frequency is 4 Hz (or 4 1/s)?

Step 1: write the equation $v = \lambda \times f$

Step 2: replace λ and f with their values
$v = 5 \text{ m} \times 4 \text{ 1/s} = 20 \text{ m/s}$

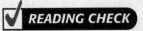 **READING CHECK**

6. Identify When the amplitudes of waves are equal, which frequency waves have the most energy?

_____

_____

Math Focus

7. Calculate What is the speed of a wave whose wavelength is 30 m and frequency is 20 Hz? Show your work.

Copyright © by Holt, Rinehart and Winston; a Division of Houghton Mifflin Harcourt Publishing Company. All rights reserved.

Section 2 Review

GLE 0707.Inq.2, GLE 0707.Inq.3, GLE 0707.Inq.5, TN
GLE 0707.11.5, GLE 0707.11.6

SECTION VOCABULARY

amplitude the maximum distance that the particles of a wave's medium vibrate from their rest position	**wavelength** the distance from any point on a wave to an identical point on the next wave
frequency the number of waves produced in a given amount of time	**wave speed** the speed at which a wave travels through a medium

1. Apply Concepts The distance between the crest and trough of an ocean wave is 1 meter. What is the amplitude of the wave?

_____

2. Identify Indicate whether the wave description should result in higher-energy wave, or a lower-energy wave.

Wave description	Wave energy
high amplitude	
low frequency	
low wavelength	

3. Apply Concepts Explain how to produce a longitudinal wave on a spring that has large energy. There are two answers; one involves wavelength and the other frequency.

_____

_____

4. Math Concepts What is the speed of a wave that has a wave length of 100 m and a frequency of 25 Hz? Show your work.

5. Apply Concepts A sound wave has a frequency of 125 Hz and a speed of 5000 m/s. What is the wavelength of the wave? Show your work.

Copyright © by Holt, Rinehart and Winston; a Division of Houghton Mifflin Harcourt Publishing Company. All rights reserved.

SECTION 3 Wave Interactions

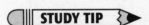

TN Tennessee Science Standards
GLE 0707.Inq.3
GLE 0707.Inq.5

BEFORE YOU READ

After you read this section, you should be able to answer these questions:

• How do waves interact with objects?

• How do waves behave when they move between two media?

• How do waves interact with other waves?

Why Do Waves Reflect?

A **reflection** occurs when a wave bounces back after hitting a barrier. All waves can be reflected. Light waves reflecting off an object allow you to see that object. For example, light waves from the sun reflecting off the moon allow you to see the moon. Sound wave can also reflect. Sound waves reflecting off a barrier are called an *echo*.

STUDY TIP

In your science notebook, define each new vocabulary word. Include sketches illustrating reflection, refraction, diffraction, and both kinds of interference.

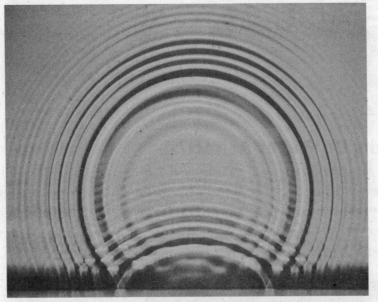

The waves in this photograph were formed by drops of water that fell into a container of water. When the waves caused by the drops of water hit one side of the container, they reflect off. The shape of the reflected waves is opposite that of the waves that struck the side of the tank.

Critical Thinking

1. Infer How does your reflection in a bathroom mirror look when you raise you right arm?

_____

_____

_____

_____

Waves are not always reflected when they hit a barrier. Sometimes they pass through a substance. When a wave passes through a substance, it is *transmitted*. Light waves transmitted through a glass window allow light to enter a room. Light waves transmitted through eyeglasses allow the wearer to see through them.

Copyright © by Holt, Rinehart and Winston; a Division of Houghton Mifflin Harcourt Publishing Company. All rights reserved.

SECTION 3 Wave Interactions *continued*

Why Do Waves Diffract?

Diffraction is the bending of waves around a barrier or through an opening. Waves usually travel in a straight line. When a wave reaches the edge of an object or an opening in a barrier, it may curve or bend. ☑

The amount of diffraction of a wave depends on its wavelength and the size of the barrier opening. Sound waves are relatively long. You can hear voices from one classroom diffract through the opening of a door into another classroom. Light waves are relatively short. You cannot see who is speaking in the other classroom.

READING CHECK

2. Describe What is diffraction?

_____

_____

_____

TAKE A LOOK

3. Describe Suppose the opening in the lower figure were made larger. What would happen to the shape of the diffracted wave?

_____

_____

If the barrier or opening is larger than the wavelength of the wave, there is only a small amount of diffraction.

If the barrier or opening is the same size or smaller than the wavelength of an approaching wave, the amount of diffraction is large.

Why Do Waves Refract?

Refraction is the bending of a wave as the wave passes from one medium to another. The wave changes speed as it passes from one material to the other. The change in speed causes the wavelength to change. The resulting wave bends and travels in a new direction. ☑

READING CHECK

4. Describe What happens to a wave because of refraction?

_____

_____

_____

This light wave is refracted as it passes into a new medium. The light wave is passing from air into water. The wave is refracted because the speed of the wave changes.

Copyright © by Holt, Rinehart and Winston; a Division of Houghton Mifflin Harcourt Publishing Company. All rights reserved.

SECTION 3 Wave Interactions *continued*

REFRACTION OF DIFFERENT COLORS

When light waves from the sun pass through a droplet of water in the air, the light is refracted. The different colors of light travel at different speeds through the drop. Therefore, the different colors are refracted by different amounts. The light is *dispersed*, or spread out, into its separate colors. The result is a rainbow. ☑

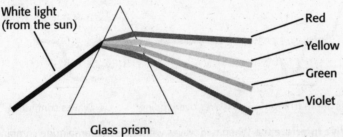

White light
(from the sun)

Red
Yellow
Green
Violet

Glass prism

White light is separated into its component colors when it passes through a prism. The red light is refracted the least. The violet light is refracted the most.

What Is Wave Interference?

All matter has volume. Therefore, objects cannot be in the same space at the same time. However, waves are made up of energy, not matter. So, more than one wave may be in the same space at the same time. Two waves can meet, share the same space, and pass through each other.

When two or more waves meet and share the same space, they overlap. **Interference** is the combination of two or more waves to form a single wave. ☑

CONSTRUCTIVE INTERFERENCE

Constructive interference occurs when the crests of one wave overlap with the crests of another wave or waves. The troughs of both waves will also overlap. The energy of the waves adds together to make a higher-energy wave. The new wave has higher crests, deeper troughs, and, therefore, higher amplitude.

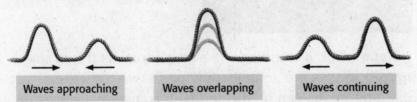

Waves approaching   Waves overlapping   Waves continuing

Constructive Interference When waves combine by constructive interference, the combined wave has a larger amplitude.

Copyright © by Holt, Rinehart and Winston; a Division of Houghton Mifflin Harcourt Publishing Company. All rights reserved.

☑ **READING CHECK**

5. Describe What does light do when it disperses?

_____

_____

TAKE A LOOK

6. Identify The order of the dispersed colors can be remembered by the mnemonic ROY G. BIV. Draw a line coming from the prism that shows the direction that orange would move in.

☑ **READING CHECK**

7. Identify What is the combination of two or more waves to form a single wave called?

_____

_____

TAKE A LOOK

8. Describe What does the medium do after the waves have overlapped and are continuing their movement?

_____

_____

_____

_____

DESTRUCTIVE INTERFERENCE

Destructive interference occurs when the crests of one wave overlap with the troughs of another wave. The energy of the new wave is less than the energy of both waves. The new wave has lower amplitude than the original waves. If a crest and trough of the same amplitude meet and cancel, the result is no wave at all. ☑

☑

READING CHECK

9. Describe What parts of a wave overlap during destructive interference?

_____

_____

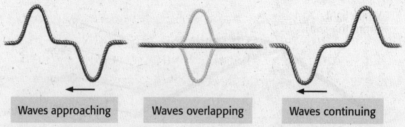

Waves approaching   Waves overlapping   Waves continuing

Destructive Interference When two waves with the same amplitude combine by destructive interference, they cancel each other out.

How Is a Standing Wave Created?

In a **standing wave**, the pattern of vibrations makes it appear as if the wave is standing still. A standing wave is caused by interference between a wave and a reflected wave. For example, pluck a guitar string. The string makes a standing wave like the top wave shown in the figure below. ☑

READING CHECK

10. Describe What causes a standing wave?

_____

_____

_____

TAKE A LOOK

11. Identify Draw two arrows in the bottom figure that show the locations of destructive interference.

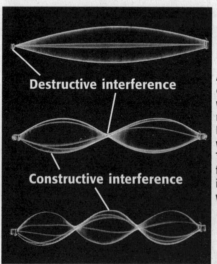

Destructive interference

Constructive interference

A rope vibrating at certain frequencies can create a standing wave. The initial wave travels down the rope and will be reflected back when it reaches the end. In a standing wave, certain parts of the wave are always at the rest position. This point is caused by destructive interference between the waves. Constructive interference can be seen at points in the wave where there is large amplitude.

Remember, a standing wave only looks as if it is standing still. Waves are actually moving in two directions. Standing waves can be formed with transverse waves or with longitudinal waves.

Copyright © by Holt, Rinehart and Winston; a Division of Houghton Mifflin Harcourt Publishing Company. All rights reserved.

SECTION 3 Wave Interactions *continued*

RESONANCE

Resonant frequencies are the frequencies at which standing waves are created. **Resonance** occurs when two objects naturally vibrate at the same frequency. The resonating object absorbs energy from the vibrating object and vibrates also. For example, when a guitar string is plucked, the wood body vibrates at the same frequency as the string. ☑

When you pluck the guitar string, you hear a musical note. The vibrating wood body makes sound waves in the air. The sound waves that reach your ear make parts of your ear vibrate, so you hear the sound of the note.

✓ READING CHECK

12. Describe When does resonance occur?

_____

_____

_____

Summary of Wave Interactions

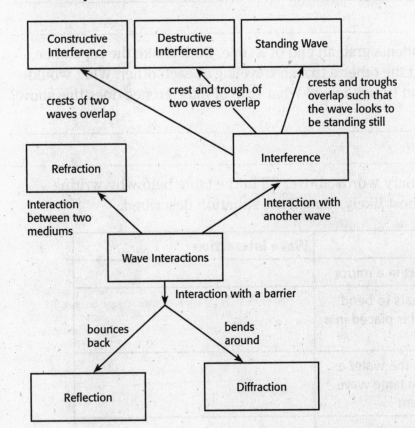

TAKE A LOOK

13. Identify What two interactions can occur when a wave strikes a barrier?

_____

_____

Copyright © by Holt, Rinehart and Winston; a Division of Houghton Mifflin Harcourt Publishing Company. All rights reserved.

Section 3 Review

GLE 0707.Inq.3, GLE 0707.Inq.5 **TN**

SECTION VOCABULARY

diffraction a change in the direction of a wave when the wave finds an obstacle or an edge, such as an opening	**refraction** the bending of a wave as the wave passes between two substances in which the speed of the wave differs
interference the combination of two or more waves that results in a single wave	**resonance** a phenomenon that occurs when two objects naturally vibrate at the same frequency; the sound produced by one object causes the other object to vibrate.
reflection the bouncing back of a wave of light, sound, or heat when the wave hits a surface that it does not go through	**standing wave** a pattern of vibration that simulates a wave that is standing still

1. Describe Suppose two students grab an end of a rope. Both shake the rope once in an upward direction to create crests traveling at each other. What would you see when the crests meet? What type of interference does this show?

_____

_____

2. Describe Suppose two students grab an end of a rope. Both shake the rope once, but one makes a crest and the other a trough traveling at each other. What would you see when the crest and trough meet? What type of interference does this show?

_____

_____

3. Identify Using the vocabulary words above, fill in the table below by writing which wave interaction most likely caused the situation described.

Wave situation	Wave interaction
The image of an object in a mirror	
A straight pencil appears to bend when the bottom half is placed in a glass of water	
Two rocks dropped in the water a meter apart produce a large wave centered between them	
A radio turned on in one classroom can be heard down the hall in a second classroom	

Copyright © by Holt, Rinehart and Winston; a Division of Houghton Mifflin Harcourt Publishing Company. All rights reserved.

Photo Credits

Abbreviations used: c-center, b-bottom, t-top, l-left, r-right, bkgd-background.

1,2,3 Harcourt; 4 (b) Peter Van Steen/Harcourt; 4 (t) Hank Morgan/Photo Researchers, Inc.; 5 Hank Morgan/Photo Researchers, Inc.; 8,9,13 Harcourt; 14 John Mitchell/Photo Researchers; 16 Sam Dudgeon/Harcourt; 15 (l,r) Fujifotos/The Image Works; 17 (t) John Mitchell/Photo Researchers; 19 Harcourt; 20 PhotoDisc/Getty Images; 21, 23, 24 Greenshoots Communications/Alamy; 25 (l) Babak A. Parviz, University of Washington; (r) Michel Baret/Photo Researchers, Inc.; 27,30 Harcourt; 33 (l) The Image Works; (inset) Leonard Lessin/Peter Arnold; 34 (tl) M.I. Walker/Photo Researchers; (tr) Steve Allen/Photo Researchers; (bl) Michael Abbey/Photo Researchers, Inc.; (br) CNRI/Photo Researchers,Inc.; 37 (t) Wolfgang Baumeister/SPL/Photo Researchers, Inc.; (b) Biophoto Associates/Photo Researchers, Inc.; 51, 56 Harcourt; 59 (b) CNRI/SPL/Photo Researchers; 60 Biophoto Associates/Photo Researchers; 61 (tl,tr,cr,cl) Ed Reschke/Peter Arnold; (bl) Biology Media/Photo Researchers; (br) Biology Media/Photo Researchers; (b) Ed Reschke/Peter Arnold, Inc.; 63 (b) National Geographic Image Collection/Ned M. Seidler; 69 (b) Norman Lightfoot/Photo Researchers, Inc.; 75 (b) CNRI/Phototake NYC; (inset) Biophoto Associates/Photo Researchers; 80 (t) Rob van Nostrand; 87 (inset) J.R. Paulson & U.K. Laemmli/University of Geneva; 89 (c) Phanie/Photo Researchers, Inc.; 93 (tr, br) Jackie Lewin/Royal Free Hospital/SPL/Photo Researchers, Inc; 97 (l) Bruce Coleman; (r) Runk/Schoenberger/Grant Heilman; 107 T. Branch/Photo Researchers; 109 (tl) Dwight R. Kuhn; (tr) Runk/Schoenberger/Grant Heilman; (b) Nigel Cattlin/Holt Studios International/Photo Researchers; 111 (br) Stephen J. Krasemann/Photo Researchers; 112 Tom Bean; 113 Everett Johnson/Jupiterimages; 115 George Bernard/SPL/Photo Researchers; 123 (bl) Jerome Wexler/Photo Researchers; (bc) Paul Hein/Unicorn Stock Photos; (br) George Bernard/Earth Scenes; 125 (b) Cathlyn Melloan/Getty Images; 126 (l,r) R.F. Evert; 171 Getty Images; 188 Sam Dudgeon/Harcourt; 189 Will & Deni McIntyre/Photo Researchers; 191 (r) M I (Spike) Walker/Alamy; (l) Andrew J. Martinez/Photo Researchers, Inc.; 193 Getty Images; 202 (tr) David Phillips/Photo Researchers; (tl) Petit Format/Nestle/Science Source/Photo Researchers; (c) Photo Lennart Nilsson/Albert Bonniers Forlag AB, A Child is Born, Dell Publishing Company; (b) Keith/Custom Medical Stock Photo; 203 Peter Van Steen/Harcourt; 210, 211 Harcourt; 219 (bl) Michael Melford/Getty Images; (br) Joseph Sohm/Visions of America/CORBIS; 220 Royalty Free/CORBIS; 222 (limestone) Breck P. Kent; (calcite) Visuals Unlimited/CORBIS; (aragonite) Breck P. Kent; (granite) Pat Lanza/Bruce Coleman; (biotite) Dr. E. R. Degginger/Color-Pic; (feldspar) Mark Schneider/Photo Researchers, Inc.; (quartz) PhotoDisc/Getty Images; 223 (bl) Dr. E. R. Degginger/Color-Pic; (br) Pat Lanza/Bruce Coleman; (c) Dorling Kindersley; (cl) Sam Dudgeon/Harcourt; (cr) Breck P. Kent; 226 (bl,cl,cr) Breck P. Kent; (br) Victoria Smith/Harcourt; 229 Royalty Free/CORBIS; 230 (cr) Joyce Photographics/Photo Researchers, Inc.; (cr,r) Sam Dudgeon/Harcourt; (l) Breck P. Kent; 231 (b) Franklin/OSF/Animals Animals/Earth Scenes; (t) Breck P. Kent; 233 George Wuerthner; 235 (bl) GC Minerals/Alamy; (bl) Carlyn Iverson/Absolute Science and Photography; (bl) Breck P. Kent; (br) Breck P. Kent/Animals Animals/Earth Scenes; (t) Jim Wark/Airphoto; 237 (shale) Ken Karp/Harcourt; (slate, phyllite) Sam Dudgeon/Harcourt; (schist) Courtesy Stan Celestian; (gneiss) Breck P. Kent 253 (cl, cr) Peter Van Steen/Harcourt; 275 Paul Chesley/Getty Images; 278 Breck P. Kent/Animals Animals/Earth Scenes; 280 (bl) B. Murton/Southhampton Oceanography Centre/SPL/Photo Researchers, Inc.; (br) Tom Bean/DRK Photo; (cl) Tui de Roy/Minden Pictures; (cr) Don Brown/Animals Animals Earth Scenes; 281 (b) Alberto Garcia/CORBIS; (cl) Dr. E.R. Degginger/Color-Pic; (cr) Tom Bean/DRK Photo; (tl) Francois Gohier/Photo Researchers, Inc.; (tr) Colin Keates/Dorling Kindersley, Courtesy of the Natural History Museum, London; 293 (bc) Andy Christiansen/Harcourt; (bl) Russell Illiq/PhotoDisc/Getty Images; (br) Mark Lewis/Getty Images; 294 (bl) James Randklev/Getty Images; (br) Myrleen Furgusson Cate/PhotoEdit; 299 Martin Harvey; 301 (l) NYC Parks Photo Archive/Fundamental Photographs; (r) Kristen Brochmann/Fundamental Photographs; 305 Bob Rowan/Progressive Image/CORBIS; 306 Mark Gibson Photography; 309 (l,r) SuperStock; 313 (t) Annie Griffiths Belt/CORBIS; 316 Harcourt; 317 (b) age fotostock/Fabio Cardosco; 321 (b) Sam Dudgeon/Harcourt; 323 (b) NASA; 332 (c) Toby Rankin/Masterfile; 335 © Michelle Bridwell/Frontera Fotos; 340 Victoria Smith/Harcourt; 341 (tl,tr) PhotoDisc/Getty Images; 347 (bl) SuperStock; (br) Zigy Kaluzny/Getty Images; 349, 350, 351, 353 Harcourt; 355 (bl) Artville/Getty Images; (br) Sam Dudgeon/Harcourt; 356 357, 358, 361, 362, 364, 365, 366, 367 Harcourt; 379 (c) Erich Schrempp/Photo Researchers, Inc.; 380 (tl,tr) Educational Development Center; (b) Richard Megna/Fundamental Photographs

Copyright © by Holt, Rinehart and Winston; a Division of Houghton Mifflin Harcourt Publishing Company. All rights reserved.